Aspergillus in the genomic era

Aspergillus
in the genomic era

edited by:

János Varga
Robert A. Samson

Wageningen Academic
Publishers

ISBN 978-90-8686-065-4

Photo front cover:
Robert A. Samson
CBS Fungal Biodiversity Centre,
Utrecht, The Netherlands

Photo back cover:
Department of Genetics
Wageningen University and
Research centre
Wageningen, The Netherlands

First published, 2008

Wageningen Academic Publishers
The Netherlands, 2008

Preface

The genus *Aspergillus* is a group of filamentous fungi which today consists of more than 250 species. The name *Aspergillus* was first applied by Micheli in Nova Plantarum Genera of 1729 to moulds with conidial heads and stalks, which reminded him of the aspergillum, a liturgical device used to sprinkle holy water. These fungi began to be recognised as active agents of decomposition processes and causes of animal and human diseases only in the middle of the ninetieth century. Today, *Aspergillus* is among the most economically important of the fungal genera. *Aspergillus* species are Janus-faced organisms in the sense that they can be both harmful and beneficial for mankind. They are responsible for a number of plant, animal and human diseases and produce several toxins including aflatoxins, ochratoxins and patulin. At the same time, Aspergilli are widely used in the food and pharmaceutical industry for the production of Oriental foods, various acids, enzymes and other compounds useful for humans. Due to its importance in biotechnology, medicine and foods, Aspergilli are in the forefront of studies dealing with various aspects of fungi. This is well illustrated by the fact that projects aiming at sequencing the genomes of nine species are in progress. In this book, some aspects of the biology of this genus are covered in chapters written by experts of the field. The chapters are organised around four main topics: phylogenetics and taxonomy of *Aspergillus*; genomics and genetics; biotechnological and agricultural importance of Aspergilli; and the clinical aspects. Regarding the phylogenetics and taxonomy, multilocus gene approaches are discussed to clarify the evolutionary history of the genus, and a new subgeneric classification of the genus *Aspergillus* is also given based on multilocus sequence data. Besides, the role of secondary metabolites, especially ochratoxins in the taxonomy is covered. Several chapters deal with the genomic and genetic aspects of Aspergilli, including the genomics of pigment biosynthesis, extracellular enzyme production, aflatoxin biosynthesis, the identification of genes taking part in sexual processes, and the variability and inheritance of mycoviruses. Regarding agricultural and biotechnological aspects, chapters deal with the biotechnological importance of Aspergilli, their role in mycotoxin contamination of food products including grapes and coffee products, and the implications of biodiversity of *Aspergillus* populations for agriculture. Finally, among the clinical aspects, the role of Aspergilli in eye infections, the mechanisms of antifungal drug resistance and their molecular identification in the clinical setting are covered.

The editors are very grateful to the authors for mostly being on time and producing such readable and authoritative contributions; to the publisher for being patient and especially to Marijn van der Gaag and Jessica van Wijngaarden for desk editing and typesetting.

János Varga & Robert A. Samson

Utrecht, The Netherlands

Table of contents

II. *Aspergillus* genomics and genetics

III. Aspergilli in the biotechnology and agriculture

IV. Clinical apects of Aspergilli

I. Phylogenetics and taxonomy of Aspergilli

A review of molecular phylogenetics in *Aspergillus*, and prospects for a robust genus-wide phylogeny

David M. Geiser[1], Robert A. Samson[2], János Varga[2,3], Antonis Rokas[4] and Sarah Melissa Witiak[1]
[1]Department of Plant Pathology, The Pennsylvania State University, University Park, PA 16802, USA
[2]CBS Fungal Biodiversity Centre, Uppsalalaan 8, NL-3584 CT Utrecht, The Netherlands
[3]Department of Microbiology, Faculty of Science and Informatics, University of Szeged, Közép fasor 52, H-6726 Szeged, Hungary
[4]Department of Biological Sciences, Vanderbilt University, VU Station B 351364, Nashville, TN 37235, USA

Abstract

Aspergillus is a diverse anamorphic genus comprising approximately 250 known species covering nine distinct teleomorph genera. Species are organised taxonomically into sections and subgenera according to anamorph morphology and teleomorph relationships. *Aspergillus* systematists have been applying DNA sequence data to address issues about species boundaries and infrageneric relationships for nearly twenty years. Most subgeneric categories have held up well under molecular phylogenetic scrutiny, but relationships among subgenera and sections have proven elusive due to an inability to resolve early diverging branches. Based on promising results in yeast species, we propose the use of up to twenty different loci for resolving relationships within the genus. Here we describe 8 loci useful for inferring phylogenetic relationships across the genus, identified and developed based on information gleaned from the eight currently available complete genome sequences in the genus. Whether or not these loci provide strong inferential power in the backbone of the *Aspergillus* phylogeny, they are promising tools for multilocus identification of species at a fine level.

Keywords: *Aspergillus*, multilocus sequence data, phylogeny, protein-coding loci, ribosomal RNA genes

1. Introduction

Over the course of the last two decades, a tremendous amount of genetic variation has been assayed and catalogued among *Aspergillus* species, yet surprisingly little progress has been made regarding the understanding of relationships among its subgeneric groups since they were outlined by Thom and Church (1926), Thom and Raper (1945) and Raper and Fennell (1965). Because of this, few of the many hypotheses about evolutionary patterns in Trichocomaceae, particularly among species with *Aspergillus* anamorphs (Malloch and Cain, 1972; Malloch, 1981, 1985), have remained untested. While part of this is a credit to the robustness of the morphological concepts outlined by previous workers, there have been surprisingly few uses of phylogenetics to develop and test evolutionary hypotheses regarding the staggering biological variation in this genus.

2. Early molecular phylogenetics

The first application of nucleic acid tools to *Aspergillus* systematics involved restriction fragment length polymorphisms in mitochondrial DNA (Kozlowski and Stepien, 1982), followed by the use of whole genomic DNA-DNA hybridisation studies (Kurtzman, 1985; Kurtzman *et al.,* 1986). While these methods represented the state of the art at the time, and DNA-DNA hybridisation showed 100% identity between an isolate of *Aspergillus flavus* and the koji species *A. oryzae*, leading those authors to reduce *A. oryzae* to varietal status under *A. flavus*, they are both technically cumbersome and limited in scope. DNA sequence based phylogenetic analyses of *Aspergillus* came along later, revealing surprisingly high levels of sequence variation. The first published study was based only on 130 nucleotides in the moderately conserved D2 region of the nuclear ribosomal large RNA subunit (LSU) gene, but revealed up to 10% sequence divergence among *Aspergillus* species (Dupont *et al.,* 1990). A subsequent study (Chang *et al.,* 1991) analysed data from three portions of the nuclear ribosomal small RNA subunit (SSU) gene totalling 558 sites, observing similarly high levels of diversity in these highly conserved regions, and concluding that the *Aspergillus* anamorph is monophyletic. Both of these studies utilised distance-based methods for phylogenetic inference, and neither incorporated a statistical test for node support.

The introduction of the polymerase chain reaction (PCR), automated DNA sequencing and user-friendly software packages such as Phylogenetic Analysis Using Parsimony (PAUP; Swofford, 2003) allowed *Aspergillus* researchers to generate larger datasets and subject them to more sophisticated analyses. Berbee *et al.* (1995) used the internal transcribed spacer (ITS) regions of the nuclear ribosomal gene repeat and SSU data to infer relationships among *Aspergillus* species, their teleomorphs and other members of the Trichocomaceae. The ITS region contained valuable information, but was largely unalignable across the genus. This work utilised statistical tests for node support and phylogenetic topology, and produced a phylogeny consistent with the subfamilies Dichlaenoideae (producing ascomata with mostly pseudoparenchymatous walls, oblate ascospores, and *Aspergillus, Paecilomyces, Merimbla, Polypaecilum* and *Penicillium* anamorphs) and Trichocomoideae (producing hyphal ascomata, non-prolate ascospores and *Penicillium* or *Paecilomyces* anamorphs) proposed by Malloch (1985). Geiser *et al.* (1996) inferred multiple, independent losses of sexual stages across the genus *Aspergillus,* based on phylogenetic analyses of sexual and putatively asexual species across the genus using the ITS and mitochondrial small subunit ribosomal RNA (mtSSU) genes, in agreement with the findings of LoBuglio *et al.* (1993) on *Penicillium* subgenus *Biverticillium.* This inference now seems far less likely in light of recent discoveries that apparently asexual *Aspergillus* species possess mating-type idiomorphs in what appears to be a heterothallic arrangement (Paoletti *et al.* 2005), consistent with cryptic sexual stages. Observations of vestiges of sexual reproduction in putatively asexual species (e.g. Hülle cells, sclerotia, hyphal masses) probably represent development of stromatic tissues that occur independently of mating and the production of ascogenous hyphae. Tamura *et al.* (2000) examined a more representative sample of taxa and countered the conclusion of Berbee *et al.* (1995) that *Aspergillus* is monophyletic. This conclusion, however, was based on the understanding that *Hemicarpenteles paradoxus* is actually an *Aspergillus* species, which is now very much in doubt (Peterson *et al.,* 2008). These authors also followed previous studies (Chang *et al.* 1991) in superimposing the major ubiquinone systems onto the molecular phylogeny, producing no obvious genus-wide patterns, but support for subgeneric groupings. Peterson (2000) took the approach of analysing LSU sequences from 215 species, and generally found weak support for basal nodes in the phylogeny, with more robust inferences associated with terminal nodes.

3. Multilocus approaches

All of the aforementioned phylogenetic analyses based on ribosomal regions produced weakly supported nodes except at terminal branches, making it difficult to test hypotheses about subgeneric relationships. Peterson (1995) used LSU and ITS sequence data to make inferences about taxa that belonged in *Aspergillus* sections *Cremei* and *Wentii,* drawing conclusions that were based on fairly weak (in some cases, <75%) bootstrap support. The same loci were used to infer relationships within *Aspergillus* sections *Circumdati* (Varga *et al.,* 2000), *Flavi* (Rigó *et al.,* 2002), *Clavati* (Varga *et al.,* 2003) and *Terrei* (Varga *et al.,* 2005). While these conclusions have held up based on additional data, this and many other studies made it clear that ribosomal genes offer limited utility. For purposes of defining species boundaries (Taylor *et al.,* 2000) and inferring historical recombination, Geiser *et al.* (2000) utilised the advanced state of *Aspergillus* molecular genetics to develop protein-coding gene based markers for the genus. This involved the identification of protein-coding genes, preferably harbouring introns, and designing primers based on sequence comparisons from two or more available *Aspergillus* species. Eleven such markers were developed for *A. flavus* and its close relatives, which showed enough variation within and between closely related taxa to reveal a cryptic species boundary and evidence for historical recombination within *A. flavus* (Geiser *et al.*, 1998b). Most, but not all markers flanked one or more introns, and others incorporated portions of the 5' untranslated regions of genes (Table 1). Unfortunately, usually being based on comparisons between no more than two species, the markers were generally not applicable beyond a narrow range of closely related taxa. Geiser *et al.* (1998a) used partial β-tubulin (*ben*A) and hydrophobin (*rod*A) gene sequences to infer phylogenetic relationships within *Aspergillus* section *Fumigati*. Based on the positions of two heterothallic species, *Neosartorya fennelliae* and *N. spathulata*, these authors furthered the position that heterothallism is derived within the genus. Later work revealing the molecular genetic basis of homothallism in Pezizomycotina (Yun *et al.*, 2000) as well as evidence suggesting *A. fumigatus* is non-clonal and heterothallic (Paoletti *et al.*, 2005; Dyer and Paoletti, 2005; Rydholm *et al.*, 2007) have rendered this inference unlikely. Instead, it is best hypothesised that any *Aspergillus* species known only from its anamorph is heterothallic. While we might think of multilocus phylogenetics as a tool for breaking up morphologically cryptic species, Feibelman *et al.* (1998) used portions of four coding genes, β-tubulin (Glass and Donaldson, 1995), calmodulin, nitrate reductase and taka-amylase (Egel *et al.*, 1994) to show that a morphologically distinct strain was identical to an isolate of *A. nomius*, and thus probably conspecific with it. Peterson *et al.* (2001), on the other hand, performed a five-locus analysis that identified the new species *A. bombycis*, employing a portion of the β-tubulin gene and two nuclear ribosomal regions, along with norsolorinic acid reductase and calmodulin. A similar set of mostly section-specific loci were used to delimit the *A. lentulus* from other taxa in section *Fumigati* (Balajee *et al.*, 2005b).

Overall, these studies firmly established a strong role for multiple protein-coding loci at the level of closely related species. Unfortunately, most of the loci in these studies have not proven applicable very far beyond that. Exceptions include the primers for β-tubulin designed by Glass and Donaldson (1995), which were previously known to be useful across the Pezizomycotina. Primers designed to amplify the 5' untranslated region of the *trp*C gene designed by Geiser *et al.* (1998b) and used to analyse *A. flavus*, *A. parasiticus* and their domesticated relatives were actually designed based on alignments of the *trp*C promoters from *A. nidulans* (Hamer and Timberlake, 1987) and *A. niger* (Kos *et al.*, 1988), yet appear to work not only in sections *Flavi* and *Nidulantes* (Geiser *et al.*, 1998c) but also broadly across the genus (D. Geiser and S.M. Witiak, unpublished results). Still, the use of a multilocus phylogenetic analysis, incorporating both ribosomal and protein-coding genes, has become a

Table 1. Protein-coding loci used for phylogenetic studies in Aspergillus.

Locus	Primer sequence	Product size (introns) Exon/intron	Breadth	Reference
Eukaryotic translation initiation factor 2 (eIF-2)	F: GAYTGYCCNGGNCAYGAYAT R: CCNGGYTTRTTNACRTCRAA	460 (1) 408/52	Genus-wide[1]	This study
Heat shock 70kDa protein, cytoplasmic, (HSP70C)	F: CAGGTCGCMATGAACCC R: TGATGGTGATCTTGTTGGTCTT	1663 (5) 1278/385	Genus-wide[1]	This study
Chaperonin complex component (TCP-1)	F: ACCATCTTRTTGCTWCCHCGHA R: AGATGGAAATYACRAACAAYGT	925 (1) 859/66	Genus-wide[1]	This study
Heat shock 70kDa protein, mitochondrial (nuclear encoded)	F: CAGGTCGCMATGAACCC R: NCCRTTNGCRTCDAYRTC	804 (0) 804/0	Genus-wide[1]	This study
Pyruvate carboxylase	F: RTCRATGAAVGTGGTCCAGCA R: GCTTCCCCATCATCATCAAGG	759 (0) 759/0	Genus-wide[1]	This study
SNF-family DNA-dependent ATPase domain- containing protein	F: GRTAYGCRGCHGCCAAGT R: GAAGTCAAASAGVGACCACA	860(0) 860/0	Genus-wide[1]	This study
Cell division control protein 42 (CDC42)	F: AARTTCCCYTCVGAATATGT R: ACCTCRTCAAARACRTCCTT	704(1) 433/271	Genus-wide[1]	This study
P-type ATPase	F: TGGYTGATGACVAAATGGAA R: GCYTCVACRTTGTTGGC	870 (1) 782/88	Genus-wide[1]	This study
Heatshock 90kDa protein (HSP90)	F: CCATCTACCTCCACGTYCTGA R: AGVAGCTGRGTGATGGACTTGAC	1198(0) 1198/0	Genus-wide[1]	This study
β-tubulin	Bena1: AGTTGTCGGACGGAAGAG Bena2: AATAGGTGCCGCTTTCTGG	561 (3) 200/361	Genus-wide[1]	Geiser *et al.*, 1996
β-tubulin	Bt1a: TTCCCCCGTCTCCACTCTTCATG Bt1b: GACGAGATCGTTCATGTTGAACTC	537 (1) 478/59	Pezizomycotina	Glass and Donaldson, 1995

Table 1. Continued.

Locus	Primer sequence	Product size (introns) Exon/intron	Breadth	Reference
β-tubulin	Bt2a: GGTAACCAAATCGGTGCTGCTTTC	495 (3)	Pezizomycotina	Glass and Donaldson, 1995
	Bt2b: ACCCTCAGTGTAGTGACCCTTGGC	287/208		
β-tubulin	Bt3-LM: GAACGTCTACTTCAACGAG	900 (2)	Fungi	Myllys *et al.*, 2002
	Bt10-LM: TCGGAAGCAGCCATCATGTTCTT			
Calmodulin	Cmd42: GGCCTTCTCCCTATTCGTAA	612	Section *Flavi*	Feibelman *et al.*, 1998
	Cmd637: CTCGCGGATCATCTCATC			
Nitrate reductase	NiaD1225: CGCTCGTCCCAACTGACCTC	2359	Section *Flavi*	Feibelman *et al.*, 1998
	NiaD3563: TTGGGGTGTATGAGCGGATGA			
Glyceraldehyde-3-phosphate dehydrogenase	Gpd1-LM: ATTGGCCGCATCGTCTTCCGCAA	1100(3)	Fungi	Myllys *et al.*, 2002
	Gpd2-LM: CCCACTCGTTGTCGTACCA	914/186		
Elongation factor 1 alpha	EF1-728F: CATCGAGAAGTTCGAGAAGG	350	Fungi	Carbone and Kohn, 1999
	EF1-986R: TACTTGAAGGAACCCTTACC	100/250		
Calmodulin	CAL-228F: GAGTTCAAGGAGGCCTTCTCCC	500	Pezizomycotina	Carbone and Kohn, 1999
	CAL-737R: CATCTTTCTGGCCATCATGG	200/350		
Calmodulin	CmdA7: GCCAAAATCTTCATCCGTAG	468 (5'NT)	Section *Flavi*	Geiser *et al.*, 1998b
	CmdA8: ATTTCGTTCAGAATGCCAGG	160/368		
Polygalacturonase	PecA1: AATCTCAGACATTTTCTCC	533 (3)	Section *Flavi*	Geiser *et al.*, 1998b
	PecA2: ACCCTCCCACTCCTTGTAG	322/211		
Acetamidase	Amds1: CCATCGGTATAGGAACTGA	550 (3)	Section *Flavi*	Geiser *et al.*, 1998b
	Amds2: AGGGTGCCACGGTATGTC	397/153		
O-methyltransferase	Omt1: GGAGTATCAGAGGATTTA	458 (3)	Section Flavi	Geiser *et al.*, 1998b
	Omt2: AGTGCTGTAATAGTCAAA	287/171		
Nitrate reductase	NiaD1: TCGTGAATGGAGAAGTGT	465 (1)	Section *Flavi*	Geiser *et al.*, 1998b
	NiaD2: GAAATTGGGGTGTATGAG	409/56		
Phosphoglycerate kinase	PecA1: ATCTCAGACAGGGGCGCC	533 (3)	Section *Flavi*	Geiser *et al.*, 1998b
	PecA2: ACCCTCCCACTCCTTGTAG	322/211		

Locus	Primer sequence	Product size (introns) Exon/intron	Breadth	Reference
Acetate regulation	FacB3: GAAAAGATCCTGTGGTTGGC	469 (5'NT)	Section *Flavi*	Geiser *et al.*, 1998b
	FacB4: CCTGGGCGATACGACTTTGG	67/402		
Glucoamylase	GlaA1: CAATCTTGAATAATATCG	499 (2)	Section *Flavi*	Geiser *et al.*, 1998b
	GlaA2: GTCCGTCGCTATGCTTGT	407/92		
Tryptophan synthesis 5' untranslated region	TrpC1: GACGGGAAATAGGCTTCC	506 (5'NT)	Genus-wide	Geiser *et al.*, 1998b,c
	TrpC3: CGCCTTGGTGGGATGGT	20/486		
Glucose-6-phosphate dehydrogenase	GsdA1: CACAATAGCACGCACTGAGG	623 (3)	Section *Flavi*	Geiser *et al.*, 1998b
	GsdA2: CTTGGGGAGGAACTTGTTGC	154/467		
Hydrophobin (*rodA*)	RodA1: GCTGGCAATGGTGTTGGCAA	461 (2)	Section *Fumigati*	Geiser *et al.*, 1998a
	RodA2: AGGGCACTGCAAGGAAAGACCC	346/116		
Putative cell surface protein	CSPf: TTGGGTGGCATTGTGCCAA	~625	*A. fumigatus*	Balajee *et al.*, 2007
	CSPr: GAGCATGACAACCCAGATACCA			
Taka-amylase A	Taa111: GGATCGATTTGCAAGGACGG		Section *Flavi*	Feibelman *et al.*, 1998; Egel *et al.*, 1994
	Taa1280: TAGAGGTCGTCCATGCTGCC			
Actin	ACT-512F: ATGTGCAAGGCCGGTTTCGC	300	Pezizomycotina	Carbone and Kohn, 1999
	ACT-783R: TACGAGTCCTTCTGGCCCAT	100/200		
Ras protein	RAS-264F: GATGAATATGATCCTACGAT	350	Pezizomycotina	Carbone and Kohn, 1999
	RAS-565R: AAATCACATTTGTTACCAAC	280/70		
RNA Polymerase II second largest subunit	See (Myllys *et al.*, 2002)		Fungi	S. Peterson, unpublished results

[1]Appears to work across diverse subgenera and sections, although not necessarily in all taxa.

recommended standard for *Aspergillus* systematics and has been applied in many *Aspergillus* sections (Hong *et al.*, 2005, 2006, 2007; Houbraken *et al.*, 2007; Perrone *et al.*, 2007; Varga *et al.*, 2007a, 2007b, 2007c; Samson *et al.*, 2007; Pildain *et al.*, 2007). Recent studies have included analyses of portions of β-tubulin, calmodulin and actin genes based on broadly designed primers (Glass and Donaldson, 1995; Carbone and Kohn, 1999), which generally provide robust phylogenetic inferences within sections. The success of the All Fungal Tree of Life (AFTOL) project (Blackwell *et al.*, 2006; James *et al.*, 2006), coupled with limited application of the genes encoding translation elongation factor 1-alpha (*tef1*) and the first and second largest subunits of RNA Polymerase II (*rpb1* and *rpb2*) in Eurotiomycetes (Geiser *et al.*, 2006), suggests that these genes are promising candidates for genus-wide phylogenetic work. However, despite its yeoman-like performance as a genus-wide marker of choice in *Fusarium* (Geiser *et al.*, 2004), *tef1* appears to have an unstable intronic structure in *Penicillium* subgenus *Eupenicillium* (Peterson *et al.*, 2004), a possible confounding factor if true in *Aspergillus* as well. Because it contains only a single intron at the 3' end of its coding sequence, *rpb2* might not be considered as a strong candidate for species-level systematics in *Aspergillus*. However, *rpb2* exons have proven to be highly phylogenetically informative in *Fusarium*, with DNA sequences alignable genus-wide yet showing variation even among closely related species (O'Donnell *et al.*, 2007). Initial genus-wide efforts to utilise *rpb2* in *Aspergillus* have been promising (S. Peterson, unpublished results)

4. Successes and shortcomings

The application of multilocus phylogenetics in *Aspergillus* has been an overwhelming success on a number of fronts. At the species level, the combination of ribosomal and protein-coding gene sequences have proven highly useful for identifying species boundaries, using principles of genealogical concordance (Taylor *et al.*, 2000). Combined analyses of multiple loci tend to yield section-level phylogenies with strongly supported internal nodes (Hong *et al.*, 2005, 2006, 2007; Houbraken *et al.*, 2007; Perrone *et al.*, 2007; Varga *et al.* 2007a,b,c; Samson *et al.* 2007; Pildain *et al.* 2007). Practically, protein-coding genes, either singly or in combination, are promising tools for correct species identification of Aspergilli from clinical sources (Balajee *et al.*, 2005a, 2007a). A particularly powerful benefit of the approach is that sequence-based loci often exhibit extensive intraspecific variation, allowing inferences of recombination (Geiser *et al.*, 1998b; Pringle *et al.*, 2005) and tracking haplotypes associated with outbreaks of infection (Balajee *et al.*, 2007b).

The establishment of a strongly supported genus-wide phylogeny, however, has proven elusive. While a four-locus (calmodulin, *rpb2*, ITS, SSU) analysis of the entire genus uncovered a variety of new insights placing species in clades and identifying synonyms, it did not produce strongly supported internal nodes in the phylogeny (Peterson *et al.,* 2008). A robust genus-wide phylogeny is highly desirable for the purposes of understanding the evolutionary basis of diversity in the genus. For example, the odd phylogenetic distribution of aflatoxin (AF) and sterigmatocystin (ST) producers in *Aspergillus,* known mostly in the divergent sections *Flavi* and *Nidulantes,* has been a mystery for some time. Population-level analyses of sequence data from the AF and ST gene clusters have revealed important insights about the evolution of this trait: the enzymatic activities required to produce AF/ST are, at least in part, encoded by modular pairs of genes, which through gene duplication and recombination have resulted in the gene clusters known to encode the biosynthetic pathway (Carbone *et al.* 2007a,b; Moore *et al.,* 2007). This raises the intriguing question of how phylogenetically distinct AF/ST clusters evolved across the genus from non AF/ST-producing ancestors, a question that is impossible to answer without knowledge of the underlying organismal phylogeny. This and many other questions regarding the basis of biological variation in the genus await a strong phylogenetic framework.

5. Genomics, and a 'many locus' approach

Early forays into multilocus analyses of *Aspergillus* were fuelled by the advanced molecular genetics in *Aspergillus* species, particularly *Emericella nidulans,* a long-standing eukaryotic genetic model system. These advances have carried over into genomics, resulting in complete genome sequences from eight species: *E. nidulans, A. flavus, A. oryzae, A. niger* (two isolates), *A. terreus, A. fumigatus, N. fischeri* and *A. clavatus* (http://www.aspergillus.org.uk/indexhome.htm?secure/sequence_info/index.php~main). These species represent a fair amount of the diversity in the genus, in addition to encompassing pairs of close relatives: two genetically distinct but conspecific isolates in *A. niger, A. flavus* and its domesticated form *A. oryzae,* and *A. fumigatus* and its close but distinct relative *N. fischeri.* These complete genomes also offer an unprecedented opportunity for the design of multiple phylogenetic markers for genus-wide phylogenetic analyses. In an analysis derived from complete genome sequences of eight yeast species, it was found that approximately twenty genes or 8,000 independent nucleotide sites were sufficient for inferring a phylogeny with consistently highly supported nodes (Rokas *et al.,* 2003). In the genus *Aspergillus,* a phylogenetic analysis of over 2,700 homologues produced a highly supported phylogeny of the eight species with completely sequenced genomes (A. Rokas, unpublished results). While the many-locus approach is not guaranteed to yield strongly supported nodes in a larger taxon set encompassing the entire genus, these results are encouraging. In an analysis of 82 *Phytophthora* species based on seven loci designed from three complete genome sequences, backbone nodes in the resulting multilocus phylogeny of 82 species were generally weak (Blair *et al.*, 2007). Terminal nodes in the phylogeny received consistently very high support across Maximum Likelihood, Parsimony and Bayesian analysis, whereas only Bayesian analysis lent strong support to early diverging branches.

Our goal is to utilise the advanced genomics in *Aspergillus* to generate a strongly supported genus-wide phylogeny. To this end, we are working to develop up to twenty loci applicable for phylogenetic analysis. Eight newly developed loci are listed in Table 1, which have been analysed along with two other previously developed loci (β-tubulin and LSU) in thirty *Aspergillus* species, with two species from *Penicillium* subgenus *Eupenicillium* used as outgroups. Sequences were concatenated across all loci, and included as many as eight missing taxa (Table 1). The resulting analyses yielded trees with consistently high Bayesian support values across the tree, but a range of parsimony bootstrap values from very weak to 100%, with strongly supported nodes tending to fall within subgenera and sections (Figure 1). The phylogenetic relationships of the eight complete genome sequences are the same as predicted by the genome-wide analyses, but different phylogenetic methods, as well as different taxon and locus sampling, disrupt that co-inference (not shown).

The lack of resolution in the *Aspergillus* tree backbone may be the due to the low numbers of genes analysed; twice as many may yield more strongly supported nodes. Furthermore, the use of outgroups from *Penicillium* subgenus *Eupenicillium* may be inappropriate. Our analyses using outgroups from *Penicillium* subgenus *Biverticillium* indicate a possible ingroup relationship between *Aspergillus* and *Penicillium* subgenera *Penicillium* and *Eupenicillium* (not shown), consistent with other inferences hinting at that (Berbee *et al.*, 1995). However, it is also possible that the twenty-locus approach to a large taxon set will not yield phylogenies of the same strength in *Aspergillus* as was observed in yeast species (Rokas *et al.*, 2003), either because more genes will be required, or because divergences in the genus occurred on a time-scale that is not compatible with the genes utilised. Methods to assess the phylogenetic informativeness of particular genes may assist in the identification of genes that are suitable to be used at the genus level (Townsend, 2007), but it may take a very large number of loci to make strong inferences across all levels in this extremely diverse genus, the origins of which likely extend back hundreds of millions of years (Galagan *et al.*, 2005).

In conclusion, the outlook remains optimistic. Even if the backbone of the *Aspergillus* phylogeny fails to gain strength, the many-locus approach yields highly supported infrasectional phylogenies, and outstanding tools for multilocus sequence typing

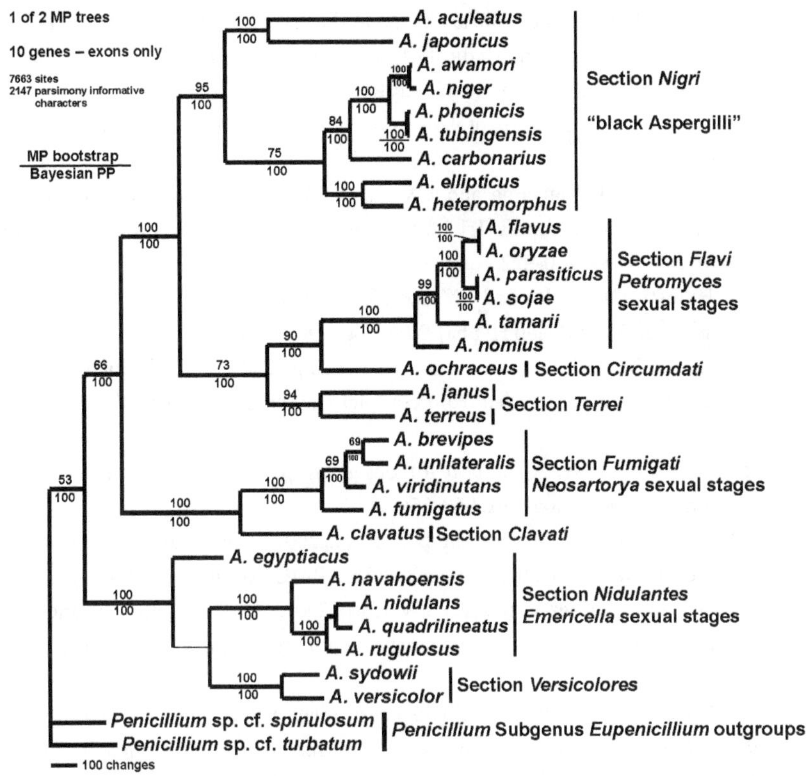

Figure 1. Phylogram representing one of two most parsimonious trees inferred based on ten gene regions: LSU, β-tubulin, and the first eight loci listed in Table 1. Parsimony bootstrap values are listed above branches, and Bayesian posterior probabilities are listed below.

and identification. As sequence data become easier and cheaper to generate, the task of compiling enough phylogenetic information to create a robust phylogenetic framework is becoming less daunting.

References

Balajee, S.A., J.L. Gribskov, E. Hanley, D. Nickle and K. A. Marr. 2005b. *Aspergillus lentulus* sp nov., a new sibling species of *A. fumigatus*. Eukaryotic Cell 4: 625-632.

Balajee, S.A., J. Gribskov, M. Brandt, J. Ito, A. Fothergill and K.A. Marr, 2005a. Mistaken identity: *Neosartorya pseudofischeri* and its anamorph masquerading as *Aspergillus fumigatus*. Journal of Clinical Microbiology 43: 5996-5999.

Balajee, S.A., L. Sigler and M.E. Brandt, 2007a. DNA and the classical way: Identification of medically important molds in the 21st century. Medical Mycology 45: 475-490.

Balajee, S.A., S.T. Tay, B.A. Lasker, S.F. Hurst and A.P. Rooney, 2007b. Characterization of a novel gene for strain typing reveals substructuring of *Aspergillus fumigatus* across North America. Eukaryotic Cell 6: 1392-1399.

Berbee, M.L., A. Yoshimura, J. Sugiyama and J.W. Taylor, 1995. Is Penicillium monophyletic? An evaluation of phylogeny in the family Trichocomaceae from 18S, 5.8S and ITS ribosomal DNA sequence data. Mycologia 87: 210-222.

Blackwell, M., D.S. Hibbett, J.W. Taylor and J.W. Spatafora, 2006. Research Coordination Networks: a phylogeny for Kingdom Fungi (Deep Hypha). Mycologia 98: 829-837.

Blair, J.E., M.D. Coffey, S.Y. Park, D.M. Geiser and S. Kang, 2007. A multi-locus phylogeny for *Phytophthora* utilizing markers derived from complete genome sequences. Fungal Genetics and Biology (in press).

Carbone, I. and L.M. Kohn, 1999. A method for designing primer sets for speciation studies in filamentous ascomycetes. Mycologia 91: 553-556.

Carbone, I., J.H. Ramirez-Prado, J.L. Jakobek and B.W. Horn, 2007b. Gene duplication, modularity and adaptation in the evolution of the aflatoxin gene cluster. BMC Evolutionary Biology 7: 111.

Carbone, I., J.L. Jakobek, J.H. Ramirez-Prado and B.W. Horn, 2007a. Recombination, balancing selection and adaptive evolution in the aflatoxin gene cluster of *Aspergillus parasiticus*. Molecular Ecology 16: 4401-4417.

Chang, J.M., H. Oyaizu and J. Sugiyama, 1991. Phylogenetic relationships among eleven selected species of *Aspergillus* and associated teleomorphic genera estimated from 18S ribosomal RNA partial sequences. Journal of General and Applied Microbiology 37: 289-308.

Dupont, J., M. Dutretre, J.F. Lafay, M.F. Roquebert and Y. Brygoo, 1990. A molecular assessment of the position of Stilbothamnium in the genus *Aspergillus*. In: Samson, R.A. and J.I. Pitt (eds.) Modern Concepts in *Penicillium* and *Aspergillus* Classification. Plenum Press, New York, pp. 335-342.

Dyer, P.S. and M. Paoletti, 2005. Reproduction in *Aspergillus fumigatus*: sexuality in a supposedly asexual species? Medical Mycology 43: S7-S14.

Egel, D.S., P.J. Cotty and K.S. Elias, 1994. Relationships among *Aspergillus* sect. *Flavi* that vary in aflatoxin production. Phytopathology 84: 906-912.

Feibelman, T.P., P.J. Cotty, M.A. Doster and T.J. Michalides, 1998. A morphologically distinct strain of *Aspergillus nomius*. Mycologia 90: 618-623.

Galagan, J.E., S.E. Calvo, C. Cuomo, L.J. Ma, J.R. Wortman, S. Batzoglou, S.I. Lee, M. Basturkmen, C.C. Spevak, J. Clutterbuck, V. Kapitonov, J. Jurka, C. Scazzocchio, M. Farman, J. Butler, S. Purcell, S. Harris, G.H. Braus, O. Draht, S. Busch, C. D'Enfert, C. Bouchier, G.H. Goldman, D. Bell-Pedersen, S. Griffiths-Jones, J.H. Doonan, J. Yu, K. Vienken, A. Pain, M. Freitag, E.U. Selker, D.B. Archer, M.A. Penalva, B.R. Oakley, M. Momany, T. Tanaka, T. Kumagai, K. Asai, M. Machida, W.C. Nierman, D.W. Denning, M. Caddick, M. Hynes, M. Paoletti, R. Fischer, B. Miller, P. Dyer, M.S. Sachs, S.A. Osmani and B.W. Birren, 2005. Sequencing of *Aspergillus nidulans* and comparative analysis with *A. fumigatus* and *A. oryzae*. Nature 438: 1105-1115.

Geiser, D.M., C. Gueidan, J. Miadlikowska, F. Lutzoni, F. Kauff, V. Hofstetter, E. Fraker, C.L. Schoch, L. Tibell, W.A. Untereiner and A. Aptroot, 2006. Eurotiomycetes: Eurotiomycetidae and Chaetothyriomycetidae. Mycologia 98: 1053-1064.

Geiser, D.M., F.M. Harbinski and J.W. Taylor, 2000. Molecular and analytical tools for characterizing *Aspergillus* and *Penicillium* species at the intra- and interspecific levels. In: Samson, R.A. and J.I. Pitt (eds) Integration of modern taxonomic methods for *Penicillium* and *Aspergillus* classification. Harwood Academic Publishers, Amsterdam, pp. 381-394.

Geiser, D.M., J.C. Frisvad and J.W. Taylor, 1998a. Evolutionary relationships in *Aspergillus* section *Fumigati* inferred from partial beta-tubulin and hydrophobin DNA sequences. Mycologia 90: 831-845.

Geiser, D.M., J.I. Pitt and J.W. Taylor, 1998b. Cryptic speciation and recombination in the aflatoxin-producing fungus *Aspergillus flavus*. Proceedings of the National Academy of Sciences USA 95: 388-393.

Geiser, D.M., J.W. Taylor, K.B. Ritchie and G.W. Smith, 1998c. Cause of sea fan death in the West Indies. Nature 394: 137-138.

Geiser, D.M., M.D. Jimenez-Gasco, S.C. Kang, I. Makalowska, N. Veeraraghavan, T.J. Ward, N. Zhang, G.A. Kuldau and K. O'Donnell, 2004. FUSARIUM-ID v. 1.0: A DNA sequence database for identifying *Fusarium*. European Journal of Plant Pathology 110: 473-479.

Geiser, D.M., W.E. Timberlake and M.L. Arnold, 1996. Loss of meiosis in *Aspergillus*. Molecular Biology and Evolution 13: 809-817.

Glass, N.L. and G.C. Donaldson, 1995. Development of primer sets designed for use with the PCR to amplify conserved genes from filamentous Ascomycetes. Applied and Environmental Microbiology 61: 1323-1330.

Hamer, J.E. and W.E. Timberlake, 1987. Functional organization of the *Aspergillus nidulans* trpC promoter. Molecular and Cellular Biology 7: 2352-2359.

Hong, S.B., H.D. Shin, J.B. Hong, J.C. Frisvad, P.V. Nielsen, J. Varga and R.A. Samson, 2007. New taxa of *Neosartorya* and *Aspergillus* in *Aspergillus* section *Fumigati*. Antonie van Leeuwenhoek (in press).

Hong, S.B., H.S. Cho, H.D. Shin, J.C. Frisvad and R.A. Samson, 2006. Novel *Neosartorya* species isolated from soil in Korea. International Journal of Systematic and Evolutionary Microbiology 56: 477-486.

Hong, S.B., S.J. Go, H.D. Shin, J.C. Frisvad and R.A. Samson, 2005. Polyphasic taxonomy of *Aspergillus fumigatus* and related species. Mycologia 97: 1316-1329.

Houbraken, J., M. Due, J. Varga, M. Meijer, J.C. Frisvad and R.A. Samson, 2007. Polyphasic taxonomy of *Aspergillus* section *Usti*. Studies in Mycology (in press).

James, T.Y., F. Kauff, C.L. Schoch, P.B. Matheny, V. Hofstetter, C.J. Cox, G. Celio, C. Gueidan, E. Fraker, J. Miadlikowska, H.T. Lumbsch, A. Rauhut, V. Reeb, A.E. Arnold, A. Amtoft, J.E. Stajich, K. Hosaka, G.H. Sung, D. Johnson, B. O'Rourke, M. Crockett, M. Binder, J.M. Curtis, C. Slot, Z. Wang, A.W. Wilson, A. Schussler, J.E. Longcore, K. O'Donnell, S. Mozley-Standridge, D. Porter, P.M. Letcher, M.J. Powell, J.W. Taylor, M.M. White, G.W. Griffith, D.R. Davies, R.A. Humber, J.B.

Morton, J. Sugiyama, A.Y. Rossman, J.D. Rogers, D.H. Pfister, D. Hewitt, K. Hansen, S. Hambleton, R.A. Shoemaker, J. Kohlmeyer, B. Volkmann-Kohlmeyer, S.A. Spotts, M. Serdani, P.W. Crous, K.W. Hughes, K. Matsuura, E. Langer, G. Langer, W.A. Untereiner, R. Lucking, B. Budel, D.M. Geiser, A. Aptroot, P. Diederich, I. Schmitt, M. Schultz, R. Yahr, D.S. Hibbett, F. Lutzoni, D.J. McLaughlin, J.W. Spatafora and R. Vilgalys, 2006. Reconstructing the early evolution of Fungi using a six-gene phylogeny. Nature 443: 818-822.

Kos, T., A. Kuijvenhoven, H.G.M. Hessing, P.H. Pouwels and C.A.M.J.J. van den Hondel, 1988. Nucleotide sequence of the *Aspergillus niger trpC* gene: structural relationship with analogous genes in other organisms. Current Genetics 13: 137-144.

Kozlowski, M. and P.P. Stepien, 1982. Restriction enzyme analysis of mitochondrial DNA of members of the genus *Aspergillus* as an aid to taxonomy. Journal of General Microbiology 128: 471-476.

Kurtzman, C.P., 1985. Classification of fungi through nucleic acid relatedness. In: Samson, R.A. and J.I. Pitt (eds.) Advances in *Penicillium* and *Aspergillus* systematics. Plenum Press, New York, pp. 233-255.

Kurtzman, C.P., M.J. Smiley, C.J. Robnett and D.T. Wicklow, 1986. DNA relatedness among wild and domesticated species in the *Aspergillus flavus* group. Mycologia 78: 955-959.

LoBuglio, K.L., J.I. Pitt and J.W. Taylor, 1993. Phylogenetic analysis of two ribosomal DNA regions indicates multiple independent losses of a sexual *Talaromyces* sexual state among asexual *Penicillium* species in subgenus *Biverticillium*. Mycologia 85: 592-604.

Malloch, D., 1981. The Plectomycete Centrum. In: Reynolds, D.R. (ed.) Ascomycete Systematics: The Luttrellian Concept. Springer-Verlag, New York, pp. 73-91.

Malloch, D., 1985. The Trichocomaceae: relationships with other ascomycetes. In: Samson, R.A. and J.I. Pitt (eds.) Advances in *Penicillium* and *Aspergillus* Systematics. Plenum Press, New York, pp. 365-382.

Malloch, D. and R.F. Cain, 1972. The Trichocomaceae: Ascomycetes with *Aspergillus, Paecilomyces* and *Penicillium* imperfect states. Canadian Journal of Botany 50: 2613-2628.

Moore, G.G., J.H. Ramirez-Prado, B.W. Horn and I. Carbone, 2007. Evidence of extensive recombination in the aflatoxin gene cluster of *Aspergillus flavus*. Phytopathology 97: S79-S79.

Myllys, L., S. Stenroos and A. Thell, 2002. New genes for phylogenetic studies of lichenized fungi: glyceraldehyde-3-phosphate dehydrogenase and beta-tubulin genes. Lichenologist 34: 237-246.

O'Donnell, K., B.A.J. Sarver, M. Brandt, D.C. Chang, J. Noble-Wang, B.J. Park, D.A. Sutton, L. Benjamin, M. Lindsley, A. Padhye, D.M. Geiser and T.J. Ward, 2007. Phylogenetic diversity and microsphere array-based genotyping of human pathogenic fusaria, including isolates from the multistate contact lens - associated US keratitis outbreaks of 2005 and 2006. Journal of Clinical Microbiology 45: 2235-2248.

Paoletti, M., C. Rydholm, E.U. Schwier, M.J. Anderson, G. Szakács, F. Lutzoni, J.P. Debeaupuis, J.P. Latge, D.W. Denning and P.S. Dyer, 2005. Evidence for sexuality in the opportunistic fungal pathogen *Aspergillus fumigatus*. Current Biology 15: 1242-1248.

Perrone, G., J. Varga, A. Susca, J.C. Frisvad, G. Stea, S. Kocsubé, B. Tóth, Z. Kozakiewicz and R.A. Samson, 2007. *Aspergillus uvarum* sp. nov., an uniseriate black *Aspergillus* species isolated from grapes in Europe. International Journal of Systematic and Evolutionary Microbiology (in press).

Peterson, S.W., 1995. Phylogenetic analysis of *Aspergillus* sections *Cremei* and *Wentii*, based on ribosomal DNA sequences. Mycological Research 99: 1349-1355.

Peterson, S.W., 2000. Phylogenetic relationships in *Aspergillus* based upon rDNA sequence analysis. In: Samson, R.A. and J.I. Pitt (eds.) Classification of *Penicillium* and *Aspergillus*: Integration of modern taxonomic methods. Harwood Publishers, Reading, UK, pp. 323-356.

Peterson, S.W., E.M. Bayer and D.T. Wicklow, 2004. *Penicillium thiersii, Penicillium angulare* and *Penicillium decaturense*, new species isolated from wood-decay fungi in North America and their phylogenetic placement from multilocus DNA sequence analysis. Mycologia 96: 1280-1293.

Peterson, S.W., J. Varga, J.C. Frisvad and R.A. Samson, 2008. Phylogeny and subgeneric taxonomy of *Aspergillus.* In: Varga, J. and R.A. Samson (eds.) *Aspergillus* in the genomic era. Wageningen Academic Publishers, Wageningen, Netherlands, pp. 33-56.

Peterson, S.W., Y. Ito, B.W. Horn and T. Goto, 2001. *Aspergillus bombycis*, a new aflatoxigenic species and genetic variation in its sibling species, *A. nomius*. Mycologia 93: 689-703.

Pildain, M.B., J.C. Frisvad, G. Vaamonde, D. Cabral, J. Varga and R.A. Samson, 2007. Two new aflatoxin producing *Aspergillus* species from Argentinean peanuts. International Journal of Systematic and Evolutionary Microbiology (in press).

Pringle, A., D.M. Baker, J.L. Platt, J.P. Wares, J.P. Latge and J.W. Taylor, 2005. Cryptic speciation in the cosmopolitan and clonal human pathogenic fungus *Aspergillus fumigatus*. Evolution 59: 1886-1899.

Raper, K.B. and D.I. Fennell, 1965. The Genus *Aspergillus*. Williams and Wilkins, Baltimore.

Rigó, K., J. Varga, B. Tóth, J. Téren, Á. Mesterházy and Z. Kozakiewicz, 2002. Evolutionary relationships within *Aspergillus* section *Flavi* based on sequences of the intergenic spacer regions and the 5.8S rRNA gene. Journal of General and Applied Microbiology 48: 9-16.

Rokas, A., B.L. Williams, N. King and S.B. Carroll, 2003. Genome-scale approaches to resolving incongruence in molecular phylogenies. Nature 425: 798-804.

Rydholm, C., P.S. Dyer and F. Lutzoni, 2007. DNA sequence characterization and molecular evolution of MAT1 and MAT2 mating-type loci of the self-compatible ascomycete mold *Neosartorya fischeri*. Eukaryotic Cell 6: 868-874.

Samson, R.A., S.B. Hong, S.W. Peterson, J.C. Frisvad and J. Varga, 2007. Polyphasic taxonomy of *Aspergillus* section *Fumigati* and its teleomorph, *Neosartorya*. Studies in Mycology (in press).

Swofford, D.L., 2003. PAUP*: Phylogenetic analysis using parsimony. version 4.0. Sinauer Associates, Sunderland, MA.

Tamura, M., K. Kawahara and J. Sugiyama, 2000. Molecular phylogeny of *Aspergillus* and associated teleomorphs in the Trichocomaceae (Eurotiales). In: Samson, R.A. and J.I. Pitt (eds.) Integration of modern taxonomic methods for *Penicillium* and *Aspergillus* classification. Harwood Academic Publishers, Amsterdam, pp. 357-372.

Taylor, J.W., D.J. Jacobson, S. Kroken, T. Kasuga, D.M. Geiser, D.S. Hibbett and M.C. Fisher, 2000. Phylogenetic species recognition and species concepts in fungi. Fungal Genetics and Biology 31: 21-32.

Thom, C. and K.B. Raper, 1945. A Manual of the Aspergilli. Williams and Wilkins, Baltimore.

Thom, C. and M.B. Church, 1926. The Aspergilli. Williams and Wilkins, Baltimore.

Townsend, J.P., 2007. Profiling phylogenetic informativeness. Systematic Biology 56: 222-231.

Varga, J., B. Tóth, K. Rigó, J. Téren, Z. Kozakiewicz, Z. and R.F. Hoekstra, 2000. Phylogenetic analysis of *Aspergillus* section *Circumdati* based on sequences of the internal transcribed spacer regions and the 5.8 S rRNA gene. Fungal Genetics and Biology 30: 71-80.

Varga, J., B. Tóth, S. Kocsubé, B. Farkas and J. Téren, 2005. Evolutionary relationships among *Aspergillus terreus* isolates and their relatives. Antonie van Leeuwenhoek 88: 141-150.

Varga, J., J.C. Frisvad and R.A. Samson, 2007b. Polyphasic taxonomy of *Aspergillus* section *Candidi* based on molecular, morphological and physiological data. Studies in Mycology (in press).

Varga, J., K. Rigó, J. Molnár, B. Tóth, S. Szencz, J. Téren, J. and Z. Kozakiewicz, 2003. Mycotoxin production and evolutionary relationships among species of *Aspergillus* section *Clavati*. Antonie van Leeuwenhoek 83: 191-200.

Varga, J., M. Due, J.C. Frisvad and R.A. Samson, 2007c. Taxonomic revision of *Aspergillus* section *Clavati* based on molecular, morphological and physiological data. Studies in Mycology (in press).

Varga, J., S. Kocsubé, B. Tóth, J.C. Frisvad, G. Perrone, A. Susca, M. Meijer and R.A. Samson, 2007a. *Aspergillus brasiliensis* sp. nov., a biseriate black *Aspergillus* species with world-wide distribution. International Journal of Systematic and Evolutionary Microbiology 57: 1925-1932.

Yun, S.H., T. Arie, I. Kaneko, O.C. Yoder and B.G. Turgeon, 2000. Molecular organization of mating type loci in heterothallic, homothallic and asexual *Gibberella/Fusarium* species. Fungal Genetics and Biology 31: 7-20.

Phylogeny and subgeneric taxonomy of *Aspergillus*

Stephen W. Peterson[1], *János Varga*[2,3], *Jens C. Frisvad*[4] *and Robert A. Samson*[2]
[1]*Microbial Genomics and Bioprocessing Research Unit, National Center for Agricultural Utilization Research, 1815 N. University Street, Peoria, IL 61604, USA*
[2]*CBS Fungal Biodiversity Centre, Uppsalalaan 8, NL-3584 CT Utrecht, The Netherlands*
[3]*University of Szeged, Faculty of Science and Informatics, Dept. of Microbiology, Szeged, H-6701 Hungary*
[4]*BioCentrum-DTU, Building 221, Technical University of Denmark, DK-2800 Kgs. Lyngby, Denmark*

Abstract

The phylogeny of the genus *Aspergillus* and its teleomorphs is discussed based on multilocus sequence data. DNA sequence analysis was used to formulate a nucleotide sequence framework of the genus and to analyse character changes in relationship to the phylogeny hypothesised from the DNA sequence analysis. The sequence data used to delineate subgeneric taxa included partial calmodulin, rDNA and RNA polymerase gene sequences. In our phylogenic structure of *Aspergillus* extrolite data of the various *Aspergillus* taxa collected from ex-type cultures and numerous other isolates are also discussed. A new subgeneric classification is proposed which includes 8 subgenera and 22 sections within the genus *Aspergillus*. Characteristics of these taxa are briefly discussed in this chapter.

Keywords: *Aspergillus*, phylogeny, classification

1. Introduction

Teleomorphs of *Aspergillus* species belong to different genera all of which are members of family Trichocomaceae of the order Eurotiales (Alexopoulos *et al.*, 1996), which has recently been assigned to the class Eurotiomycetes (Geiser *et al.*, 2006). This order is monophyletic as proved by several molecular studies (Geiser and LoBuglio, 2001). The subgeneric taxonomy of *Aspergillus* was for many years informal, relying on the groups concept of Thom and Church (1926), Thom and Raper (1945) and Raper and Fennell (1965). Raper and Fennell (1965) described 18 species groups within the *Aspergillus* (form) genus based mainly on cultural and morphological features. These groups were treated as sections belonging to six subgenera by Gams *et al.*

(1985). Phylogenetic studies of ribosomal RNA gene sequences led to the suggestion of three subgenera with a total of 15 sections and the so-called '*Warcupiella* group', a treatment currently accepted by most *Aspergillus* researchers (Peterson, 2000).

An important addition to taxonomy was initiated by a publication on phylogenetic systematics (Hennig, 1950, 1966) in which monophyly is required for all taxa, and outgroup analysis is used to determine the polarity of character change. Hennig's system was based on relations, not shared properties (Mayr, 1974; Rieppel, 2007), so traditional characters used in *Aspergillus* taxonomy were less suited for this type of analysis. More recently DNA sequence analysis has become possible in fungi, and sequence difference based characters are well suited for cladistic analysis. We have used DNA sequence analysis to formulate a nucleotide sequence framework of the genus *Aspergillus* and analyse character changes in relationship to the phylogeny hypothesised from the DNA sequence analysis. The extrolite data of the various *Aspergillus* taxa were collected over many years from the studies of the ex-type cultures and numerous isolates and analysed by the methods described by Frisvad *et al.* (2004a,b).

2. Taxonomic outline of the genus *Aspergillus* based on phylogenetic analysis of multilocus sequence data

A sample of species was chosen to represent the diversity of *Aspergillus* and to represent the 'groups' or sections and subgenera of prior authors. Those species are listed in Table 1. The cultures were cultivated and DNA was isolated using the technique of Serra and Peterson (2007). Parts of the calmodulin gene (CF), the internal transcribed spacer region (ITS) and part of the 28 S rDNA (ID), and part of the RNA polymerase beta (RPB2) gene was amplified using the primers and protocols of Peterson (2008). DNA sequences were obtained using Applied Biosystems (Foster City, CA) fluorescent dye terminators and ABI sequencers (ABI 3100, ABI 3730). DNA sequencing was bidirectional in all cases. Edited sequences were aligned using CLUSTALW (Thompson *et al.*, 1994) and then examined using a text editor. At the calmodulin locus, all intronic sequences were removed from the alignment, at the ID locus the ITS1 and ITS2 sequences were removed leaving the 5.8S and 28S partial sequences, while at the RPB2 locus all nucleotides were used because there are no introns in this fragment of the gene. Parsimony trees were generated using PAUP* (Swofford, 2002). Initially each locus was evaluated individually using PAUP* in maximum parsimony search. Random addition order (500 replications) was used with NNI branch swapping. The resulting tree(s) were used as starting point for heuristic search with addition order 'asis' and TBR branch swapping. A combined dataset was analysed using weighted parsimony. Weights for each locus

Table 1. Species assignments to sections. Parenthetical names of sections are previous assignments taken from the literature, sect. Versicolores *is abbreviated (Vers).*

Genus	Species	Authority	Section
Section Aspergillus			
Eurotium	*amstelodami*	Mangin	Aspergillus
Eurotium	*carnoyi*	Malloch & Cain	Aspergillus
Eurotium	*chevalieri*	Mangin	Aspergillus
Eurotium	*cristatum*	(Raper & Fennell) Malloch & Cain	Aspergillus
Eurotium	*echinulatum*	Delacr.	Aspergillus
Eurotium	*herbariorum*	Link	Aspergillus
Eurotium	*intermedium*	Blaser	Aspergillus
Eurotium	*leucocarpum*	Hadlok & Stolk	Aspergillus
Eurotium	*niveoglaucum*	(Thom & Raper) Malloch & Cain	Aspergillus
Aspergillus	*proliferans*	G. Sm.	Aspergillus
Eurotium	*repens*	de Bary	Aspergillus
Eurotium	*rubrum*	Konig Spieckermann & Bremer	Aspergillus
Eurotium	*tonophilum*	Ohtsuki	Aspergillus
Eurotium	*umbrosum*	(Bain. & Sart.) Malloch & Cain	Aspergillus
Eurotium	*xerophilum*	Samson & Mouch.	Aspergillus
Section Bispori			
Aspergillus	*bisporus*	Kwon-Chung & Fennell	Bispori (Cervini)
Section Candidi			
Aspergillus	*campestris*	M. Chr.	Candidi
Aspergillus	*candidus*	Link	Candidi
Section Cervini			
Aspergillus	*cervinus*	Massee	Cervini
Aspergillus	*kanagawaensis*	Nehira	Cervini
Aspergillus	*nutans*	McLennan & Ducker	Cervini
Aspergillus	*parvulus*	G. Sm.	Cervini
Section Circumdati			
Aspergillus	*auricomus*	(Guégen) Saito	Circumdati
Aspergillus	*bridgeri*	M. Chr.	Circumdati
Aspergillus	*cretensis*	Frisvad & Samson	Circumdati
Aspergillus	*elegans*	Gasperini	Circumdati
Aspergillus	*flocculosus*	Frisvad & Samson	Circumdati
Aspergillus	*fresenii*	Subram.	Circumdati
Aspergillus	*insulicola*	Montem. & A.R. Santiago	Circumdati
Aspergillus	*melleus*	Yukawa	Circumdati

Table 1. Continued.

Genus	Species	Authority	Section
Section Circumdati (Continued)			
Aspergillus	*muricatus*	Udagawa, Uchiy. & Kamiya	Circumdati
Aspergillus	*neobridgeri*	Frisvad & Samson	Circumdati
Aspergillus	*ochraceopetaliformis*	Bat. & Maia	Circumdati
Aspergillus	*ochraceus*	G. Wilh.	Circumdati
Aspergillus	*ostianus*	Wehmer	Circumdati
Aspergillus	*petrakii*	Vörös	Circumdati
Aspergillus	*persii*	A.M. Corte & Zotti	Circumdati
Aspergillus	*pseudoelegans*	Frisvad & Samson	Circumdati
Aspergillus	*robustus*	M. Chr. & Raper	Circumdati
Aspergillus	*roseoglobosus*	Frisvad & Samson	Circumdati
Aspergillus	*sclerotiorum*	G.A. Huber	Circumdati
Aspergillus	*steynii*	Frisvad & Samson	Circumdati
Aspergillus	*sulphureus*	(Fresen.) Thom & Church	Circumdati
Aspergillus	*westerdijkiae*	Frisvad & Samson	Circumdati
Section Clavati			
Neocarpenteles	*acanthosporus*	(Udagawa & Takada)	Clavati
Aspergillus	*clavatonanicus*	Bat., H. Maia & Alecrim	Clavati
Aspergillus	*clavatus*	Desm.	Clavati
Aspergillus	*giganteus*	Wehmer	Clavati
Aspergillus	*longivesica*	L. H. Huang & Raper	Clavati
Aspergillus	*rhizopodus*	J.N. Rai, Wadhwani & S.C. Agarwal	Clavati
Dichotomomyces	*cejpii*	(Milko) D.B. Scott	Clavati
Section Cremei			
Aspergillus	*brunneouniseriatus*	Suj. Singh & B. K. Bakshi	Cremei (Ornati)
Aspergillus	*dimorphicus*	B. S. Mehrotra & R. Prasad	Cremei (Circumdati)
Aspergillus	*flaschentraegeri*	Stolk	Cremei
Aspergillus	*gorakhpurensis*	Kamal & Bhargava	Cremei
Aspergillus	*itaconicus*	Kinoshita	Cremei
Aspergillus	*pulvinus*	Kwon-Chung & Fennell	Cremei
Aspergillus	*stromatoides*	Raper & Fennell	Cremei
Aspergillus	*wentii*	Wehmer	Cremei
Chaetosartorya	*cremea*	(Kwon-Chung & Fennell) Subram.	Cremei
Chaetosartorya	*chrysella*	(Kwon-Chung & Fennell) Subram.	Cremei
Chaetosartorya	*stromatoides*	B. J. Wiley & E. G. Simmons	Cremei

Table 1. Continued.

Genus	Species	Authority	Section
Section Flavi			
Aspergillus	*avenaceus*	G. Sm.	Flavi
Aspergillus	*bombycis*	S. W. Peterson et al	Flavi
Aspergillus	*caelatus*	B. W. Horn	Flavi
Aspergillus	*coremiiformis*	Bartoli & Maggi	Flavi
Aspergillus	*flavus*	Link	Flavi
Aspergillus	*lanosus*	Kamal & Bhargava	Flavi (Circumdati)
Aspergillus	*leporis*	States & M. Chr.	Flavi
Aspergillus	*nomius*	Kurtzman B.W. Horn & Hesselt.	Flavi
Aspergillus	*parasiticus*	Speare	Flavi
Aspergillus	*pseudotamarii*	Y. Ito et al	Flavi
Aspergillus	*subolivaceus*	Raper & Fennell	Flavi
Aspergillus	*tamarii*	Kita	Flavi
Petromyces	*albertensis*	J. P. Tewari	Flavi
Petromyces	*alliaceus*	Malloch & Cain	Flavi
Section Flavipedes			
Aspergillus	*aureofulgens*	Luppi Mosca	Flavipedes
Aspergillus	*flavipes*	(Bain. & Sart.) Thom & Church	Flavipedes
Aspergillus	*janus*	Raper & Thom	Flavipedes
Aspergillus	*brevijanus*	(Raper & Thom) Peterson	Flavipedes
Fennellia	*flavipes*	B. J. Wiley & E. G. Simmons	Flavipedes
Fennellia	*nivea*	(B. J. Wiley & E. G. Simmons) Samson	Flavipedes
Section Fumigati			
Aspergillus	*brevipes*	G. Sm.	Fumigati
Aspergillus	*duricaulis*	Raper & Fennell	Fumigati
Aspergillus	*fumigatus*	Fresen.	Fumigati
Aspergillus	*fumigatus* v. *ellipticus*	Raper & Fennell	Fumigati
Aspergillus	*lentulus*	Balajee & K. A. Marr	Fumigati
Aspergillus	*unilateralis*	Thrower	Fumigati
Aspergillus	*viridinutans*	Ducker & Thrower	Fumigati
Neosartorya	*aurata*	(Warcup) Malloch & Cain	Fumigati
Neosartorya	*aureola*	(Fennell & Raper) Malloch & Cain	Fumigati
Neosartorya	*fennelliae*	Kwon-Chung & S.J. Kim	Fumigati
Neosartorya	*fischeri*	(Wehmer) Malloch & Cain	Fumigati
Neosartorya	*glabra*	(Fennell & Raper) Kozak.	Fumigati
Neosartorya	*laciniosa*	S.B. Hong, Frisvad & Samson	Fumigati

Table 1. Continued.

Genus	Species	Authority	Section
Section Fumigati (Continued)			
Neosartorya	*pseudofischeri*	S. W. Peterson	Fumigati
Neosartorya	*quadricincta*	(J.L. Yuill) Malloch & Cain	Fumigati
Neosartorya	*spinosa*	(Raper & Fennell) Kozak.	Fumigati
Neosartorya	*stramenia*	(Novak & Raper) Malloch & Cain	Fumigati
Neosartorya	*spathulata*	Takada & Udagawa	Fumigati
Neosartorya	*tatenoi*	Y. Horie, Miyaji, Koji Yokoy., Udagawa & Camp.-Takagi	Fumigati
Section Nidulantes			
Aspergillus	*aeneus*	Sappa	Nidulantes
Aspergillus	*asperescens*	Stolk	Nidulantes (Vers)
Aspergillus	*aureolatus*	Muntanjola-Cvetkovic & Bata	Nidulantes
Aspergillus	*caespitosus*	Raper & Thom	Nidulantes
Aspergillus	*crustosus*	Raper & Fennell	Nidulantes
Aspergillus	*eburneocremeus*	Sappa	Nidulantes
Aspergillus	*multicolor*	Sappa	Nidulantes
Aspergillus	*protuberus*	Muntanjola-Cvetkovic	Nidulantes (Vers)
Aspergillus	*recurvatus*	Raper & Fennell	Nidulantes
Aspergillus	*speluneus*	Raper & Fennell	Nidulantes (Vers)
Aspergillus	*sydowii*	(Bain. & Sart.) Thom & Church	Nidulantes (Vers)
Aspergillus	*varians*	Wehmer	Nidulantes (vers)
Aspergillus	*versicolor*	(Vuill.) Tiraboschi	Nidulantes (Vers)
Emericella	*astellata*	(Fennell & Raper) Y. Horie	Nidulantes
Emericella	*aurantiobrunnea*	(G. A. Atkins et al) Malloch & Cain	Nidulantes
Emericella	*bicolor*	M. Chr. & States	Nidulantes
Emericella	*desertorum*	Samson & Mouch.	Nidulantes
Emericella	*fruticulosa*	(Raper & Fennell) Malloch & Cain	Nidulantes
Emericella	*navahoensis*	M. Chr. & States	Nidulantes
Emericella	*nidulans*	(Eidam.) G. Winter	Nidulantes
Emericella	*purpurea*	Samson & Mouch.	Nidulantes
Emericella	*quadrilineata*	(Thom & Raper) C. R. Benjamin	Nidulantes
Emericella	*rugulosa*	(Thom & Raper) C. R. Benjamin	Nidulantes
Emericella	*striata*	(Rai, Tewari & Mukerji) Malloch & Cain	Nidulantes
Emericella	*unguis*	Malloch & Cain	Nidulantes
Emericella	*variecolor*	Berk. & Br.	Nidulantes
Emericella	*violacea*	(Fennell & Raper) Malloch & Cain	Nidulantes

Table 1. Continued.

Genus	Species	Authority	SECTION
Section Nigri			
Aspergillus	*aculeatus*	Iizuka	NIGRI
Aspergillus	*brasiliensis*	Varga et al	NIGRI
Aspergillus	*carbonarius*	(Bain.) Thom	NIGRI
Aspergillus	*ellipticus*	Raper & Fennell	NIGRI
Aspergillus	*heteromorphus*	Batista & Maia	NIGRI
Aspergillus	*ibericus*	R. Serra, J. Cabañes & G. Perrone	NIGRI
Aspergillus	*japonicus*	Saito	NIGRI
Aspergillus	*niger*	Van Tieghem	NIGRI
Aspergillus	*tubingensis*	Mosseray	NIGRI
Section Ochraceorosei			
Aspergillus	*ochraceoroseus*	Bartoli & Maggi	OCHRACEOROSEI (CIRC)
Section Ornati			
Sclerocleista	*ornata*	(Raper, Fennell & Tresner) Subram.	ORNATI
Sclerocleista	*thaxteri*	Subram.	ORNATI
Penicillium			
Aspergillus	*crystallinus*	Kwon-Chung & Fennell	PEN (VERSICOLORES)
Aspergillus	*malodoratus*	Kwon-Chung & Fennel	PEN (VERSICOLORES)
Hemicarpenteles	*paradoxus*	Sarbhoy & Elphick	PEN (ORNATI)
Section Raperi			
Aspergillus	*ivoriensis*	Bartoli & Maggi	RAPERI
Aspergillus	*raperi*	Stolk	RAPERI (ORNATI)
Section Restricti			
Aspergillus	*caesiellus*	Saito	RESTRICTI
Aspergillus	*conicus*	Blochwitz	RESTRICTI
Aspergillus	*gracilis*	Bain.	RESTRICTI
Aspergillus	*penicillioides*	Speg.	RESTRICTI
Aspergillus	*restrictus*	G. Sm.	RESTRICTI
Aspergillus	*vitricola*	Ohtsuki	RESTRICTI
Eurotium	*halophilicum*	C. M. Chr. et al	RESTRICTI (ASPERGILLUS)
Section Silvati			
Aspergillus	*silvaticus*	Fennell & Raper	SILVATI (VERSICOLOR)
Section Sparsi			
Aspergillus	*anthodesmis*	Bartoli & Maggi	SPARSI (WENTII)
Aspergillus	*biplanus*	Raper & Fennell	SPARSI
Aspergillus	*conjunctus*	Kwon-Chung & Fennell	SPARSI (USTI)

Table 1. Continued.

Genus	Species	Authority	Section
Section Sparsi (Continued)			
Aspergillus	*diversus*	Raper & Fennell	Sparsi
Aspergillus	*funiculosus*	G. Sm.	Sparsi
Aspergillus	*panamensis*	Raper & Thom	Sparsi (Usti)
Aspergillus	*sparsus*	Raper & Thom	Sparsi
Section Terrei			
Aspergillus	*allahabadii*	Mehrotra & Agnihotri	Terrei
Aspergillus	*ambiguus*	Sappa	Terrei (Versicolores)
Aspergillus	*carneus*	Blochwitz	Terrei
Aspergillus	*microcysticus*	Sappa	Terrei (Versicolores)
Aspergillus	*niveus*	Blochwitz	Terrei
Aspergillus	*niveus* var. *indicus*	Lal & A.K. Sarbhoy	Terrei
Aspergillus	*terreus*	Thom	Terrei
Aspergillus	*terreus* var. *africanus*	Fennell & Raper	Terrei
Aspergillus	*terreus* var. *aureus*	Thom & Raper	Terrei
Section Usti			
Aspergillus	*amylovorus*	Panasenko ex Samson	Usti
Aspergillus	*deflectus*	Fennell & Raper	Usti
Aspergillus	*egyptiacus*	Moubasher & Moustafa	Usti
Aspergillus	*elongatus*	Rai & Agarwal	Usti (Versicolores)
Aspergillus	*granulosus*	Raper & Thom	Usti
Aspergillus	*kassunensis*	Baghdadi	Usti (Nidulantes)
Aspergillus	*lucknowensis*	J. N. Rai et al	Usti (Versicolores)
Aspergillus	*pseudodeflectus*	Samson & Mouchacca	Usti (Versicolores)
Aspergillus	*puniceus*	Kwon-Chung & Fennell	Usti
Aspergillus	*subsessilis*	Raper & Fennell	Usti (Nidulantes)
Aspergillus	*ustus*	(Bain.) Thom & Church	Usti
Emericella	*heterothallica*	(Kwon et al) Malloch & Cain	Usti (Nidulantes)
Section Warcupi			
Warcupiella	*spinulosa*	(Warcup) Subram.	Warcupi (Ornati)
Section Zonati			
Aspergillus	*zonatus*	Kwon & Fennell	Zonati (Flavi)
Aspergillus	*clavatoflavus*	Raper & Fennell	Zonati (Flavi)
Penicilliopsis	*clavariiforme*	Solms	Zonati

were calculated as the number of phylogenetically informative sites for the locus with the fewest informative sites divided by the number of informative sites at each other locus. The weighted parsimony trees were generated as for single locus data sets and bootstrap values were determined using PAUP* with 'asis' addition order and TBR branch swapping for 1000 replications. Bayesian analysis was performed using MrBayes 3.1.2 (Huelsenbeck and Ronquist, 2001). The 5.8S rDNA and 28S rDNA were designated separate character sets, and the protein coding regions of CF and RPB2 were divided into 1st, 2nd and 3rd position character sets. The model for analysis was GTR with gamma and invariants. Each character set was independent in the analysis. MCMC analysis was performed for up to 10^7 generations until the chains had converged, as assessed by the log likelihood values graphically presented in MrBayes using the sump command. One thousand initial trees were removed from analysis as the 'burnin' sample. Trees were prepared for publication using TREEVIEW (Page, 1996) and CorelDrawXIII.

Based on parsimony analysis of the CF, RPB2 and ID sequence data sets, sixteen monophyletic groups were identified among the *Aspergillus* species that correspond largely with the sections of *Aspergillus* (Figure 1, Gams *et al.*, 1985; Peterson, 2000). Probability statistics strongly support most of the sections, but the relationships of the sections that might be viewed as subgenera are not strongly supported in all cases.

3. Subgenus *Nidulantes*

Sections *Nidulantes, Sparsi, Ochraceorosei* and *Usti* form a monophyletic group along with *A. raperi* and *A. ivoriensis, Aspergillus silvaticus* and *Aspergillus bisporus.* Section *Nidulantes* contains species with *Emericella* teleomorphs plus anamorphic species that have previously been assigned to section *Nidulantes* or section *Versicolores.* Several species in this section produce sterigmatocystins, while *E. astellata* and *E. venezuelensis* also produce aflatoxin B_1 (Frisvad and Samson, 2004; Frisvad *et al.,* 2004a).

Section *Usti* (Figure 2, Table 1) includes *A. granulosus, A. pseudodeflectus, A. lucknowensis* and *A. elongatus* from section *Versicolores,* and *A. puniceus, A. ustus* and *A. deflectus* that were originally assigned to section *Usti* (Raper and Fennell, 1965). This section also includes *E. heterothallica* the only *Emericella* species outside of the monophyletic group section *Nidulantes.* Both *E. heterothallica* and *A. ustus* produce the metabolite NF8054X, a 18, 22-cyclosterol derivative (Mizuno *et al.,* 1995), and several species produce sterigmatocystin, versicolorins and asperthecin (Rabie *et al.,* 1977; Frisvad, 1985). *A. panamensis* and *A. conjunctus,* two other species placed in the *A. ustus* group by Raper and Fennell (1965), were found to belong to section *Sparsi* based on multilocus sequence data, and are excluded from section *Usti.*

Figure 1. Maximum parsimony tree of representative isolates of the Aspergillus *sections. Numbers at node are the bootstrap statistic followed by the Bayesian posterior probability.*

Basal in the tree to sections *Nidulantes* and *Usti* are several species that do not form a monophyletic group as a whole. *A. raperi* and *A. ivoriensis* form a distinct monophyletic group with strong support in both the bootstrap and Bayesian statistics (Figure 3). In order to maintain monophyly in the subgeneric taxonomy, these species are being designated section *Raperi. Aspergillus silvaticus* and *A. bisporus* is each strongly supported as a separate monophyletic group. *A. silvaticus* has been found to produce silvaticamide, a toxic compound (Yukio and Mikio, 1985), and arugosin E (Kawahara *et al.,* 1988) which were found to be structurally similar to arugosins produced by *E. rugulosa,* shamixanthons produced by *E. variecolor,* cycloisoemericellin produced by *E. striata* and emericellin produced by *E. nidulans* (Kawahara *et al.,* 1988), and dithiosilvatin and silvathione, dioxopiperazine derivatives structurally related to emethallicin produced by *E. heterothallica* (Kawahara *et al.,* 1987, 1989). Extrolites

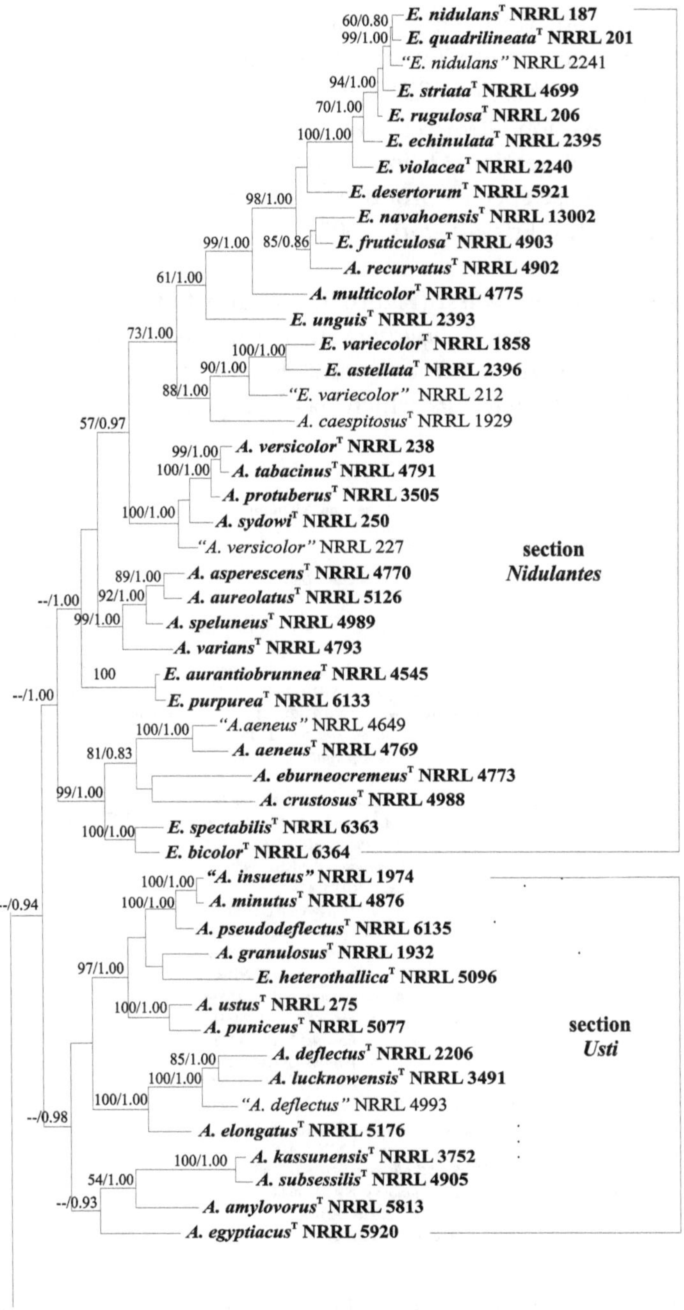

Figure 2. Maximum parsimony tree of Aspergillus *sections* Nidulantes *and* Usti. *Numbers at node are the bootstrap statistic followed by the Bayesian posterior probability.*

of *A. bisporus* have not yet been examined, this species was found to be able to cause nephritis in mice (Kwon-Chung and Fennell, 1971). These two species will be designated monospecific sections *Silvati* and *Bispori* respectively. *A. ochraceoroseus*, the type species of section *Ochraceorosei* is also in this clade (Frisvad *et al.*, 2005). The other member of section *Ochraceorosei*, *A. rambellii*, was not included in this study. Both of these species produce aflatoxin B1, sterigmatocystin and 3-O-methylsterigmatocystin (Frisvad *et al.*, 2005).

The monophyletic section *Sparsi* (Figure 3) contains four species assigned to this section previously: *A. sparsus, A. biplanus, A. diversus* and *A. funiculosus*. In addition it includes two species, *A. panamensis* and *A. conjunctus* previously assigned to section *Usti*, and *A. anthodesmis* which has been placed in the *A. wentii* group

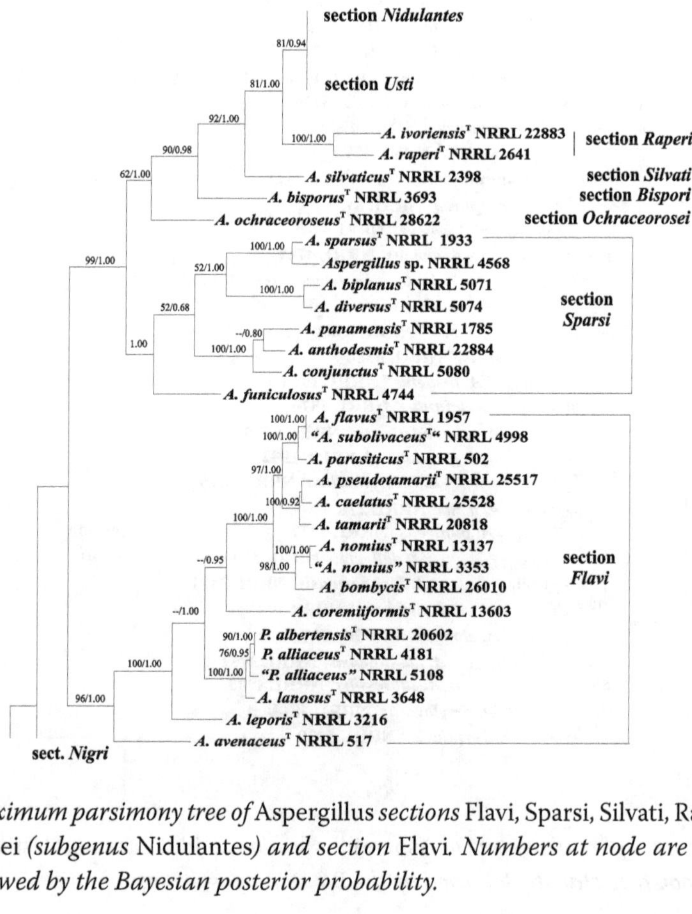

Figure 3. Maximum parsimony tree of Aspergillus *sections* Flavi, Sparsi, Silvati, Raperi, Bispori, Ochraceorosei *(subgenus* Nidulantes*) and section* Flavi. *Numbers at node are the bootstrap statistic followed by the Bayesian posterior probability.*

previously (Raper and Fennell, 1965). *A. panamensis* produces cyclogregatin and tetronic acid derivatives (Anke *et al.*, 1980, 1988), while *A. funiculosus* has been found to produce ethericins A and B (König *et al.*, 1978, 1980) and the antimicrobial substance funicin (Nakamura *et al.*, 1983).

4. Subgenus *Circumdati*

Section *Flavi* (Figure 3) includes *Aspergillus flavus*, a species producing the carcinogenic mycotoxin aflatoxin. In order for sect *Flavi* to be monophyletic, *A. clavatoflavus* and *A. zonatus* (placed in the group by Raper and Fennell, 1965) have been excluded from this section. These two species are distant from other *Aspergillus* species (Figure 7). High resolution studies (Peterson, 2008; Frisvad *et al.*, 2005) have established the species in the section and answered important questions of synonymy. Several species in this section produce aflatoxins, aspergillic acid, cyclopiazonic acid and kojic acid (Frisvad and Samson, 2000; Pildain *et al.*, 2007).

In some analyses section *Nigri* is a sister group to section F*lavi*, but in this phylogenetic tree it is ancestral to section *Flavi* (Figure 4). Section *Nigri* includes 4 species, *A. carbonarius, A. niger, A. lacticoffeatus* and *A. sclerotioniger* that produce ochratoxin A, and are responsible for ochratoxin contamination of agricultural products including grapes, spices and coffee (Samson *et al.*, 2004). The uniseriate and biseriate species from section *Nigri* form monophyletic sister group in this section. Peterson (2008) found eleven monophyletic lineages in this section, which are shown in Figure 3. Most species in section *Nigri* produce citric acid, but the species also produce different combinations of pyranonigrins, asperazine, naphtho-γ-pyrones, malformins, secalonic acids, antafumicins, and kotanins (Samson *et al.*, 2004), most of which have not been found outside section *Nigri*.

Section *Circumdati* (Figure 4) has recently been studied intensively and revised taxonomically (Frisvad *et al.*, 2004b). There are 22 lineages in the section that are shown in the figure. The placement of *A. robustus* in relation to other *Aspergillus* species was questioned by Peterson (2000) who could not place it in any particular group. In Figure 3, *A. robustus* is the most outlying of the species in the monophyletic group that corresponds to section *Circumdati*, and there is strong statistical support for its placement in the section. Eight species consistently produce large amounts of ochratoxin A: *A. cretensis, A. flocculosus, A. pseudoelegans, A. roseoglobulosus, A. westerdijkiae, A. sulphureus,* and *Neopetromyces muricatus*. Two species produce large or small amounts of ochratoxin A, but less consistently: *A. ochraceus* and *A. sclerotiorum*. In view of ochratoxin contamination of various agricultural products, the most important species are *A. ochraceus, A. westerdijkiae* and *A. steynii*. Nearly

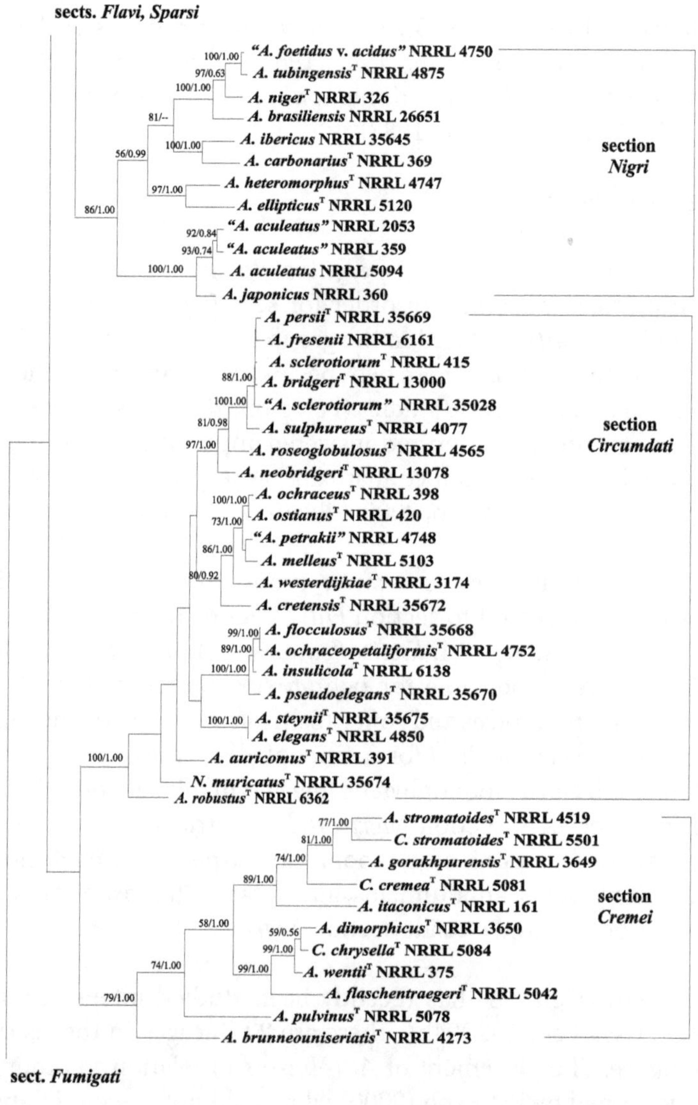

Figure 4. Maximum parsimony tree of Aspergillus *sections* Cremei, Circumdati *and* Nigri. *Numbers at node are the bootstrap statistic followed by the Bayesian posterior probability.*

all species produce penicillic acids, xanthomegnins, melleins and ochratoxins, except *A. robustus* (Frisvad and Samson, 2000; Frisvad *et al.*, 2004b).

Even though species in each section in subgenus *Circumdati* produce section specific metabolites, certain metabolites are present in all sections, including aflavinines, ochratoxins or in two of the sections: aspergillic acids have been found in section *Circumdati* and *Flavi* (Frisvad and Samson, 2000).

Section *Cremei* (Figure 4) as reported by Peterson (1995) is a monophyletic group containing eleven lineages. It is a sister group to section *Circumdati*. Species in section *Cremei* produce secondary metabolites such as wentilacton (Dorner *et al.*, 1980), emodin (Wells *et al.*, 1975) and citraconic anhydride plus bianthrones (Assante *et al.*, 1980) and are chemically and physiologically somewhat different from the other sections in subgenus *Circumdati*.

5. Subgenus *Fumigati*

Sections *Fumigati, Cervini* and *Clavati* form a monophyletic group together with *Dichotomomyces cejpii* (Figure 5). These sections have previously been assigned to subgenera *Fumigati* and *Clavati* (Gams *et al.*, 1985). Phylogenetically, the position of section *Cervini* as outgroup to sections *Fumigati* and *Clavati* is strongly supported by Bayesian analysis, with insignificant bootstrap support. These three sections should be combined into a single subgenus, subgenus *Fumigati*. Section *Clavati* was found to be closely related to *Dichotomomyces cejpii*. Species in *Neosartorya* and *Neocarpenteles* have anamorphs with green conidia and share the production of tryptoquivalins, while *Dichotomomyces* was found to be able to produce gliotoxin (Seigle-Murandi *et al.*, 1990), which is also produced by *A. fumigatus* and some *Neosartorya* species. Tryptoquivalins, tryptoquivalons, pseurotins are produced by members of both sections *Clavati* and *Fumigati* (Hong *et al.*, 2005), but each section also contain producers of metabolites unique to that section. In section *Cervini, A. cervinus* was found to be able to produce the quinol derivative terremutin and 3,6-dihydroxy-2,5-toluquinone (Elsohly *et al.*, 1974), and 4R*,5S*-dihydroxy-3-methoxy-5-methylcyclohex-2-enone and 6-methoxy-5-dihydropenicillic acid (He *et al.*, 2004), while *A. parvulus* produces parvulenone (Chao *et al.*, 1979), naphthalenone (Bartman and Campbell, 1979) and asparvenone derivatives (Bos *et al.*, 1997).

6. Subgenus *Terrei*

Sections *Terrei* and *Flavipedes* are sister groups with 77% bootstrap and 100% posterior probability support for their relative positions (Figure 6). However, the assignment of the different species to either of these sections is revised compared to previous assignments (Raper and Fennell, 1965; Kozakiewicz, 1989). Section *Terrei* includes *A. terreus* and its varieties as well as *A. niveus, A. carneus, A. niveus* var.

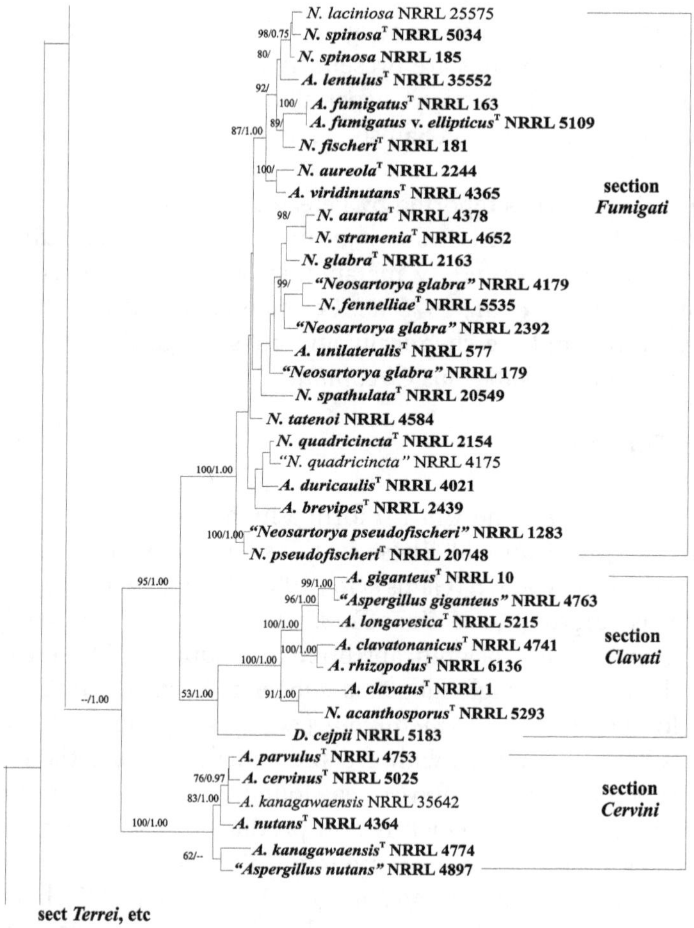

Figure 5. Maximum parsimony tree of Aspergillus *subgenus* Fumigati *including sections* Fumigati, Clavati *and* Cervini*. Numbers at node are the bootstrap statistic followed by the Bayesian posterior probability.*

indicus, A. allahabadii, A. ambiguus and *A. microcysticus.* The first three species have previously been placed in section *Flavipedes* and the last three species were placed in section *Versicolores* (Raper and Fennell, 1965; Samson, 1979). Section *Flavipedes* includes species with *Fennellia* teleomorphs, several described and undescribed anamorphic species from section *Flavipedes* and *A. janus* and *A. brevijanus,* two species from section *Versicolores.* The close and well supported relationship of these sections suggests that they be placed in a subgenus apart from other species in *Aspergillus.* There is no monophyletic relationship of these two sections with the

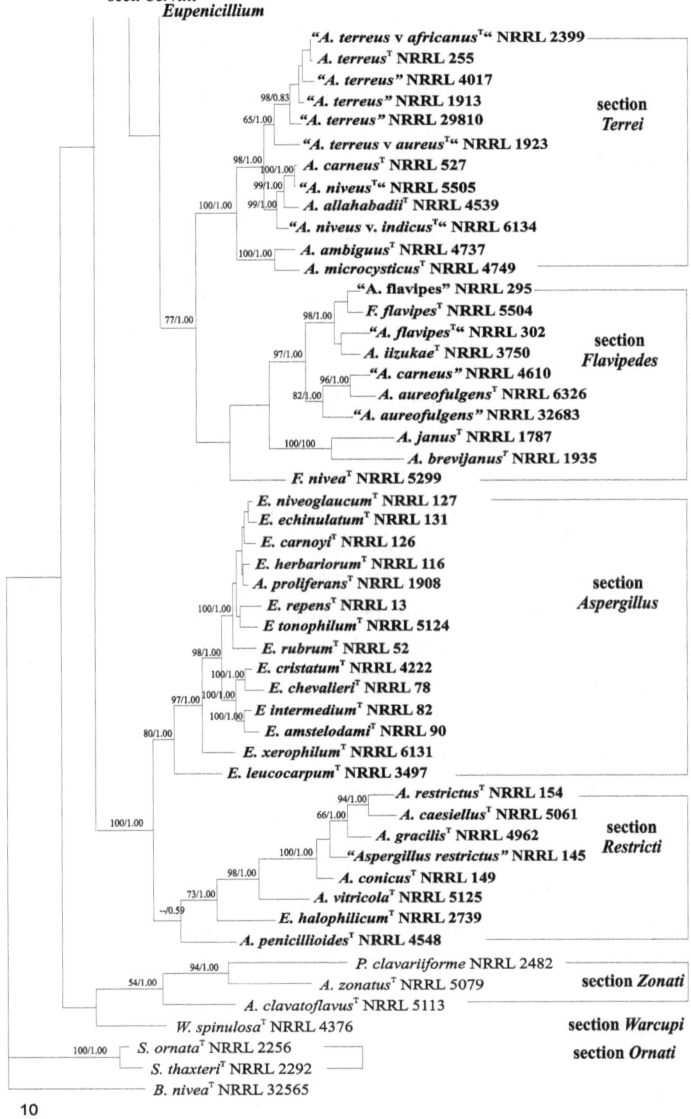

Figure 6. Maximum parsimony tree of Aspergillus *subgenera* Aspergillus *and* Terrei, *and the basal sections. Numbers at node are the bootstrap statistic followed by the Bayesian posterior probability.*

other sections in subgenus *Nidulantes*. To maintain monophyly these two sections have been united in a new subgenus. The species in section *Terrei* produce different combinations of citrinin, citreoviridin, asperphenamate and gregatins (Frisvad, unpublished), supporting the cladification suggested above.

7. Subgenus *Aspergillus*

Subgenus *Aspergillus* (Figure 6) is composed of sections *Aspergillus* and *Restricti* (Gams *et al.*, 1985). In Figure 6 there is very strong statistical support for this arrangement. Section *Aspergillus* contains all except one species with *Eurotium* teleomorphs and form a monophyletic group. *E. halophilicum* appears in section *Restricti*, but with limited statistical support. Section *Restricti* contains anamorphic species that are osmophilic and those species form a monophyletic group with *E. halophilicum*. Other than the one exception just noted, the groups as proposed by Raper and Fennell (1965) and Gams *et al.* (1985) are supported by the phylogenetic analysis. The phylogenetic analysis is also supported by profiles of secondary metabolites and physiology of the individual species: nearly all species in *Eurotium* produce echinulins, neoechinulins, flavoglaucins and auroglaucins, secondary metabolites which have only been found in this subgenus (Turner and Aldridge, 1983; Frisvad, unpublished). Furthermore all species in subgenus *Aspergillus* have much lower water activity growth optima than the other Aspergilli (Pitt and Hocking, 1997)

Basal in the tree (Figure 6) are two groups with members that were earlier placed in section *Flavi* (*A. zonatus, A. clavatoflavus*) or section *Ornati* (*S. ornata, S. thaxteri, W. spinulosa*; Raper and Fennell, 1965; Gams *et al.*, 1985), together with *Penicilliopsis clavariiforme*. *A. zonatus, A. clavatoflavus* and *P. clavariiformis* form a monophyletic clade and are proposed to belong to section *Zonati*. The species of *Sclerocleista* form a monophyletic group and constitute section *Ornati*. *Warcupiella spinulosa* appears to be a monospecific section, section *Warcupi*.

8. Subgenus *Candidi*

Forming a sister group to the *Eupenicillium* lineage is section *Candidi* (Figure 7). This section was treated earlier as part of subgenus *Circumdati* (Gams *et al.* 1985). Members of this section produce candidusins, terphenyllin and chlorflavonin related compounds in species-specific combinations (Rahbaek *et al.*, 2000). Interestingly, both *H. paradoxus* (Frisvad, unpublished) and *A. candidus* produce xanthocillin derivatives, which are also produced by some *Eupenicillium* and related *Penicillium* species including *Penicillium chrysogenum* (Rothe, 1954; Hagedorn and Tönjes, 1957) and *E. egyptiacum* (Vesonder, 1979).

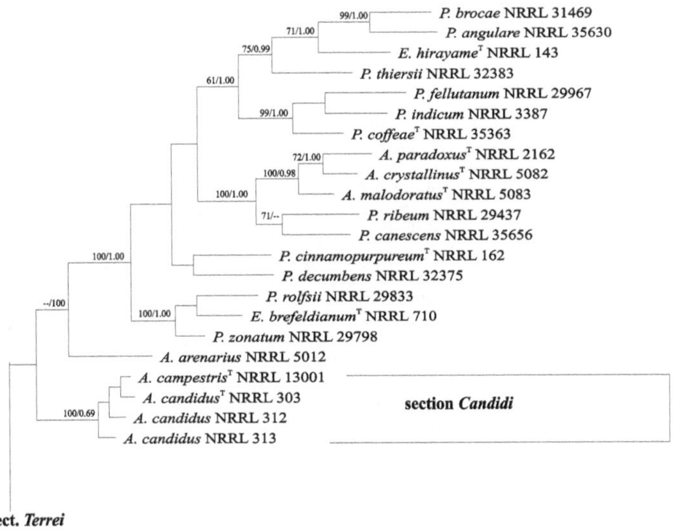

Figure 7. Maximum parsimony tree of Aspergillus *subgenus* Candidi *and excluded taxa. Numbers at node are the bootstrap statistic followed by the Bayesian posterior probability.*

9. Species excluded from the *Aspergillus* genus

Four of the *Aspergillus* species, *A. malodoratus, A. crystallinus, A. paradoxus* and *A. arenarius* are on a branch with *Penicillium* species (Figure 7) from the *Eupenicillium* clade. The disposition of these species will be published separately.

10. Subgeneric arrangements

Based on phylogenetic analysis of the sequence data, we propose to divide the genus *Aspergillus* into 8 subgenera (Table 2). Subgenus *Aspergillus* is composed of sections *Aspergillus* and *Restricti* and is a valid monophyletic group as reported by Gams *et al.* (1985). Sections *Terrei* and *Flavipedes* were placed in subgenus *Nidulantes* but do not form a monophyletic group with section *Nidulantes.* Subgenus *Terrei* is proposed to include these two sections. Sections *Cervini, Fumigati* and *Clavati* form a monophyletic group, but the sections have been placed in two subgenera. We propose to place all three sections in subgenus *Fumigati.* Sections *Cremei, Circumdati, Nigri* and *Flavi* do not have statistical support for their relationships to each other. Gams *et al.* (1985) placed them in subgenus *Circumdati* and our data do not contradict that arrangement. We accept subgenus *Circumdati* with the four sections *Circumdati, Flavi, Nigri* and *Cremei.* Subgenus *Nidulantes* contains sections

Table 2. Overview of the infrageneric classification of the Aspergillus *genus based on multilocus sequence data.*

Subgenus	Section	Teleomorph
Aspergillus	*Aspergillus*	*Eurotium*
	Restricti	*Eurotium*
Fumigati	*Fumigati*	*Neosartorya*
	Clavati	*Neocarpenteles, Dichotomomyces*
	Cervini	-
Circumdati	*Circumdati*	*Neopetromyces*
	Nigri	-
	Flavi	*Petromyces*
	Cremei	*Chaetosartorya*
Terrei	*Terrei*	-
	Flavipedes	*Fennellia*
Nidulantes	*Nidulantes*	*Emericella*
	Usti	*Emericella*
	Sparsi	
	Raperi	
	Silvati	
	Ochraceorosei	
	Bispori	
Ornati	*Ornati*	*Sclerocleista*
Warcupi	*Warcupi*	*Warcupiella*
	Zonati	*Penicilliopsis*
Candidi	*Candidi*	-

Sparsi, Usti, Nidulantes, Raperi, Silvati, Bispori and *Ochraceorosei* and we propose these sections as the constituents of a modified subgenus *Nidulantes*.

11. Conclusions

The sections or 'groups' of *Aspergillus* have become more reflective of true relationships with each revision of the genus (Thom and Church, 1926; Thom and Raper, 1945; Raper and Fennell, 1965). The membership in groups has changed as knowledge increased and as the defining characters of groups were better understood. Several of the subgenera proposed by Gams *et al.* (1985) are shown here to be monophyletic, but the other subgenera have been revised to reflect monophyly in all taxa. This new arrangement of subgenera and sections is supported by certain morphological and physiological features, and especially by different section specific secondary metabolites. Each section is a polythetic class based on phenotypic features (Beckner, 1959), but interestingly these classes agree with the clades based on nucleotide sequences.

Four of the newly proposed sections in *Aspergillus* are presently monospecific, and some other new sections contain two or three species. Further studies with more isolates belonging to these sections are needed to clarify their taxonomic status.

Acknowledgements

The authors are grateful for skillful technical assistance provided by Dr. Bruno Mannarelli, Ms. Jennifer Scoby and Ms. Jennifer Steele. The mention of firm names or trade products does not imply that they are endorsed or recommended by the U.S. Department of Agriculture over other firms or similar products not mentioned.

References

Alexopoulos, C.J., C.W. Mims and M. Blackwell, 1996. Introductory mycology. John Wiley & Sons, New York.

Anke, H., H. Schwab and H. Achenbach, 1980. Tetronic acid derivatives from *Aspergillus panamensis*. Production, isolation, characterization and biological activity. Journal of Antibiotics 33: 931-939.

Anke, H., I. Casser, M. Schrage and W. Steglich, 1988. Cyclogregatin, a new metabolite from *Aspergillus panamensis*. Journal of Antibiotics 41: 1681-1684.

Assante, G., L. Camarda and G. Nasini, 1980. Secondary mould metabolites. IX. Structure of a new bianthrone and of three new secoanthraquinones from *Aspergillus wentii* Wehmer. Gazzetta Chimica Italiana 110: 629-631.

Bartman, C.D. and I.M. Campbell, 1979. Naphthalenone production in *Aspergillus parvulus*. Canadian Journal of Microbiology 25: 130-137.

Beckner, M., 1959. The biological way of thought. Columbia University Press, New York.

Bos M., R. Canesso, N. Inoue-Ohga, A. Nakano, Y. Takehana and A.J. Sleight, 1997. O-methylasparvenone, a Nitrogen-free serotonin antagonist. Bioorganic and Medicinal Chemistry 5: 2165-2171.

Chao, P.D., P.L. Jr. Schiff, D.J. Slatkin and J.E. Knapp, 1979. Metabolites of Aspergilli. Part IV. New naphthalenones and 6-ethyl-7-methoxyjuglone from *Aspergillus parvulus*. Journal of Chemical Research Synopses 1979: 236.

Dorner, J.W., R.J. Cole, J.P. Springer, R.H. Cox, H. Cutler and D.T. Wicklow, 1980. Isolation and identification of two new biologically active norditerpene dilactones from *Aspergillus wentii*. Phytochemistry 19: 1157-1161.

Elsohly, H.N., D.J. Slatkin, P.L.Jr. Schiff and J.E. Knapp, 1974. Metabolites of *Aspergillus cervinus* Massee (Moniliaceae). Journal of Pharmacological Sciences 63: 1632-1633.

Frisvad, J.C., 1985. Secondary metabolites as an aid to *Emericella* classification. In: Samson, R.A. and J.I. Pitt (eds.) Advances in *Penicillium* and *Aspergillus* systematics. Plenum Press, New York, pp. 437-443.

Frisvad, J.C. and R.A. Samson, 2000. *Neopetromyces* gen. nov., and an overview of teleomorphs of *Aspergillus* subg. *Circumdati*. Studies in Mycology 45: 201-207.

Frisvad, J.C. and R.A. Samson, 2004. *Emericella venezuelensis*, a new species with stellate ascospores producing sterigmatocystin and aflatoxin B_1. Systematic and Applied Microbiology 27: 672-680.

Frisvad, J.C., J.M. Frank, J.A.M.P. Houbraken, A.F.A. Kuijpers and R.A. Samson, 2004b. New ochratoxin A producing species of *Aspergillus* section *Circumdati*. Studies in Mycology 50: 23-43.

Frisvad, J.C., P. Skouboe and R.A. Samson, 2005. Taxonomic comparison of three different groups of aflatoxin producers and a new efficient producer of aflatoxin B_1, sterigmatocystin and 3-O-methylsterigmatocystin, *Aspergillus rambellii* sp. nov. Systematic and Applied Microbiology 28: 442-453

Frisvad, J.C., R.A. Samson and J. Smedsgaard, 2004a. *Emericella astellata*, a new producer of aflatoxin B_1, B_2 and sterigmatocystin. Letters in Applied Microbiology 38: 440-445.

Gams, W., M. Christensen, A.H.S. Onions, J.I. Pitt and R.A. Samson, 1985. Infrageneric taxa of *Aspergillus*. In Samson, R.A. and J.I. Pitt (eds.) Advances in *Penicillium* and *Aspergillus* systematics. Plenum Press, New York, pp. 55-62.

Geiser, D.M. and K.F. LoBuglio, 2001. The monophyletic Plectomycetes: Ascosphaerales, Onygenales and Eurotiales. In: Lemke, P., D.J. McLaughlin and E. McLaughlin (eds.) The Mycota. Vol. VII, Systematics and evolution. Springer-Verlag, Berlin, pp. 201-219.

Geiser, D.M., C. Gueidan, J. Miadlikowska, F. Lutzoni, F. Kauff, V. Hofstetter, E. Fraker, C.L. Schoch, L. Tibell, W.A. Untereiner and A. Aptroot, 2006. Eurotiomycetes: Eurotiomycetidae and Chaetothyriomycetidae. Mycologia 98: 1053-1064.

Hagedorn, I. and H. Tönjes, 1957. Konstitutionsaufklärung von xanthocillin, einem neuen Antibiotikum. Pharmazie 12: 567-580.

He, J., E.M. Wijeratne, B.P. Bashyal, J. Zhan, C.J. Seliga, M.X. Liu, E.E. Pierson, L.S. Pierson, H.D. VanEtten and A.A.L. Gunatilaka, 2004. Cytotoxic and other metabolites of *Aspergillus* inhabiting the rhizosphere of Sonoran desert plants. Journal of Natural Products 67: 1985-1991.

Hennig, W., 1950. Grundzüge einer Theorie der phylogenetischen Systematik. Deutcher Zentralverlag, Berlin.

Hennig, W., 1966. Phylogenetic systematics. University of Illinois Press, Urbana, Il.

Hong, S.B., S.J. Go, H.D. Shin, J.C. Frisvad and R.A. Samson, 2005. Polyphasic taxonomy of *Aspergillus fumigatus* and related species. Mycologia 97: 1316-1329.

Huelsenbeck, J.P. and F. Ronquist, 2001. MRBAYES: Bayesian inference of phylogeny. Bioinformatics 17: 754-755.

Kawahara, N. K. Nozawa, S. Nakajima and K. Kawai, 1988. Studies on fungal products. Part 15. Isolation and structure determination of arugosin E from *Aspergillus silvaticus* and cycloisoemericellin from *Emericella striata.* Journal of the Chemical Society. Perkin Transactions 1 1988: 907-911.

Kawahara, N., K. Nozawa, S. Nakajima and K. Kawai, 1987. Studies on fungal products. Part 13. Isolation and structures of dithiosilvatin and silvathione, novel dioxopiperazine derivatives from *Aspergillus silvaticus.* Journal of the Chemical Society. Perkin Transactions 1 1987: 2099-2101.

Kawahara, N., S. Nakajima, Y. Yamazaki and K. Kawai, 1989. Structure of a novel epidithiodioxopiperazine, emethallicin A, a potent inhibitor of histamine release from *Emericella heterothallica.* Chemical and Pharmaceutical Bulletin 37: 2592-2595.

König, W.A., K.P. Pfaff, W. Loeffler, D. Schanz and H. Zähner, 1978. Stoffwechselprodukte von Kiroorganismen, 171. Ethericin A; Isolierung, Charakterisierung und Strukturaufklärung eines neuen, antibiotisch wirksamen Diphenylethers. Justus Liebigs Annalen der Chemie 1978: 1289-1296.

König, W.A., R. Krause, W. Loeffler and D. Schanz, 1980. Metabolic products of microorganisms. 196 The structure of ethericin B, a new diphenylether antibiotic. Journal of Antibiotics 33: 1270-1273.

Kozakiewicz, Z., 1989. *Aspergillus* species on stored products. Mycological Papers 161: 1-188.

Kwon-Chung, K.J. and D.I. Fennell, 1971. A new pathogenic species of *Aspergillus.* Mycologia 63: 478-489.

Mayr, E., 1974. Cladistic analysis or cladistic classification. Zeitschrift für Zoologie, Systematrik und Evolution 12: 94-128.

Mizuno, R., N. Kawahara, K. Nozawa, M. Yamazaki, S. Nakajima and K. Kawai, 1995. Stereochemistry of an 18,22-Cyclosterol, Mer-NF8054X, from *Emericella heterothallica* and *Aspergillus ustus.* Chemical and Pharmaceutical Bulletin 43: 9-11.

Nakamura, M., K. Fukuyama, T. Tsukihara, Y. Katsube and T. Hamasaki, 1983. Structure of funicin, antimicrobial substance from *Aspergillus funiculosus,* $C_{17}H_{18}O_{5}$. Acta Crystallographica Section C. Crystal Structure Communications 39: 268-270.

Page, R.D.M., 1996. TREEVIEW: An application to display phylogenetic trees on personal computers. Computer Applications in the Biosciences 12: 357-358.

Peterson, S.W., 1995. Phylogenetic analysis of *Aspergillus* sections *Cremei* and *Wentii,* based on ribosomal DNA sequences. Mycological Research 99: 1349-1355.

Peterson, S.W., 2000. Phylogenetic relationships in *Aspergillus* based upon rDNA sequence analysis. In: Samson, R.A. and J.I. Pitt (eds.) Classification of *Penicillium* and *Aspergillus*: Integration of modern taxonomic methods. Harwood Publishers, Reading, UK, pp. 323-356.

Peterson, S.W., 2008. Phylogenetic analysis of the genus *Aspergillus.* Mycologia (in press).

Pildain, M.B., J.C. Frisvad, G. Vaamonde, D. Cabral, J. Varga and R.A. Samson, 2007. Two new aflatoxin producing *Aspergillus* species from Argentinean peanuts. International Journal of Systematic and Evolutionary Microbiology (in press).

Pitt, J.I. and A.D. Hocking, 1997. Fungi and food spoilage. 2nd ed. Blackie Academic and Professional, London.

Rabie, C.J., M. Steyn and G.C. van Schalkwyk, 1977. New species of *Aspergillus* producing sterigmatocystin. Applied and Environmental Microbiology 33: 1023-1025.

Rahbaek, L., J.C. Frisvad and C. Christophersen, 2000. An amendment of *Aspergillus* section *Candidi* based on chemotaxonomical evidence. Phytochemistry 53: 581-586.

Raper, K.B. and D.I. Fennell, 1965. The genus *Aspergillus*. Williams and Wilkins, Baltimore.

Rieppel, O., 2007. Species: kinds of individuals or individuals of a kind. Cladistics 23: 373-384.

Rothe, W., 1954. Das neue Antibiotikum Xanthocillin. Deutsche Medizinische Wochenschrift 79: 1080-1081.

Samson, R.A., 1979. Compilation of Aspergilli described since 1965. Studies in Mycology 18: 1-38.

Samson, R.A., J.A.M.P. Houbraken, A.F.A. Kuijpers, J.M. Frank and J.C. Frisvad, 2004. New ochratoxin A or sclerotium producing species in *Aspergillus* section *Nigri*. Studies in Mycology 50: 45-61.

Seigle-Murandi, F., S. Krivobok, R. Steiman and D. Marzin, 1990. Production, mutagenicity, and immunotoxicity of gliotoxin. Journal of Agricultural and Food Chemistry 38: 1854-1856.

Serra, R. and S.W. Peterson, 2007. *Penicillium astrolabium* and *Penicillium neocrassum*, two new species isolated from grapes and their phylogenetic placement in the *P. olsonii* and *P. brevicompactum* clade. Mycologia 99: 78-87.

Swofford, D.L., 2002. PAUP*. Phylogenetic Analysis Using Parsimony (*and Other Methods). Version 4. Sunderland, Massachusetts, Sinauer Associates.

Thom, C. and K.B. Raper, 1945. Manual of the Aspergilli. Williams and Wilkins, Baltimore.

Thom, C. and M.B. Church, 1926. The Aspergilli. Williams and Wilkins, Baltimore.

Thompson, J.D., D.G. Higgins and T.J. Gibson, 1994. CLUSTALV: Improving the sensitivity of progressive multiple sequence alignments through sequence weighting, position-specific gap penalties and weight matrix choice. Nucleic Acids Research 22: 4673-4680.

Turner, W.B. and D.C. Aldridge, 1983. Fungal Metabolites II. Academic Press, London.

Vesonder, R.F., 1979. Xanthocillin, a metabolite of *Eupenicillium egyptiacum* NRRL 1022. Journal of Natural Products 42: 232-233.

Wells, J.M., R.K. Cole and J.W. Kirksey, 1975. Emodin, a toxic metabolite of *Aspergillus wentii* isolated from weevil-damaged chestnuts. Applied Microbiology 30: 26-28.

Yukio, M. and Y. Mikio, 1985. Biosynthesis of silvaticamide, a toxin from *Aspergillus silvaticus*. Chemical and Pharmaceutical Bulletin 33: 4296-4298.

Ochratoxin A in profiling and speciation

F. Javier Cabañes and M. Rosa Bragulat
Veterinary Mycology Group, Department of Animal Health and Anatomy, Autonomous University of Barcelona, E-08193 Bellaterra, Barcelona, Spain

Abstract

More than 20 species have been cited as ochratoxin A-producing fungi in the genus *Aspergillus*. However, few of them are known to be regularly the source of ochratoxin A (OTA) contamination of foods. OTA contamination of food was until recently believed to be produced only by *Aspergillus ochraceus* and by *Penicillium verrucosum*, which affect mainly dried stored foods and cereals respectively, in different regions of the world. Different methods have been developed to detect OTA producing fungi. Traditional mycological methods are time consuming and require taxonomical and chromatography expertise. Different molecular diagnostic methods for an early detection of ochratoxigenic fungi, using mainly PCR techniques, have also been proposed. However, OTA production is also included as a character for taxonomical purposes in classification and also for identification. As is well known in taxonomy, one difficulty in devising identification schemes is that the results of characterisation tests may vary depending on different conditions. We can find some confusing or controversial data about the ability to produce OTA by some species. Some examples found in the literature about these issues are discussed.

Keywords: detection methods, mycotoxins, ochratoxin A, ochratoxin-producing species, speciation, taxonomy

1. Introduction

Ochratoxin A (OTA) is a nephrotoxic mycotoxin classified as a human renal carcinogen (group 2B). It also has teratogenic, carcinogenic and immunosuppressive properties. Natural occurrence of OTA has been reported from temperate to tropical climates in different food commodities such as cereals, coffee, dried fruits, wine and grape juice. In the European diet, cereals and cereal-based products have been identified as the first major source of human exposure to OTA (Anonymous, 1998). Human exposure to OTA is most likely coming from low level contamination of a wide range of different foods. Elimination of OTA in humans is extremely slow, since the toxin has the longest half-life known for living mammals. As a result of this exposure there is a high incidence of OTA in human blood and breast milk (Clark and Snedeker 2006, Petzinger and Weidenbach 2002). Regarding this toxic potential, the European Union has established maximum OTA levels for different food products (Commission Regulation (EC) No 123/2005).

Some guidance values for this mycotoxin have been recently recommended for cereals, cereal products intended for animal feed and complete and complementary feedingstuffs for pigs and poultry (Commission Recommendation (EC) No 576/2006). Commercial mixed feeds mainly contain mixtures of cereals. Swine, like poultry, are exposed to OTA through their feed which is composed of cereals such as barley, maize, oats, wheat that are susceptible to contamination by this mycotoxin (Dragacci *et al.*, 1999). Contamination of animal feeds with OTA may result in the presence of residues in edible offal and blood products. However, higher concentrations of OTA may occur in certain local specialties such as blood puddings and sausages prepared with pig blood serum (EFSA, 2004). At present it is not established maximum levels for OTA in meat and meat products in the European Community. However, the consideration of setting a maximum level for OTA for edible offal and blood products is under discussion. In Denmark, since 1978, the contamination of pig meat with OTA has been assessed indirectly by the inspection of pigs' kidneys for the presence of macroscopic lesions of porcine nephropathy (Jorgensen and Petersen, 2002).

2. Ochratoxin A producing fungi

OTA producing fungi are included in the genera *Aspergillus* and *Penicillium* (Table 1). OTA was originally described in 1965 (Van der Merwe *et al.*, 1965) as a metabolite of *A. ochraceus* consisting of a polyketide derived from a dihydromethyl-isocoumarin linked to phenylalanine (Figure 1). Some years after, it was detected from strains of *P. viridicatum* and some other species of *Penicillium*, being today classified in *Penicillium verrucosum* (Pitt, 1987). Recently, *P. nordicum*, formed with some strains isolated mainly from fermented meat and cheese, split from the latter species (Larsen *et al.*, 2001; Castella *et al.*, 2002). At present, these two species are the only OTA producers known and accepted in this genus (Frisvad *et al.*, 2004b).

In the case of the genus *Aspergillus*, other OTA producing fungi were described in subsequent years, belonging mainly to the section *Circumdati*, such as *A. melleus, A. ostianus, A. petrakii, A. persii, A. sclerotiorum, A. sulphureus* or *Neopetromyces muricatus. Petromyces alliaceus* and *Petromyces albertensis*, previously considered in the above mentioned section, are now included in section *Flavi*, based on rDNA studies (Peterson, 2000) and supported by phenotypic data (Frisvad and Samson, 2000).

Table 1. Ochratoxin A producing species.

Genus	Subgenus	Section[1]	Species	Reference
Aspergillus	*Circumdati*	*Circumdati*	*A. melleus*	Ciegler, 1972
		Circumdati	*A. muricatus*	Frisvad and Samson, 2000
		Circumdati	*A. ochraceus*	Van der Merwe *et al.*, 1965
		Circumdati	*A. ostianus*	Ciegler, 1972
		Circumdati	*A. petrakii*	Ciegler, 1972
		Circumdati	*A. sclerotiorum*	Ciegler, 1972
		Circumdati	*A. sulphureus*	Ciegler, 1972
		Circumdati	*A. persii*	Zotti and Corte, 2002
		Circumdati	*A. westerdijkiae*	Frisvad *et al.*, 2004a
		Circumdati	*A. steynii*	Frisvad *et al.*, 2004a
		Circumdati	*A. cretensis*	Frisvad *et al.*, 2004a
		Circumdati	*A. flocculosus*	Frisvad *et al.*, 2004a
		Circumdati	*A. pseudoelegans*	Frisvad *et al.*, 2004a
		Circumdati	*A. roseoglobulosus*	Frisvad *et al.*, 2004a
		Flavi	*A. alliaceus*	Ciegler, 1972
		Flavi	*A. albertensis*	Varga *et al.*, 1996
		Nigri	*A. carbonarius*	Horie, 1995; Wicklow *et al.*, 1996; Téren *et al.*, 1996
		Nigri	*A. niger*	Abarca *et al.*, 1994
		Nigri	*A. lacticoffeatus*	Samson *et al.*, 2004
		Nigri	*A. sclerotioniger*	Samson *et al.*, 2004
Penicillium	*Penicillium*	*Penicillium*	*P. verrucosum*	Pitt, 1987
		Penicillium	*P. nordicum*	Larsen *et al.*, 2001; Castellá *et al.*, 2002

[1]According to Gams *et al.* (1985) and Stolk and Samson (1985).

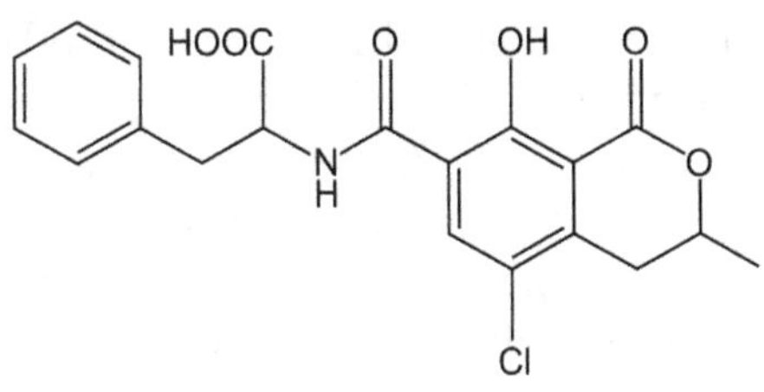

Figure 1. Chemical structure of ochratoxin A.

There is a long list of species known to produce OTA in the genus *Aspergillus*, but few of them are known to contaminate foods with this mycotoxin. OTA contamination of food was until recently believed to be caused only by *A. ochraceus* and by *P. verrucosum*, which affect mainly dried stored foods and cereals respectively, in different regions of the world. It is generally assumed that *P. verrucosum* produces OTA in temperate and cold climates, whereas *A. ochraceus* is more commonly associated with warmer and tropical climates. Nevertheless, such fungi are not always the source of OTA in some commodities. Recent surveys have clearly shown that some species belonging to the black aspergilli, including the *A. niger* aggregate and *A. carbonarius*, are sources of OTA in food commodities such as wine, grapes, dried vine fruits and coffee (Cabañes *et al.*, 2002; Abarca *et al.*, 2003; Taniwaki, 2006). *Petromyces alliaceus* (*A. alliaceus*) has been cited as a possible source for the OTA contamination occasionally observed in figs (Bayman *et al.*, 2002).

Recently, new OTA-producing species have been described in section *Circumdati*: *A. cretensis, A. flocculosus, A. pseudoelegans, A. roseoglobulosus, A. steynii, A. westerdijkiae* (Frisvad *et al.*, 2004a) and in section *Nigri*: *A. lacticoffeatus* and *A. sclerotioniger* (Samson *et al.*, 2004). With the exception of *A. steynii* and *A. westerdijkiae*, the rest of the new species are apparently rare, may not be important in foods or few strains have been studied. These two new species have often been identified as *A. ochraceus* and they seem to be more frequently isolated and are better OTA producers than *A. ochraceus sensu stricto.* In fact the original OTA producing strain (NRRL 3174) (Van der Merwe *et al.*, 1965) has been now identified as *A. westerdijkiae* (Frisvad *et al.*, 2004a).

3. Methods for detecting OTA producing fungi

Different methods have been developed to detect OTA producing fungi. Traditional mycological methods are time consuming and require taxonomical and chromatography expertise. Direct microscopic examination of the samples, isolation and culture techniques followed by chemical characterisation of OTA [e.g. thin layer chromatography (TLC), high-performance liquid chromatography (HPLC)] produced by fungal isolates are included in these methods. However, molecular diagnostic methods for an early detection of ochratoxigenic fungi, using mainly PCR techniques, have been also proposed (O'Callagham *et al.*, 2003; Geisen *et al.*, 2004). Two main approaches have been used to detect the presence of these fungi. They are based on the detection of species-specific DNA fragments of OTA producing species (e.g. rDNA genes, anonymous specific sequences, calmodulin) and the detection of DNA sequences of genes involved in OTA biosynthesis (e.g. polyketide synthase genes). One of the goals of these techniques is to differentiate between toxigenic and non-toxigenic strains belonging to species known to produce OTA. To date, one of the problems is that little is known about the genes involved in the OTA biosynthesis. These molecular aspects have been recently reviewed by Niessen *et al.* (2005) and O'Callaghan and Dobson (2005).

Some differential culture media have been used for facilitating the detection of OTA producing species. *Penicillium verrucosum* shows a brown to reddish- brown reverse in Dichloran Yeast Extract Sucrose 18% Glycerol agar (DYSG agar) (Frisvad *et al.*, 1992) and are distinguished from other penicillia, which show reverse colourations in shades of yellow. But, not all the strains belonging to this ochratoxigenic species are mycotoxin producers. If we determine the presence of OTA producing species using this culture medium we have to take into account that this technique does not differentiate between toxigenic and non-toxigenic strains belonging to *P. verrucosum*. However, this medium has been successfully used to correlate the *P. verrucosum* infestation and OTA content in cereals (Lund and Frisvad, 2003).

For *Aspergillus* spp. the medium Coconut Cream Agar (CCA), a culture medium previously designed for detection of aflatoxigenic fungi (e.g. *A. flavus*, *A. parasiticus*), has been also used for detecting OTA producers by Heenan *et al.* (1998). Nearly all of the cultures of *A. carbonarius* tested in CCA by these authors produced a strong green-blue fluorescence in the reverse when observed under long wave UV light. Twenty six of the 27 cultures testing positive for OTA by TLC were positive on CCA, obtaining in this case a good correlation. This medium has been also used in confirmation of non OTA producing mutants in the study of the biosynthetic pathway for OTA in *A. ochraceus* (O'Callagham *et al.*, 2003).

Numerous analytical methods have been described to detect toxigenic fungi based on toxin production in natural or synthetic substrates followed by extraction, purification and detection by TLC or HPLC. In general, these methods are time-consuming and are not suitable when many fungal isolates need to be screened. Some simpler techniques have been developed, particularly the agar plug method (Filtenborg *et al.*, 1983) described the agar plug sampling TLC technique for toxigenic moulds in pure culture. In these techniques, the agar plugs are removed from the cultures with the aid of a cork borer and placed on TLC plates. The secondary metabolites are extracted on the TLC plate with a suitable solvent.

A simple and clean screening method for detecting specifically ochratoxigenic fungi in pure cultures, based on HPLC determination of OTA in the extract obtained from agar plugs cut from plate cultures has been also developed (Bragulat *et al.*, 2001). This method facilitates the quantification of the OTA level produced by the strains tested. It has several advantages over other conventional methods which use large amounts of natural substrates or culture media and subsequently are quite costly in terms of solvents and amount of work. The simplicity of the method makes it very useful when many fungal isolates need to be screened. Another advantage is that the extracts are very clean and the HPLC chromatograms show no interference. Chromatographic techniques are usually also used for confirming the detection of OTA by other methods.

Complementing these conventional methods, various immunological techniques have been also adapted to detect and quantify OTA in food, providing user friendly formats (e.g. ELISA (enzyme-linked immunosorbent assay) kits, lateral flow devices) making easy the detection of the mycotoxin produced by the isolates. Unfortunately commercially available ELISA test kits for OTA are validated only for some substrates (e.g. corn) and they may need further chromatographic confirmation. Other matrix similar compounds can interact with the specific antibodies and cause production of false positive results, due to cross-reaction. We can find in the literature some examples of false positive OTA producers belonging to *Aspergillus* spp. (e.g. *A.sydowii, A. terreus, A. ustus*) or *Penicillium* spp. (e.g. *P. corylophilum, P. implicatum, P. montanense*) using ELISA techniques (Ueno *et al.*, 1991). Techniques using anti-OTA antibodies and other novel technologies for detecting OTA in food and beverages have been recently reviewed by Visconti and De Girolamo (2005).

4. Ochratoxin A in profiling and speciation

OTA production is also included as a character for taxonomical purposes in classification (e.g. secondary metabolite profiles for describing or defining species) and also for identification (e.g. synoptic key to species, illustrated manuals of identification).

Penicillium taxonomy is not easy for the inexperienced, and compared to *Aspergillus* it is a more diverse genus in terms of numbers of species and range of habitats (Pitt and Hocking, 1997). However, in this genus OTA is only produced by *P. verrucosum* and *P. nordicum*, despite some reports on OTA production by other species (Frisvad *et al.*, 2004b, Frisvad and Samson, 2004). Both *P. verrucosum* and *P. nordicum* are slowly growing species belonging to the series *Verrucosa*, subgenus *Penicillium*. Species in this subgenus are among the most frequently isolated moulds and some of the most difficult to identify of all fungi. Recently, Frisvad and Samson (2004) pointed out that a polyphasic approach, including the study of extrolites, is necessary to classify and identify species of *Penicillium* subgenus *Penicillium*. These authors keyed a total of 58 species in this subgenus. Due to the size and complexity of the genus *Penicillium*, detection of OTA production in isolates of *Penicillium* spp. provides a practical way for the identification of *P. verrucosum* and *P. nordicum*, and their differentiation of the rest *Penicillium* species.

These OTA producing species have common morphological characteristics such as very similar colony diameters on many media or rough stipes. Larsen *et al.* (2001) found that based on differences in secondary metabolite profiles, colony reverse colour on YES and habitat preference, *P. nordicum* could be separated from *P. verrucosum*. All meat-derived and most cheese-derived isolates are usually *P. nordicum* and all plant-derived isolates are *P. verrucosum*. Most of the isolates of this latter species produced both OTA and citrinin and had a characteristic dark brown reverse colour on YES, whereas almost all the *P. nordicum* strains tested had a pale, creamy or dull yellow reverse colour and produced more OTA than *P. verrucosum* but are non citrinin producers. On the other hand, not all the strains belonging to an ochratoxigenic species are OTA producers. For example, the consistency in OTA production among *P. verrucosum* isolates seems to be high. However, percentages of OTA producers cited by different authors in this species ranges from 66 to >95% (Frisvad and Filtenborg, 1989; Frisvad *et al.*, 2005; Pitt, 1987).

As far as the genus *Aspergillus* is concerned, OTA has been also included as a character for their speciation (Frisvad *et al.*, 2004a; Samson *et al.*, 2004; Serra *et al.*, 2006). Ochratoxin A production consistency varies in the reported OTA producers in section *Circumdati* (Frisvad *et al.*, 2004a). The trace production of OTA in *A. melleus, A. ostianus, A. petrakii* and *A. persii* may be inconsistent. In *A. ochraceus sensu stricto* and *A. sclerotiorum* several isolates do not produce this mycotoxin. In most cases, 100% of the isolates belonging to the rest of the mentioned species (e.g. *A. steynii, A. westerdijkiae*) produce OTA.

In section *Nigri*, some members of the *A. niger* aggregate (*A. niger* species complex) and *A. carbonarius*, are OTA producers. Recently, *A. lacticoffeatus* and *A. sclerotioniger*, two new OTA-producing species which are close to *A. niger* and *A. carbonarius* respectively, have been described in section *Nigri* (Samson *et al.*, 2004). These two new species are known from only very few isolates. In this section, the ability of the uniseriate species (*A. japonicus, A. aculeatus*) to produce OTA has been mentioned (Dalcero *et al.*, 2002; Battilani *et al.*, 2003; Ponsone *et al.*, 2007), but this fact needs to be confirmed, since by the moment they are not considered as OTA producing species (Teren *et al.*, 1996; Parenicova *et al.*, 2001). Recently, none of the 66 strains having uniseriate conidiophores which had been isolated from grapes of different origins were able to produce OTA (Bau *et al.*, 2005b).

The reported percentages of ochratoxigenic isolates in *A. carbonarius* ranges from 25 to 100%, whereas in the *A. niger* aggregate those percentages are lower, ranging from 0.6% to 50% (Abarca *et al.*, 2004). However, the reported percentage of OTA producing strains in *A. carbonarius* from grapes achieves the 100% in some studies (Cabañes *et al.*, 2002; Bau *et al.*, 2005a; Leong *et al.*, 2004), confirming its consistency in producing this mycotoxin. Bau *et al.* (2005b) analysed 123 strains morphologically identified as *A. carbonarius* involved in the OTA contamination of grapes from Europe and Israel. From these strains, only four strains were not able to produce OTA. These strains showed a different RAPD pattern and the section of DNA sequenced differed from the sequence of the other *A. carbonarius* strains. Afterwards, these OTA-non producing strains were described as a new species named *A. ibericus* (Serra *et al.*, 2006). These strains, moreover, have characters that allow morphological distinction from the other species in the section, particularly the spore size, which allows separation of the species from the two most common biseriate species in section *A. carbonarius* and *A. niger* and its aggregate species. The validation of this new taxon was further supported by analysis of the ITS-5.8S rDNA and calmodulin gene sequences and by analysis of the amplified fragment length polymorphism (AFLP) patterns. *Aspergillus ibericus* can be distinguished from *A. sclerotioniger* based on cultural traits, conidia size and by the secondary metabolite profile, as *A. sclerotioniger* produces OTA.

Species included in the *A. niger* aggregate are difficult to distinguish one from each other and molecular methods and secondary metabolites are usually necessary to identify them. The taxonomy of this group, also referred as *A. niger* species complex, is not fully resolved as the number of accepted species depends on the methodology used. So far there has not been complete agreement between phenotypical and molecular data. For this reason, misidentification is easy in this group. Recently, Samson *et al.* (2004) provisionally accepted eight species in this complex (*A. brasiliensis, A. costaricensis, A. foetidus, A. lacticoffeatus, A. niger, A. piperis, A. tubingensis* and *A. vadensis*) and among them only some strains of *A. niger* and *A. lacticoffeatus* were considered OTA producers. However, some strains of *A. awamori, A. usamii* and *A. foetidus* (Téren *et al.*, 1996; Abarca *et al.*, 2004) have been also cited as OTA producers, but their identity has been questioned (Samson *et al.*, 2004).

5. Some problems in determining OTA production

As is well known in taxonomy, one difficulty in devising identification schemes is that the results of characterisation tests may vary depending on different conditions such as the growth conditions (e.g. incubation temperature, water activity, pH), the length of incubation period, the composition of the medium, the method of detection (e.g.

TLC, HPLC, ELISA) and the criteria used to define a positive or negative reaction, in this case, a mycotoxin production. Perhaps, for these reasons we can find some confusing or controversial data about the ability to produce OTA by some species in the literature. Therefore, the results of characterisation tests obtained by one laboratory often do not match exactly those obtained by another laboratory, although the results within each laboratory may be quite consistent. The blind acceptance of an identification scheme without reference to the particular conditions employed by those who devised the scheme can lead to error, and, unfortunately, such conditions are not always specified.

A list of logical and clear stated recommendations have been recently published to avoid incorrect reporting of fungal species producing particular mycotoxins (Frisvad *et al.*, 2006). However, in general, we should consider that a mycotoxin producing species should produce consistently a mycotoxin and not only possible traces (e.g. ppt). The reliability of analytical results is essential when data are used in food surveillance studies. The use of validated methods is generally recommended, although it is no guarantee of accurate results (Visconti and De Girolamo, 2005). This is also valid for defining OTA producers. However, in our case we have not to detect a very low quantity of a mycotoxin in a food; we have to detect and define a mycotoxin producer, that, in theory, it should produce an easy detectable quantity of this product.

A weak production should be clearly distinguishable from no production. Basic rules of assessing the quality and useability of analytical methods and detection limits should be followed for determining the identity of an analyte. For example, a spot, using TLC, or a peak using HPLC obtained from a fungal culture with/at the same Rf value (retardation factor, TLC) or Rt (retention time HPLC) matching of that a mycotoxin standard should be confirmed, in order to avoid incorrect reporting of a species producing a mycotoxin.

The concept of limit of detection should be properly and carefully used. Detection limits can be usually calculated for the instrument used for measurement (instrument detection limit), for the analytical method (method detection limit) or as a sample-specific quantification limit (sample quantification limit). Some papers reporting OTA production by some fungal species at detected concentrations close or below the limit of detection of this mycotoxin in the sample have been recently reported. Battilani *et al.* (2003) cited that 16 out of 63 isolates of *Aspergillus* having uniseriate conidial heads produced OTA at <1ppb concentrations, giving a detection limit of 0.7 ppb. Similarly, Ponsone *et al.* (2007) reported that 28 out of 107 isolates of these species produced OTA between 1.2-8.6 ppm, giving a detection limit of 1 ppm. The quantification of analytes at concentrations close to the limit of detection can result in quite large errors, because of contribution to the measurement from noise. The closer the mycotoxin concentration determined is to the detection limit, the greater the possibility of false positives.

Recently, some strains of *A. tubingensis* have also been cited as OTA producers (Medina *et al.*, 2005; Perrone *et al.*, 2006). OTA production by these fungal species were detected at concentrations close or below the limit of quantification obtained for this mycotoxin (Perrone *et al.*, 2006). Medina *et al.* (2005) did not cite the limit of quantification for OTA in their paper, but the identity of OTA in YES-5% bee pollen cultures was determined using LC-ion trap MS. These authors claimed that the fact of failing to detect OTA previously in cultures of *A. tubingensis* might be due to culture medium and/or incubation time and pointed out that bee pollen stimulate biosynthesis of OTA. Accensi *et al.* (1999) divided the *A. niger* aggregate into two groups, designated types N and T, based on ITS-5.8S rDNA RFLP of the type strains of *A. niger* and *A. tubingensis*. All the OTA-positive isolates belonging to the *A. niger* aggregate characterised by this method tested in our laboratory were of type N, whereas type T strains were not able to produce OTA (Abarca *et al.*, 2003; Accensi *et al.*, 2001; Cabañes *et al.*, 2002). In a recent study, 20 out of the 173 *A. niger* aggregate strains from grapes of different European countries and Israel produced OTA (Bau *et al.*, 2006). All the OTA producing species belonged to the N type. More recently, Leong *et al.* (2007) reported that all the Australian isolates of *A. niger* displaying the type T (n: 52) were non ochratoxigenic and the only three OTA producing isolates showed the RFLP pattern type N.

Although, in black aspergilli other different OTA producing species than *A. niger*, *A. carbonarius*, *A. lacticoffeatus* and *A. sclerotioniger* may exist, OTA production by uniseriate species and *A. tubingensis* should be confirmed. Improvements in the speciation of the *A. niger* species complex would also facilitate a better understanding of the OTA-producing species in this group.

Acknowledgements

The authors acknowledge the financial support of the Spanish Government (AGL2004-07549-C05-03) and the Generalitat de Catalunya (2005SGR00684).

References

Abarca, M.L., F. Accensi, J. Cano and F.J. Cabañes, 2004. Taxonomy and significance of black aspergilli. Antonie van Leeuwenhoek 86: 33-49.

Abarca, M.L., F. Accensi, M.R. Bragulat, G. Castellá and F.J. Cabañes, 2003. *Aspergillus carbonarius* as the main source of ochratoxin A contamination in dried vine fruits from the Spanish market. Journal of Food Protection 66: 504-506.

Abarca, M.L., M.R. Bragulat, G. Castellá and F.J. Cabañes, 1994. Ochratoxin A production by strains of *Aspergillus niger* var. *niger*. Applied and Environmental Microbiology 60: 2650-2652.

Accensi, F., J. Cano, L. Figuera, M.L. Abarca and F.J. Cabañes, 1999. New PCR method to differentiate species in the *Aspergillus niger* aggregate. FEMS Microbiology Letters 180: 191-196.

Accensi, F., M.L. Abarca, J. Cano, L. Figuera and F.J. Cabañes, 2001. Distribution of ochratoxin a producing strains in the *A. niger* aggregate. Antonie van Leeuwenhoek 79: 365-370.

Anonymous, 1998. Position paper on ochratoxin A, Codex Alimentarius Commission, Codex Committee on Food Additives and Contaminants, Thirty-first session, The Hague, The Netherlands, CX/FAC 99/14.

Battilani, P., A. Pietri, T. Bertuzzi, L. Languasco, P. Giorni and Z. Kozakiewicz, 2003. Occurrence of ochratoxin A-producing fungi in grapes grown in Italy. Journal of Food Protection 66: 633-636.

Bau, M., G. Castellá, M.R. Bragulat and F.J. Cabañes, 2005b. DNA based characterization of ochratoxin A-producing and non producing *Aspergillus carbonarius* strains from grapes, Research in Microbiology 156: 375-381.

Bau, M., G. Castellá, M.R. Bragulat and F.J. Cabañes, 2006. RFLP characterization of *Aspergillus niger* aggregate species from grapes from Europe and Israel. International Journal of Food Microbiology 111: S18-S21.

Bau, M., M.R. Bragulat, M.L. Abarca, S. Mínguez and F.J. Cabañes, 2005a. Ochratoxigenic species from Spanish wine grapes. International Journal of Food Microbiology 98: 125-130.

Bayman, P., J.L. Baker, M.A. Doster, T.J. Michailides and N.E. Mahoney, 2002. Ochratoxin production by the *Aspergillus ochraceus* group and *Aspergillus alliaceus*. Applied and Environmental Microbiology 68: 2326-2329.

Bragulat, M.R., M.L. Abarca and F.J. Cabañes, 2001. An easy screening method for fungi producing ochratoxin A in pure culture. International Journal of Food Microbiology 71: 139-144.

Cabañes, F.J., F. Accensi, M.R. Bragulat, M.L. Abarca, G. Castellá, S. Mínguez and A. Pons, 2002. What is the source of ochratoxin A in wine? International Journal of Food Microbiology 79: 213-215.

Castellá, G., T.O. Larsen, F.J. Cabañes, H. Schmidt, A. Alboresi, L. Niessen, P. Färber and R. Geisen, 2002. Molecular characterization of ochratoxin A producing strains of the genus *Penicillium*. Systematic and Applied Microbiology 25: 74-83.

Ciegler, A., 1972. Bioproduction of ochratoxin A and penicillic acid by members of the *Aspergillus ochraceus* group. Canadian Journal of Microbiology 18: 631-636.

Clark, H.A. and S.M. Snedeker, 2006. Ochratoxin A: its cancer risk and potential for exposure. Journal of Toxicology and Environmental Health B9: 265-296.

Commission of the European Communities, 2005. Commission Regulation (EC) No 123/2005 of 26 January 2005 amending Regulation (EC) No 466/2001 as regards ochratoxin A. Official Journal of the European Communities L25: 3-5.

Commission of the European Communities, 2006. Commission Recommendation (EC) No 576/2006 of 17 August 2006 on the presence of deoxynivalenol, zearalenone, ochratoxin A, T-2 and HT-2 and fumonisins in products intended for animal feeding. Official Journal of the European Communities L229: 3-5.

Dalcero, A., C. Magnoli, C. Hallak, S.M. Chiachiera, G. Palacio and C.A.R. Rosa, 2002. Detection of ochratoxin A in animal feeds and capacity to produce this mycotoxin by *Aspergillus* section *Nigri* in Argentina. Food Additives and Contaminants 19: 1065-1072.

Dragacci, S., F. Grosso, R. Bire, J.M. Fremy and S. Coulon, 1999. A French monitoring programme for determining ochratoxin A occurrence in pig kidneys. Natural Toxins 7: 167-173.

EFSA, 2004. Opinion of the Scientific Panel on Contaminants in Food Chain on a request from the Commission related to ochratoxin A (OTA) as an undesirable substance in animal feed. The EFSA Journal 101: 1-36.

Filtenborg, O., J. C. Frisvad and J.A. Svendsen, 1983. Simple screening method for molds producing intracellular mycotoxins in pure cultures. Applied and Environmental Microbiology 45: 581-585.

Frisvad J.C. and O. Filtenborg, 1989. Terverticillate penicillia: chemotaxonomy and mycotoxin production. Mycologia 81: 837-861.

Frisvad, J.C. and R.A. Samson, 2000. *Neopetromyces* gen. nov. and an overview of teleomorphs of *Aspergillus* subgenus *Circumdati*. Studies in Mycology 45: 201-207.

Frisvad, J.C. and R.A. Samson, 2004. Polyphasic taxonomy of *Penicillium* subgenus *Penicillium*. A guide to identification of food and air-borne terverticillate Penicillia and their mycotoxins. Studies in Mycology 49: 1-173.

Frisvad J.C., F. Lund and S. Elmholt, 2005. Ochratoxin A producing *Penicillium verrucosum* isolates from cereals reveal large AFLP fingerprinting variability. Journal of Applied Microbiology 98: 684-692.

Frisvad J.C., J.M. Frank, J.A.M.P. Houbraken, A.F.A. Kuijpers and R.A. Samson, 2004a. New ochratoxin A producing species of *Aspergillus* section *Circumdati*. Studies in Mycology 50: 23-43.

Frisvad J.C, J. Smedsgaard, T.O. Larsen and R.A. Samson, 2004b. Mycotoxins, drugs and other extrolites produced by species in *Penicillium* subgenus *Penicillium*. Studies in Mycology 49: 201-242.

Frisvad J.C., K.F. Nielsen, R.A. Samson, 2006. Recommendations concerning the chronic problem of misidentification of mycotoxigenic fungi associated with foods and feeds. In: A.D. Hocking, J.I. Pitt, R.A. Samson and U. Thrane (eds.) Advances in Food Mycology. Advances in Experimental Medicine and Biology Vol.571, Springer Science, New York, 33-46.

Frisvad J.C., O. Filtenborg, F. Lund and U. Thrane, 1992. New selective media for the detection of toxigenic fungi in cereal products, meat and cheese. In: R.A. Samson, A.D. Hocking, J.I. Pitt and A.D. King (eds.) Modern Methods in Food Mycology. Elsevier, Amsterdam, pp 275-284.

Gams, W., M. Christensen, A.H.S. Onions, J.I. Pitt and R.A. Samson, 1985. Infrageneric taxa of *Aspergillus*. In: R.A. Samson and J.I. Pitt (eds.) Advances in *Aspergillus* and *Penicillium* systematics. Plenum Press, New York, pp. 55-62.

Geisen, R., Z. Mayer, A. Karolewiez and P. Farber, 2004. Development of a real-time PCR system for detection of *P. nordicum* and for monitoring ochratoxin A production in foods based on the ochratoxin polyketide synthase gene. Systematic and Applied Microbiology 27: 501-507.

Heenan C.N., K.J. Shaw and J.I. Pitt, 1998. Ochratoxin A production by *Aspergillus carbonarius* and *A. niger* isolates and detection using coconut cream agar. Journal of Food Mycology 1:67-72.

Horie, Y., 1995. Productivity of ochratoxin A of *Aspergillus carbonarius* in *Aspergillus* section *Nigri*. Nippon Kingakukai Kaiho 36: 73-76.

Jorgensen, K. and A. Petersen, 2002. Content of ochratoxin in paired kidney and meat samples from healthy Danish slaughter pigs. Food Additives and Contaminants 19: 562-567.

Larsen, T.O., A. Svendsen and J. Smedsgaard, 2001. Biochemical characterization of ochratoxin A-producing strains of the genus *Penicillium*. Applied and Environmental Microbiology 67: 3630-3635.

Leong, S.L., A.D. Hocking and E.S. Scott, 2007. *Aspergillus* species producing ochratoxin A: isolation from vineyard soils and infection of Semillon bunches in Australia. Journal of Applied Microbiology 102: 124-133.

Leong, S.L., A.D. Hocking and J.I. Pitt, 2004. Occurrence of fruit rot fungi (*Aspergillus* section *Nigri*) on some drying varieties of irrigated grapes. Australian Journal of Grape and Wine Research 10: 83-88.

Lund, F. and J.C. Frisvad, 2003. *Penicillium verrucosum* in wheat and barley indicates presence of ochratoxin A. Journal of Applied Microbiology 95: 1117-1123.

Medina, A., R. Mateo, L. López-Ocaña, F.M. Valle-Algarra and M. Jiménez, 2005. Study of Spanish grape mycobiota and ochratoxin A production by isolates of *Aspergillus tubingensis* and other members of *Aspergillus* section *Nigri*. Applied and Environmental Microbiology 71: 4696-4702.

Niessen, L., H. Schmidt, E. Mühlencoert, P. Farber, A. Karolewiez and R. Geisen, 2005. Advances in the molecular diagnosis of ochratoxin A-producing fungi. Food Additives and Contaminants 22: 324-334.

O'Callaghan, J. and A.D.W. Dobson, 2005. Molecular characterization of ochratoxin A biosynthesis and producing fungi. Advances in Applied Microbiology 58: 227-243.

O'Callaghan, J., M.X. Caddick, A.D.W. Dobson, 2003. A polyketide synthase gene required for ochratoxin A biosynthesis in *Aspergillus ochraceus*. Microbiology 148: 3485-3491.

Parenicova, L., P. Skouboe, J. Frisvad, R.A. Samson, L. Rossen, M. Ten Hoor-Suykerbuyk and J. Visser, 2001. Combined molecular and biochemical approach identifies *Aspergillus japonicus* and *Aspergillus aculeatus* as two species. Applied and Environmental Microbiology 67: 51-527.

Perrone, G., G. Mulè, A. Susca, P.Battilani, A. Pietri and A. Logrieco, 2006. Ochratoxin A production and amplified fragment length polymorphism analysis of *Aspergillus carbonarius, Aspergillus tubingensis,* and *Aspergillus niger* strains isolated from grapes in Italy. Applied and Environmental Microbiology 72: 680-685.

Peterson, S.W. 2000. Phylogenetic relationships in *Aspergillus* based on rDNA sequence analysis. In: R.A. Samson and J.I. Pitt (eds.) Integration of modern taxonomic methods for *Penicillium* and *Aspergillus* systematics. Harwood Academic Publishers, Amsterdam, pp. 323-355.

Petzinger, E. and A. Weidenbach, 2002. Mycotoxins in the food chain: the role of ochratoxins. Livestock Production Science 76: 245-250.

Pitt, J.I., 1987. *Penicillium viridicatum, Penicillium verrucosum* and production of ochratoxin A. Applied and Environmental Microbiology 53: 266-269.

Pitt, J.I. and A.D. Hocking, 1997. Fungi and food spoilage. Blackie Academic and Professional, London.

Ponsone, M.L., M. Combina, A. Dalcero and S. Chulze, 2007. Ochratoxin A and ochratoxigenic *Aspergillus* species in Argentinean wine grapes cultivated under organic and non-organic systems. International Journal of Food Microbiology 114: 131-135.

Samson, R.A., J.A.M.P. Houbraken, A.F.A. Kuijpers, J.M. Frank and J.C. Frisvad, 2004. New ochratoxin A or sclerotium producing species of *Aspergillus* section *Nigri*. Studies in Mycology 50: 45-61.

Serra, R., F.J. Cabañes, G. Perrone, G. Castellá, A. Venâncio, G. Mulè and Z. Kozakiewicz, 2006. *Aspergillus ibericus*: a new species of the section *Nigri* isolated from grapes. Mycologia 98: 295-306.

Stolk, A.C. ans R.A. Samson, 1985. A new taxonomic scheme for *Penicillium* anamorphs. In: R.A. Samson and J.I. Pitt (eds.) Advances in *Aspergillus* and *Penicillium* systematics. Plenum Press, New York, pp. 163-191.

Taniwaki, M.H., 2006. An update on ochratoxigenic fungi and ochratoxin A in coffee. In: A.D. Hocking, J.I. Pitt, R.A. Samson and U. Thrane U (eds.) Advances in Food Mycology. Advances in Experimental Medicine and Biology Vol. 571, Springer Science, New York, pp. 189-202.

Téren, J., J. Varga, Z. Hamari, E. Rinyu and F. Kevei, 1996. Immunochemical detection of ochratoxin A in black *Aspergillus* strains. Mycopathologia 134: 171-176.

Ueno, Y., O. Kawamura, Y. Sugiura, K. Horiguchi, M. Nakajima, K. Yamamoto and S. Sato, 1991. Use of monoclonal antibodies, enzyme-linked immunosorbent assay and immunoaffinity column chromatography to determine ochratoxin A in porcine sera, coffee products and toxin-producing fungi. In: M. Castegnaro, R. Plestina, G. Dirheimer, I.N. Chernozemsky and H. Bartsch (eds.) Mycotoxins, endemic nephropahy and urinary tract tumours. IARC-Sci.-Publ., Lyon, pp. 71-75.

Van der Merwe K.J., P.S. Steyn, L. Fourie, D.B. Scott and J.J. Theron, 1965. Ochratoxin A, a toxic metabolite produced by *Aspergillus ochraceus* Wilh. Nature 205: 1112-1113.

Varga, J., E. Kevei, E. Rinyu, J. Téren and Z. Kozakiewicz, 1996. Ochratoxin production by *Aspergillus* species. Applied and Environmental Microbiology 62: 4461-4464.

Visconti, A. and A. de Girolamo, 2005. Fitness for purpose- ochratoxin A analytical developments. Food Additives and Contaminants (Suppl. 1): 37-44.

Wicklow, D.T., P.F. Dowd, A.A. Alfatafta and J.B. Gloer, 1996. Ochratoxin A: an antiinsectan metabolite from the sclerotia of *Aspergillus carbonarius* NRRL 369. Canadian Journal of Microbiology 42: 1100-1103.

Zotti, M. and A.M. Corte, 2002. *Aspergillus persii:* a new species in section *Circumdati.* Mycotaxon 83: 269-278.

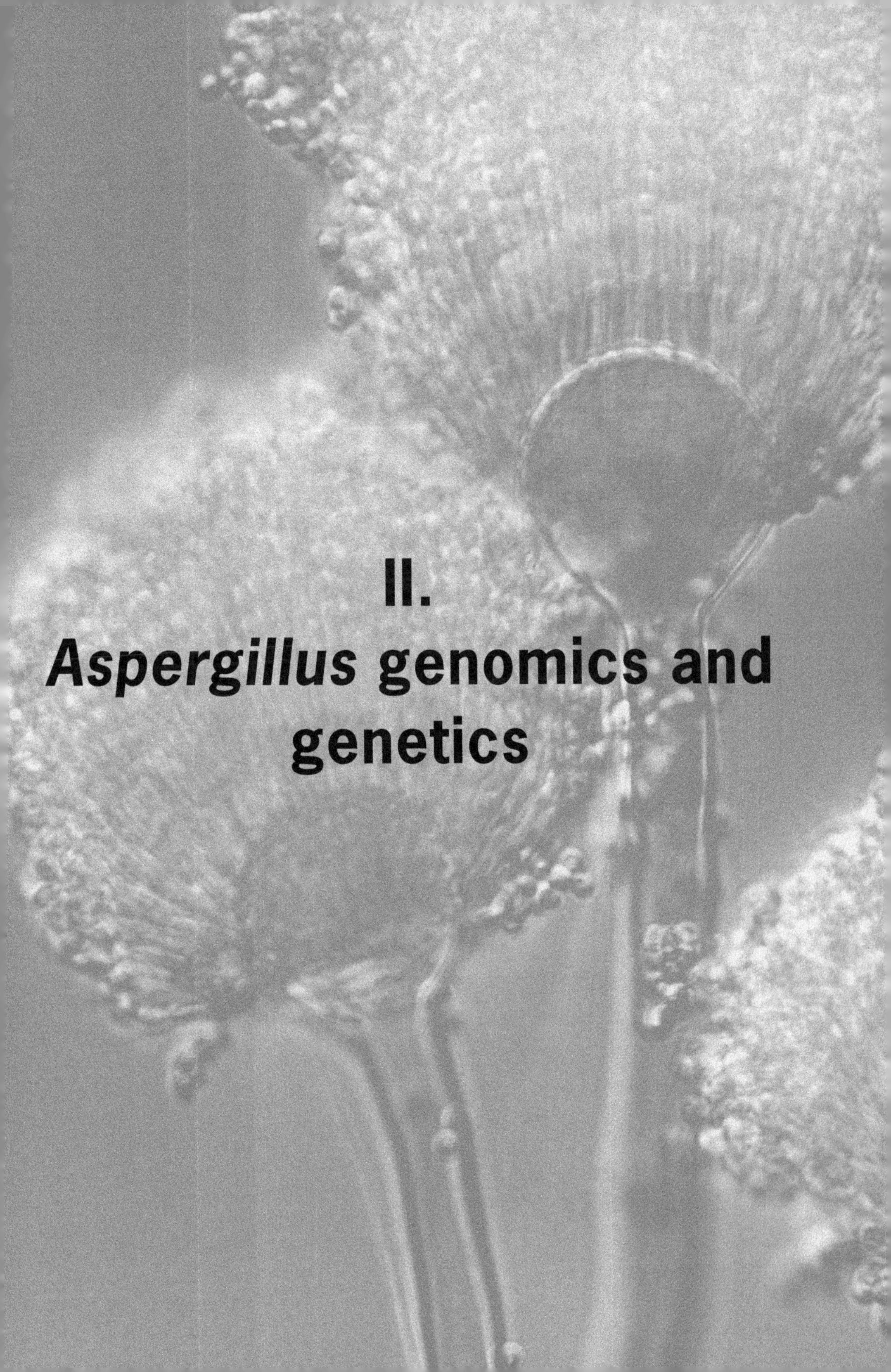

II.
Aspergillus genomics and genetics

Aspergillus genomics and DHN-melanin conidial pigmentation

Scott E. Baker
Fungal Biotechnology Team, Chemical and Biological Process Group, Energy and Environment Directorate, Pacific Northwest National Laboratory, 902 Battelle Blvd., Richland, Washington 99354, USA

Abstract

The genome sequencing projects spread across the phylogenetic diversity of the genus *Aspergillus* have brought an enormous amount of gene sequence data to a research community with a rich history of genetics and biochemistry. Included in this history is a body of research focused on conidial pigments, in particular DHN-melanin. The DHN-melanin biosynthetic pathway has been characterised biochemically and genetically in Aspergilli and other ascomycete fungi, and while there are differences in the polyketide backbone produced in *Aspergillus* compared to some other fungi, a large part of the pathway to the base subunit of the melanin polymer, 1,8-dihydroxynapthalene, is conserved. The genome sequence resources currently available confirm that the genes described at the genetic and/or biochemical level in a given organism are conserved across numerous Aspergilli.

Keywords: DHN-melanin, polyketide synthase, pigment, biosynthesis, genomics, *A. fumigatus, A. nidulans, A. oryzae, A. niger*

1. Introduction

Fungi commonly possess characteristic pigments that are used as a character for distinguishing between species. The Aspergilli are a particularly rich source of pigmentation genes and gene models; genome sequences of *Aspergillus nidulans, Aspergillus fumigatus, Aspergillus oryzae, Aspergillus flavus, Aspergillus clavatus, Aspergillus terreus, Neosartorya fischeri* and *Aspergillus niger*, are published and/or publicly available. The conidial pigments of these and other Aspergilli have been studied for decades at both the biochemical and genetic levels resulting in an impressive body of scientific research. Indeed, the prior research in the area of *Aspergillus* species pigmentation combined with an explosion of genome sequencing projects makes it possible to associate genes and gene models with pigment biosynthetic pathways.

Of fungal pigments, melanin is the best studied. Melanin is the generic name for dark brown or black pigments produced not only by fungi, but other microbes, animals, and plants. Within the kingdom Fungi, there are a number of known biosynthetic pathways for melanin production. A large number of ascomycete fungi produce melanin through a pathway where 1,8-dihydroxynapthalene (1,8-DHN) is the precursor for polymerisation. DHN-melanin is produced from a polyketide backbone that is subsequently modified by reduction and dehydration. While not essential for growth, fungal melanins are known to play important roles in protection from environmental and competitive stresses, infection of plants and animals, and development including asexual and sexual reproduction (Wheeler and Bell, 1988; Langfelder *et al.*, 2003). DHN-melanin is integral to the wide assortment of conidial colours that occur in *Aspergillus* species.

2. Biochemistry and genetics of DHN-melanin production

2.1. Biosynthesis

Genetic mutants were integral to the early characterisation of the DHN-melanin pathway. Albino microsclerotia (*alm*) mutants were co-cultured with brown microsclerotia (*brm*) mutants of *Verticillium dahliae* (Bell *et al.*, 1976). It was found that a diffusible factor, shown to be scytalone, a precursor to 1,8-DHN, from the *brm* mutants was able to restore wildtype melanisation to the albino microsclerotia (Bell *et al.*, 1976). Subsequent research showed that scytalone or 1,8-DHN could be fed to a variety of albino mutants from ascomycetes (Wheeler and Bell, 1988; Langfelder *et al.*, 2003).

Two similar pathways have been discovered for the production of 1,8-DHN. The main difference between the two pathways is the way in which 1,3,6,8-tetrahydroxynapthahalene (1,3,6,8-THN) is synthesised. In many fungi, 1,3,6,8-THN is direct pentaketide product of a polyketide synthase (Wheeler and Bell, 1988). However, in *A. fumigatus* and *A. nidulans*, the melanin pathway associated polyketide synthase has been shown to produce a heptaketide which must subsequently be 'shortened' to form 1,3,6,8-THN (Tsai *et al.*, 2001; Fujii *et al.*, 2004). In *A. fumigatus* it has been shown that the conversion of the heptaketide *yA* product is catalysed by hydrolytic action by the protein product of *aygl* (Fujii *et al.*, 2004). From 1,3,6,8-THN two sets of reduction followed by dehydration produce 1,8-DHN which is subsequently polymerised to form melanin (Figure 1). In many cases the reduction steps can be inhibited by the compound tricyclazole; sensitivity of pigmentation to tricyclazole has been used as an indicator of a DHN-melanin pathway (Wheeler, 1983; Wheeler and Bell, 1988). Copper oxidases are thought to be involved in the

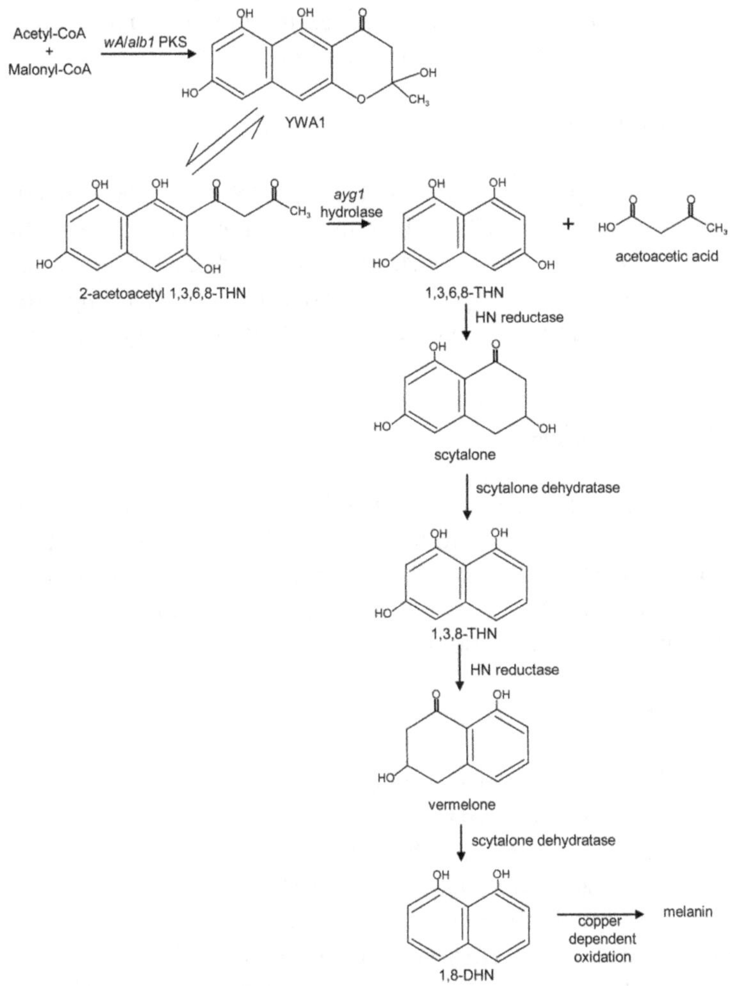

Figure 1. The 1,8-DHN biosynthetic pathway in Aspergilli (Tsai et al.*, 2001; Langfelder* et al.*, 2003; Fujii* et al.*, 2004).*

polymerisation of 1,8-DHN into melanin (Clutterbuck, 1972; Wheeler and Bell, 1988; Brown *et al.*, 1993). The 1,8-DHN-based melanin polymers are likely elaborated by covalent interactions with proteins, carbohydrates and other compounds (Ray and Eakin, 1975).

Research conducted primarily in *A. fumigatus* and *A. nidulans* has formed the basis of associating gene function with biochemical reaction in the Aspergilli. This work has formed the foundation upon which an integrated genetic and biochemical mechanism for DHN-melanin production has been built.

2.2. Polyketide synthase

Many fungi are known produce a 1,3,6,8-THN pentaketide DHN-melanin starting backbone (Wheeler and Bell, 1988). In contrast, there is experimental evidence that two Aspergilli, *A. fumigatus* and *A. nidulans* produce a heptaketide backbone called naphtopyrone YWA1. When *A. nidulans wA* and *A. fumigatus pksP/alb1* were expressed in *A. oryzae,* the naphtopyrone YWA1 heptaketide was produced (Watanabe *et al.*, 1999, 2000).

2.3. aygl, aYWA1 hydrolase

Genetic and biochemical evidence in *A. fumigatus* indicate that the napthapyrone YWA1 is hydrolytically shortened to form 1,3,6,8-THN and acetoacetic acid by the protein product of *aygl* (Tsai *et al.*, 2001). Ayglp is thought to work on the open form of YWA1 (2-acetoacetyl 1,3,6,8-THN) which is in equilibrium with the closed napthapyrone form (Fujii *et al.*, 2004).

2.4. HN reductase

The reduction of 1,3,6,8-THN to scytalone and the reduction of 1,3,8-THN to vermelone are catalysed by hydroxynaphtalene (HN) reductases. Across a number of ascomycetes, HN reductases have been shown to be inhibited by tryicyclazole. Thus, there is a strong association pigment production inhibition by tricyclazole with the presence of a DHN-melanin pathway. It has therefore been assumed that melanin producing fungi which are insensitive to tricyclazole such as *A. nidulans* and *A. niger,* do not use the DHN-melanin pathway (Wheeler and Bell, 1988). Alternatively, it is possible that tricyclazole insensitive fungi produce tricyclazole resistant HN reductases.

2.5. Scytalone dehydrogenase

As noted previously, scytalone is the diffusible compound that came to define the commonality of the DHN-melanin pathway. From study of the *A. fumigatus* DHN-pathway it is known that both scytalone and vermelone are substrates for scytalone dehydrogenase (Tsai *et al.,* 1999). However, the genome sequences of several Aspergilli reveal the presence of at least one hydrolase with similarity to the A. *fumigatus* scytalone dehydrogenase.

2.6. Copper dependent oxidases

Copper dependent oxidases, such as the laccase encoded by the *A. nidulans yA* gene, are thought to be involved in the polymerisation of DHN to form the melanin polymer (Clutterbuck, 1972, Wheeler and Bell, 1988). For example, it has been shown that 2,4-dithiopyrimidine which chelates Cu^{2+} inhibits aspergillin production in *A. niger* and in the conidia of laccase deficient *Aspergillus parasiticus* strains, an orange pigment intermediate accumulates (Ray and Eakin, 1975, Brown *et al.*, 1993).

Interestingly, deletion studies of *A. fumigatus abr2* indicate that laccase activity on 1,8-DHN is needed for wildtype pigmentation and conidial surface formation, but not for reactive oxygen species protection or virulence in a mouse infection model (Sugareva *et al.*, 2006). Encoded in an individual fungal genome, there are multiple polyketide synthases (PKS) that when subjected to phylogenetic analysis group together with known YWA1 or 1,3,6,8-THN producing PKS encoding genes (Kroken *et al.*, 2003; Varga *et al.*, 2003). It is attractive to speculate that the products of the non-YWA1 or non-1,3,6,8-THN may produce compounds that are incorporated into the DHN-melanin polymer to give each organism its unique conidial colour.

3. Conidial DHN-melanin pigments and genomics

3.1. Aspergillus fumigatus

Aspergillus fumigatus conidia are grayish-green in colour (Klich, 2002). Due to the body of recent molecular genetic studies and the clustering of its genes on chromosome 2 (Table 1), *A. fumigatus* would appear to have the simplest conidial pigmentation system of the Aspergilli studied thus far. However, this is likely not to be the case. A single tricyclazole sensitive HN reductase is found in the *A. fumigatus* melanin gene cluster (Tsai *et al.*, 1999). However, evidence from gene deletion and inhibition studies has shown that a second tricyclazone sensitive HN reductase is present in *A. fumigatus* that can catalyse 1,3,8-THN reduction to vermelone (Tsai *et al.*, 1999). The gene encoding this second HN reductase is not co-located with the well characterised pigmentation gene cluster (Tsai *et al.*, 1999).

A combination reverse genetic and biochemical approach has been used to characterise the conidial pigmentation pathway in *A. fumigatus* (Tsai *et al.*, 1997, 1999, 2001; Langfelder *et al.*, 1998; Fujii *et al.*, 2004; Sugareva, *et al.*, 2006). In contrast to *A. nidulans*, six *A. fumigatus* DHN-melanin synthesis genes are clustered. Each gene has been deleted by targeted gene replacement. *pksP/alb1* was the first *A. fumigatus* pigment gene to be characterised genetically and was found to play a

Table 1. DHN-melanin biosynthetic pathway genes with GenBank accession numbers.

Organism	Polyketide synthase	YWA1 hydrolase	HN reductase	Scytalone dehydratase	Laccase	Multi-copper oxidase
A. fumigatus[1]	*pksP/alb*	*ayg1*	*arp2*	*arp1*	*abr2*	*abr1*
	CAA76740	AAF03354	AAF03314	AAC49843	AAF03349	AAF03353
A. nidulans[2]	*wA/*		*chaA* (?)	*chaA* (?)	*yA*	
	CAA46695	EAA61462	EAA66019	EAA66018	EAA58164	EAA61461
	II	IV	VIII	VIII	I	IV
A. oryzae[3]						
	BAE61567	BAE55407	BAE59519	no gene model	BAE61606	BAE61568
	13	2	10	12	13	13
A. niger[4]	CAL00851	CAK42096	multiple candidates	CAK96768	multiple candidates	CAK42098
	1	1		8		1

[1]All six *A. fumigatus* genes listed are clustered on chromosome 2.

[2]*chaA* predicted to be associated with gene models based on physical map, genome sequence, linkage map (http://www.gla.ac.uk/ibls/molgen/Aspergillus/index.html); EAA66018 and EAA66019 are adjacent to each other; EAA61461 and EAA61462 are adjacent to each other. Roman numerals indicate linkage group.

[3]Genes BAE61567 and BAE61568 are adjacent to one another. The HN reductase is located in what is considered to be a non-functional aflatoxin biosynthetic gene cluster (Machida *et al.*, 2005). Numbers indicate supercontigs.

[4]Genes are from *A. niger* strain CBS 513.88 (Pel *et al.*, 2007); CAK42096 and CAK42098 are co-located; CAL00851 is located on the opposite chromosome arm from CAK42096 and CAK42098. Numbers indicate chromosome location. 'Multiple candidates' indicates that a clear ortholog could not be determined by BLAST-based sequence analysis or by co-localisation in the genome with other candidate DHN-melanin biosynthetic genes.

role in pathogenicity in a mouse model of *A. fumigatus* infection (Langfelder *et al.*, 1998; Tsai *et al.*, 1998). In addition *pksP/alb1* mutants had defects in spore surface morphology and increased sensitivity to reactive oxygen species (Langfelder *et al.*, 1998; Tsai *et al.*, 1998).

3.2. Aspergillus nidulans

Aspergillus nidulans has been a model genetic fungus for over 60 years with a number of conidial colour mutants having been isolated. Wildtype conidia are green in colour (Klich, 2002). Two *A. nidulans* genes in particular, *wA* and *yA* have been studied in great detail - *wA* mutant conidia lack pigment while *yA* mutant conidia are yellow in colour and deficient in copper laccase activity (Pontecorvo *et al.*, 1953b; Clutterbuck, 1972; Mayorga and Timberlake, 1990; Tilburn *et al.*, 1990). While *A. nidulans* conidial pigmentation would appear to be more complex than *A. fumigatus* based on number of conidial pigmentation mutants and the fact that they are not clustered, the strong genetic (sexual and parasexual) infrastructure of the *A. nidulans* system has made the isolation and characterisation of pigment colour mutants technically more straight forward (Pontecorvo *et al.*, 1953b; Kafer, 1958, 1961; Clutterbuck, 1965, 1972; Kurtz and Champe, 1981).

Based on correlations between the genome sequence, physical map and linkage map, a number of classical genetic conidial colour mutants have been putatively associated with gene models (classical mutants matched to the most current *A. nidulans* gene models have been listed on the Glasgow *Aspergillus nidulans* gene list website http://www.gla.ac.uk/ibls/molgen/Aspergillus/index.html) (Galagan *et al.*, 2005). Indeed, for each of the six genes in the *A. fumigatus* DHN-melanin cluster, a candidate *A. nidulans* gene ortholog can be found (Table 1).

3.3. Aspergillus oryzae

While work has been done on DOPA-melanin, there is a paucity of research on conidial pigmentation/DHN-melanin in *A. oryzae* (Langfelder *et al.*, 2003). The conidia of *A. oryzae* can vary from grayish yellow to olive brown (Klich, 2002). Interestingly, *A. oyzae* has been used as an expression host for the characterisation of the product (napthapyrone YWA1) of both the *A. nidulans wA* and *A. fumigatus pksP/alb1* genes (Watanabe *et al.*, 1999, 2000).

The release and analysis of the genome sequence for *A. oryzae* occurred in 2005 (Machida *et al.*, 2005). The genome sequence reveals that genes for DHN-melanin synthesis are present (Table 1). Interestingly the best candidate HN reductase for

1,8-DHN synthesis (based on *A. fumigatus arp2* as a query) is found in the *A. oryzae* aflatoxin cluster. The *A. oyzae* aflatoxin cluster is considered non-functional with regard to aflatoxin production (Machida *et al.*, 2005). Interestingly, tricyclazole inhibits aflatoxin production in *A. flavus* (Wheeler *et al.*, 1989). The *A. flavus* tricyclazole sensitve versicolorin reductase amino acid sequence is 97% identical to its *A. oryzae* ortholog (GenBank accession BAE59519). An intriguing possibility is that the *A. oryzae* and *A. flavus* versicolorin reductase genes may be involved in 1,8-DHN biosynthesis. While the current gene models for *A. oryzae* lack a scytalone dehydratase encoding gene, BLAST-based sequence analysis indicates that a candidate gene ortholog is present.

3.4. Aspergillus niger

Aspergillus niger conidia are dark brown to black in colour (Klich, 2002). Conidial colour mutants have been generated in *A. niger* for use in development of parasexual or mitotic recombination linkage maps (Pontecorvo *et al.*, 1953a). These colour mutants include *fwnA* (fawn coloured conidia), *brnA* (brown coloured conidia) and *olvA* (olive coloured conidia) all found in linkage group I (Bos *et al.*, 1988, 1994; Debets *et al.*, 1993).

Aspergillin, the black pigment of *A. niger* conidia, has been the subject of limited biochemical characterisation. Aspergillin is thought to be a polymer composed of DHN melanin and a hexahydroxyl pentacyclic quinoid (HPQ) compound (Ray and Eakin, 1975). Chelation of copper ions was found to inhibit the biochemical reactions that form aspergillin from the combination of DHN-melanin and the HPQ compound (Ray and Eakin, 1975). The copper ion dependency of aspergillin formation could be indicative of a role for a copper dependent laccase similar to the *A. nidulans yA* gene that is thought to play a role in the polymerisation of DHN-melanin (Clutterbuck, 1972; Wheeler and Bell, 1988).

Three strains of *A. niger* have been the subject of genome projects (Baker, 2006; Pel *et al.*, 2007). In early 2007, the detailed analysis of one of these strains, CBS 513.88 was published (Pel *et al.*, 2007). BLAST-based sequence analyses of the *A. niger* strain CBS 513.88 and strain ATCC 1015 genomes indicates that there are strong candidate gene orthologs for the six clustered and characterised *A. fumigatus* DHN-melanin biosynthetic genes (Table 1).

4. Conclusions and future directions

Conidial DHN-melanin pigments are more than just compounds that make the Aspergilli such beautiful organisms – they are the product of a fascinating biosynthetic pathway that is developmentally regulated. In addition, DHN-melanin in Aspergilli is important in pathogenicity and in resistance to chemical and environmental stress. Research in *Aspergillus* pigment production led to the discovery of YWA1, a heptaketide napthapyrone, that serves as a starting point in the biosynthesis to 1,8-DHN (Tsai *et al.*, 2001; Fujii *et al.*, 2004). The presence of YWA1 hydrolase *aygl* orthologs in the genomes of *A. nidulans, A. oryzae* and *A. niger* indicate that a heptaketide 'shortening' route to DHN-melanin could be common across the genus *Aspergillus* (Table 1).

Because pigmentation in both *A. niger* and *A. nidulans* is resistant to tricyclazole inhibition, it has been asserted that these fungi did not make DHN-melanin (Wheeler and Bell, 1988). However, the presence of scytalone dehydratase and other orthologs of DHN-melanin synthesis could be taken as evidence that tricyclazole resistant HN reductases are encoded with the genomes of *A. niger* and *A. nidulans*. BLAST-based homology searches indicate the presence of several candidate HN reductases in both *A. niger* and *A. nidulans* that are attractive candidates for future research.

Putative regulatory genes for melanin synthesis with similarity to the *Magnaporthe grisea pig1* and *Colletotrichum lagenarium cmr1* melanin regulatory genes have been noted in the annotations of *A. fumigatus* and *A. niger* (Nierman *et al.*, 2005; Pel *et al.*, 2007). However, detailed studies of this gene are not currently published, so only sequence similarity can be used at this point to establish orthology with *pig1* and *cmr1*.

Aspergillus conidial pigmentation research in the 'genomics era' should prove to be exciting. For example, the genome sequences of *A. nidulans* and *A. niger* will aid in the characterisation of previously isolated colour mutant genes at the sequence level. In fact, some predictions linking classical mutants to genome sequence derived gene models for *A. nidulans* have already been made (Table 1). Furthermore, the genome sequences and the transcriptomic and proteomic tools that they enable will provide insight into many facets of conidial pigmentation in Aspergilli. These insights may include an expanded list of DHN-melanin pathway regulation genes as well as specific genes involved in the polymerisation of 1,8-DHN into DHN-melanin based pigments. The 1,8-DHN biosynthetic pathway was characterised well before the *Aspergillus* genomics era; it is in the post-genomics era that the genetics and biochemistry of 1,8-DHN polymerisation into the myriad of DHN-melanin based conidial pigments will prove a fruitful area of study for *Aspergillus* researchers.

Acknowledgements

Thanks to Brooke Smitha for her work on aspergillin, Kenneth S. Bruno and Mikael R. Andersen for critical readings of the text and helpful scientific discussions on conidial pigmentation and Kirsten J. Holmboe for crucial logistical support.

References

Baker, S.E., 2006. *Aspergillus niger* genomics: past, present and into the future. Medical Mycology 44 Suppl 1: S17-S21.

Bell, A.A., J.E. Puhalla, W.J. Tolmsoff and R.D. Stipanovic, 1976. Use of mutants to establish (+)-scytalone as an intermediate in melanin biosynthesis by *Verticillium dahliae*. Canadian Journal of Microbiology 22: 787-799.

Bos, C.J., A.J. Debets, K. Swart, A. Huybers, G. Kobus and S.M. Slakhorst, 1988. Genetic analysis and the construction of master strains for assignment of genes to six linkage groups in *Aspergillus niger*. Current Genetics 14: 437-443.

Bos, C.J., S.M. Slakhorst, A.J.M. Debets and K. Swart, 1994. Linkage group analysis in *Aspergillus niger*. Applied Microbiology and Biotechnology 38: 742-745.

Brown, D.W., F.M. Hauser, R. Tommasi, S. Corlett and J.J. Salvo, 1993. Structural elucidation of a putaive conidial pigment intermediate in *Aspergillus parasiticus*. Tetrahedron Letters 34: 419-422.

Clutterbuck, A.J., 1965. A fawn conidia mutant in *Aspergillus nidulans*. Aspergillus Newsletter 6: 12.

Clutterbuck, A.J., 1972. Absence of laccase from yellow-spored mutants of *Aspergillus nidulans*. Journal of General Microbiology 70: 423-435.

Debets, F., K. Swart, R.F. Hoekstra and C.J. Bos, 1993. Genetic maps of eight linkage groups of *Aspergillus niger* based on mitotic mapping. Current Genetics 23: 47-53.

Fujii, I., Y. Yasuoka, H.F. Tsai, Y.C. Chang, K.J. Kwon-Chung and Y. Ebizuka, 2004. Hydrolytic polyketide shortening by ayg1p, a novel enzyme involved in fungal melanin biosynthesis. Journal of Biological Chemistry 279: 44613-44620.

Galagan, J.E., S.E. Calvo, C. Cuomo, L.J. Ma, J.R. Wortman, S. Batzoglou, S.I. Lee, M. Basturkmen, C.C. Spevak, J. Clutterbuck, V. Kapitonov, J. Jurka, C. Scazzocchio, M. Farman, J. Butler, S. Purcell, S. Harris, G.H. Braus, O. Draht, S. Busch, C. D'Enfert, C. Bouchier, G.H. Goldman, D. Bell-Pedersen, S. Griffiths-Jones, J.H. Doonan, J. Yu, K. Vienken, A. Pain, M. Freitag, E.U. Selker, D.B. Archer, M.A. Penalva, B.R. Oakley, M. Momany, T. Tanaka, T. Kumagai, K. Asai, M. Machida, W.C. Nierman, D.W. Denning, M. Caddick, M. Hynes, M. Paoletti, R. Fischer, B. Miller, P. Dyer, M.S. Sachs, S.A. Osmani and B.W. Birren, 2005. Sequencing of *Aspergillus nidulans* and comparative analysis with *A. fumigatus* and *A. oryzae*. Nature 438: 1105-1115.

Kafer, E., 1958. An 8-chromosome map of *Aspergillus nidulans*. Advances in Genetics 9: 105-145.

Kafer, E., 1961. The processes of spontaneous recombination in vegetative nuclei of *Aspergillus nidulans*. Genetics 46: 1581-609.

Klich, M.A., 2002. Identification of common *Aspergillus* species. Centraalbureau voor Schimmelcultures, Utrecht, 116p.

Kroken, S., N.L. Glass, J.W. Taylor, O.C. Yoder and B.G. Turgeon, 2003. Phylogenomic analysis of type I polyketide synthase genes in pathogenic and saprobic ascomycetes. Proceedings of the National Academy of Sciences USA 100: 15670-15675.

Kurtz, M.B. and S.P. Champe, 1981. Dominant spore color mutants of *Aspergillus nidulans* defective in germination and sexual development. Journal of Bacteriology 148: 629-638.

Langfelder, K., B. Jahn, H. Gehringer, A. Schmidt, G. Wanner and A.A. Brakhage, 1998. Identification of a polyketide synthase gene (*pksP*) of *Aspergillus fumigatus* involved in conidial pigment biosynthesis and virulence. Medical Microbiology and Immunology 187: 79-89.

Langfelder, K., M. Streibel, B. Jahn, G. Haase and A.A. Brakhage, 2003. Biosynthesis of fungal melanins and their importance for human pathogenic fungi. Fungal Genetics and Biology 38: 143-158.

Machida, M., K. Asai, M. Sano, T. Tanaka, T. Kumagai, G. Terai, K. Kusumoto, T. Arima, O. Akita, Y. Kashiwagi, K. Abe, K. Gomi, H. Horiuchi, K. Kitamoto, T. Kobayashi, M. Takeuchi, D.W. Denning, J.E. Galagan, W.C. Nierman, J. Yu, D.B. Archer, J.W. Bennett, D. Bhatnagar, T.E. Cleveland, N.D. Fedorova, O. Gotoh, H. Horikawa, A. Hosoyama, M. Ichinomiya, R. Igarashi, K. Iwashita, P.R. Juvvadi, M. Kato, Y. Kato, T. Kin, A. Kokubun, H. Maeda, N. Maeyama, J. Maruyama, H. Nagasaki, T. Nakajima, K. Oda, K. Okada, I. Paulsen, K. Sakamoto, T. Sawano, M. Takahashi, K. Takase, Y. Terabayashi, J.R. Wortman, O. Yamada, Y. Yamagata, H. Anazawa, Y. Hata, Y. Koide, T. Komori, Y. Koyama, T. Minetoki, S. Suharnan, A. Tanaka, K. Isono, S. Kuhara, N. Ogasawara and H. Kikuchi, 2005. Genome sequencing and analysis of *Aspergillus oryzae*. Nature 438: 1157-1161.

Mayorga, M.E. and W.E. Timberlake, 1990. Isolation and molecular characterization of the *Aspergillus nidulans wA* gene. Genetics 126: 73-79.

Nierman, W.C., A. Pain, M.J. Anderson, J.R. Wortman, H.S. Kim, J. Arroyo, M. Berriman, K. Abe, D.B. Archer, C. Bermejo, J. Bennett, P. Bowyer, D. Chen, M. Collins, R. Coulsen, R. Davies, P.S. Dyer, M. Farman, N. Fedorova, N. Fedorova, T.V. Feldblyum, R. Fischer, N. Fosker, A. Fraser, J.L. Garcia, M.J. Garcia, A. Goble, G.H. Goldman, K. Gomi, S. Griffith-Jones, R. Gwilliam, B. Haas, H. Haas, D. Harris, H. Horiuchi, J. Huang, S. Humphray, J. Jimenez, N. Keller, H. Khouri, K. Kitamoto, T. Kobayashi, S. Konzack, R. Kulkarni, T. Kumagai, A. Lafon, J.P. Latge, W. Li, A. Lord, C. Lu, W.H. Majoros, G.S. May, B.L. Miller, Y. Mohamoud, M. Molina, M. Monod, I. Mouyna, S. Mulligan, L. Murphy, S. O'Neil, I. Paulsen, M.A. Penalva, M. Pertea, C. Price, B.L. Pritchard, M.A. Quail, E. Rabbinowitsch, N. Rawlins, M.A. Rajandream, U. Reichard, H. Renauld, G.D. Robson, S. Rodriguez de Cordoba, J.M. Rodriguez-Pena, C.M. Ronning, S. Rutter, S.L. Salzberg, M. Sanchez, J.C. Sanchez-Ferrero, D. Saunders, K. Seeger, R. Squares, S. Squares, M. Takeuchi, F. Tekaia, G. Turner, C.R. Vazquez de Aldana, J. Weidman, O. White, J. Woodward, J.H. Yu, C. Fraser, J.E. Galagan, K. Asai, M. Machida, N. Hall, B. Barrell and D.W. Denning, 2005. Genomic sequence of the pathogenic and allergenic filamentous fungus *Aspergillus fumigatus*. Nature 438: 1151-1156.

Pel, H.J., J.H. de Winde, D.B. Archer, P.S. Dyer, G. Hofmann, P.J. Schaap, G. Turner, R.P. de Vries, R. Albang, K. Albermann, M.R. Andersen, J.D. Bendtsen, J.A. Benen, M. van den Berg, S. Breestraat, M.X. Caddick, R. Contreras, M. Cornell, P.M. Coutinho, E.G. Danchin, A.J. Debets, P. Dekker, P.W. van Dijck, A. van Dijk, L. Dijkhuizen, A.J. Driessen, C. d'Enfert, S. Geysens, C. Goosen, G.S. Groot, P.W. de Groot, T. Guillemette, B. Henrissat, M. Herweijer, J.P. van den Hombergh, C.A. van den Hondel, R.T. van der Heijden, R.M. van der Kaaij, F.M. Klis, H.J. Kools, C.P. Kubicek, P.A. van Kuyk, J. Lauber, X. Lu, M.J. van der Maarel, R. Meulenberg, H. Menke, M.A. Mortimer, J. Nielsen, S.G. Oliver, M. Olsthoorn, K. Pal, N.N. van Peij, A.F. Ram, U. Rinas, J.A. Roubos, C.M. Sagt, M. Schmoll, J. Sun, D. Ussery, J. Varga, W. Vervecken, P.J. van de Vondervoort, H. Wedler, H.A. Wosten, A.P. Zeng, A.J. van Ooyen, J. Visser and H. Stam, 2007. Genome sequencing and analysis of the versatile cell factory *Aspergillus niger* CBS 513.88. Nature Biotechnology 25: 221-231.

Pontecorvo, G., J.A. Roper and E. Forbes, 1953a. Genetic recombination without sexual reproduction in *Aspergillus niger*. Journal of General Microbiology 8: 198-210.

Pontecorvo, G., J.A. Roper, L.M. Hemmons, K.D. Macdonald and A.W. Bufton, 1953b. The genetics of *Aspergillus nidulans*. Advances in Genetics 5: 141-238.

Ray, A.C. and R.E. Eakin, 1975. Studies on the biosynthesis of aspergillin by *Aspergillus niger*. Applied Microbiology 30: 909-915.

Sugareva, V., A. Hartl, M. Brock, K. Hubner, M. Rohde, T. Heinekamp and A.A. Brakhage, 2006. Characterisation of the laccase-encoding gene *abr2* of the dihydroxynaphthalene-like melanin gene cluster of *Aspergillus fumigatus*. Archives of Microbiology 186: 345-355.

Tilburn, J., F. Roussel and C. Scazzocchio, 1990. Insertional inactivation and cloning of the wA gene of *Aspergillus nidulans*. Genetics 126: 81-90.

Tsai, H.F., I. Fujii, A. Watanabe, M.H. Wheeler, Y.C. Chang, Y. Yasuoka, Y. Ebizuka and K.J. Kwon-Chung, 2001. Pentaketide melanin biosynthesis in *Aspergillus fumigatus* requires chain-length shortening of a heptaketide precursor. Journal of Biological Chemistry 276: 29292-29298.

Tsai, H.F., M.H. Wheeler, Y.C. Chang and K.J. Kwon-Chung, 1999. A developmentally regulated gene cluster involved in conidial pigment biosynthesis in *Aspergillus fumigatus*. Journal of Bacteriology 181: 6469-77.

Tsai, H.F., R.G. Washburn, Y.C. Chang and K.J. Kwon-Chung, 1997. *Aspergillus fumigatus* arp1 modulates conidial pigmentation and complement deposition. Molecular Microbiology 26: 175-183.

Tsai, H.F., Y.C. Chang, R.G. Washburn, M.H. Wheeler and K.J. Kwon-Chung, 1998. The developmentally regulated *alb1* gene of *Aspergillus fumigatus*: its role in modulation of conidial morphology and virulence. Journal of Bacteriology 180: 3031-3038.

Varga, J., K. Rigo, S. Kocsube, B. Farkas and K. Pal, 2003. Diversity of polyketide synthase gene sequences in *Aspergillus* species. Research in Microbiology 154: 593-600.

Watanabe, A., I. Fujii, H. Tsai, Y.C. Chang, K.J. Kwon-Chung and Y. Ebizuka, 2000. *Aspergillus fumigatus alb1* encodes naphthopyrone synthase when expressed in *Aspergillus oryzae*. FEMS Microbiology Letters 192: 39-44.

Watanabe, A., I. Fujii, U. Sankawa, M.E. Mayorga, W.E. Timberlake and Y. Ebizuka, 1999. Re-identification of *Aspergillus nidulans wA* gene to code for a polyketide synthase of napthopyrone. Tetrahedron Letters 40: 91-94.

Wheeler, M.H., 1983. Comparisons of fungal melanin biosynthesis in ascomycetous, imperfect and basidiomycetous fungi. Transactions of the British Mycological Society 81: 29-36.

Wheeler, M.H. and A.A. Bell, 1988. Melanins and their importance in pathogenic fungi. Current Topics in Medical Mycology 2: 338-387.

Wheeler, M.H., D. Bhatnagar and M.G. Rojas, 1989. Chlobenthiazone and tricyclazole inhibition of aflatoxin biosynthesis by *Aspergillus flavus*. Pesticide Biochemistry and Physiology 35: 315-323.

A genomic look at physiology and extracellular enzymes of *Aspergillus* in relation to utilization of plant matter

Ronald P. de Vries
Microbiology, Department of Biology, Utrecht University, Padualaan 8, 3584 CH Utrecht, The Netherlands

Abstract

Utilisation of plant matter by filamentous fungi such as *Aspergillus* requires extracellular degradation of the polymeric compounds, uptake of the monomeric products and metabolic conversions to allow growth of the fungus. Fungi organise efficient use of plant matter though a set of transcriptional regulators which ensure that the right extracellular enzyme mixture is produced at the right time and that the related metabolic pathways are active. This field has been subject of study for many decades due to the high potential of the extracellular enzymes for industrial applications. Since the availability of *Aspergillus* genome sequences, the amount of data on these systems has grown exponentially allowing more global approaches to understanding fungal physiology. In this chapter the available knowledge of plant polysaccharide degradation is compared with insights form the *Aspergillus* genome to point to conserved and differentiated aspects between different *Aspergillus* species. In addition, the suitability of genes encoding plant cell wall degrading enzymes for use in phylogeny is discussed.

Keywords: plant cell wall degradation, gene regulation, metabolic pathways

1. Introduction

Although some *Aspergillus* species can be opportunistic pathogens of plants and animals, this genus mainly has a saprophytic lifestyle. In nature it grows predominantly on dead and decaying plant matter, as this is the most commonly available carbon source. The majority of plant material consists of the plant cell walls which themselves are largely made up of polysaccharides with varying amounts of lignin and proteins (McNeill *et al.*, 1984). In addition to plant cell walls crude plant matter also contains storage polysaccharides, making the polysaccharide fraction the most abundant carbon source available to *Aspergillus* and other fungi. Since polymeric compounds cannot be taken up by the fungus, these polysaccharides need to be degraded extracellular. In order to do this, fungi produce diverse enzyme cocktails that are tailored to the polysaccharides that are available to the fungus (De Vries and Visser, 2001). The production of these enzymes is controlled by a complex system of transcriptional regulators that allow the fungus to efficiently adapt to changing carbon sources available in the environment (De Vries, 2003).

The recent availability of several genome sequences of this genus has allowed a more detailed analysis of the ability of the different species to degrade polysaccharides and grow on the resulting monomeric components, as well as the regulatory systems related to these processes. These resources also allow detailed comparisons between close and more distantly related species, providing additional insights into fungal evolution.

In this chapter, the ability of different *Aspergillus* species to degrade and utilise plant components will be discussed with respect to genome content, growth, physiology and regulation.

2. Plant polysaccharides

Plant polysaccharides can be divided in two groups based on their location, namely plant cell wall polysaccharides and storage polysaccharides. The first group consist of cellulose, hemicellulose and pectin, while the second group comprises compounds such as starch, inulin and various gums.

2.1. Cellulose

Cellulose is a linear polymer of β-1,4-linked D-glucose residues that can be organised in bundles called microfibrils (Kolpak and Blackwell 1976). Degradation of cellulose requires the combined action of β-1,4-endoglucanases, β-1,4-glucosidases and cellobiohydrolases (De Vries and Visser, 2001). The first class of enzymes cleaves the polymer into short oligosaccharides, while the second class cleaves single D-glucose residues from these oligosaccharides and from cellobiose. Two types of cellobiohydrolases have been described, either cleaving cellobiose from the reducing end or the non-reducing end of cellulose (De Vries *et al.*, 2005b).

2.2. Hemicellulose

Hemicellulose is a collective name for various heterogeneous polysaccharides. The most commonly found hemicelluloses in plants are xyloglucan, xylan and galacto(gluco)mannan.

Xyloglucan has the same backbone as cellulose (β-1,4-linked-D-glucose), but contains α-1,6-linked D-xylose residues on the backbone as well as acetyl residues (Vincken *et al.* 1997). Attached to these D-xylose residues α-1,2-linked D-fucose, α-1,2-linked L-arabinose and α- and β-1,2-linked D-galactose residues can be found. Degradation of xyloglucan therefore not only requires specialised endoglucanases (often referred to as xyloglucanases) but also xyloglucan acetyl esterases, α-1,4-D-xylosidases, α-arabinofuranosidases and α- and β-galactosidases (De Vries and Visser, 2001).

Xylan is a polymer with a β-1,4-linked D-xylose backbone, which can be decorated with acetyl-groups at O2 or O3, α-1,2-linked 4-O-methyl-glucuronic acid, α-1,2- or 1,3-linked L-arabinose, and α-1,4-linked and β-1,4- or 1,5-linked D-galactose (Montgomery *et al.*, 1956; Eda *et al.*, 1976; Wilkie, 1979; Brillouet and Joseleau, 1987; Ebringerová *et al.*, 1990; Schooneveld-Bergmans *et al.*, 1998; Ebringerová and Heinze, 2000; Huisman *et al.*, 2000). In addition, ferulic acid can be linked to O5 of terminal L-arabinose residues (Smith and Hartley, 1983; Saulnier *et al.*, 1995; Wende and Fry, 1997). Xylans rich in L-arabinose or 4-O-methyl glucuronic acid are often referred to as arabinoxylans and glucuronoxylans, and are most commonly found in cereals and hardwoods, respectively. Degradation of the backbone of xylan involves β-1,4-endoxylanases and β-1,4-D-xylosidases, while the side groups are removed by acetylxylan esterases, arabinoxylan arabinofuranohydrolases, α-L-arabinofuranosidases, α-glucuronidases, feruloyl esterases and α- and β-galactosidases (De Vries and Visser, 2001).

Galacto(gluco)mannan consists of a backbone of β-1,4-D-mannose residues that can be interrupted by β-1,4-linked D-glucose residues in the case of galactoglucomannan (Timell, 1967; Dea and Morrison, 1975; Dey, 1978). Acetyl residues at O2 or O3 and α-1,6-linked and β-1,2-linked D-galactose residues can be attached to the main chain. Degradation of these polymers requires the combined action of β-1,4-endomannanases, β-1,4-D-mannosidases, α-1,4-D-galactosidases, acetylmannan esterases, β-D-galactosidases and β-D-glucosidases (De Vries and Visser, 2001).

2.3. Pectin

Pectin also consists of a mixture of polymeric compounds (Perez *et al.*, 2000). Homogalacturonan is a major component of pectin and is often referred to as the pectin smooth region. It consists of a linear polymer of α-1,4-linked D-galacturonic acid residues that can be decorated with methyl (at O6) or acetyl (at O2 or O3) residues. Another major component is rhamnogalacturonan I, which has a backbone of α-1,4-linked D-galacturonic acid residues interrupted by α-1,2-linked L-rhamnose residues. Long side chains can be attached to O4 of the rhamnose residues consisting of α-1,5-linked L-arabinose residues (arabinan) or β-1,4-linked D-galactose residues (galactan) or combination of the two. These side chains can be branched and can contain feruloyl residues on terminal L-arabinose (O2) or D-galactose (O6) residues. Two other structures can be found in pectin. Xylogalacturonan is a modified homogalacturonan with D-xylose residues attached to the main chain. Rhamnogalacturonan II is an oligosaccharide of approximately 30 residues consisting of a homogalacturonan backbone with 4 side chains consisting of uncommon sugars such as 2-O-methyl-L-fucose and 3-deoxy-D-manno-2-octulosonic acid (Mazeau and Perez, 1998). Degradation of pectin is achieved by the combined action of endo- and exopolygalacturonases, endo- and exorhamnogalacturonases, xylogalacturonases, pectin lyases, pectate lyases, rhamnogalacturonan lyases, pectin methyl esterases, pectin acetyl esterases, rhamnogalacturonan acetyl esterases, feruloyl esterases, endo- and exoarabinanases, endo- and exogalactanases, α-L-arabinofuranosidases, β-galactosidases and α-rhamnosidases (De Vries and Visser, 2001).

2.4. Starch and inulin

Starch and inulin are both storage polysaccharides. Starch is a branched polysaccharides consisting entirely of D-glucose and contains two main structures (Hoover and Sosulski, 1991). Amylose consists almost completely of linear α-1,4-linked D-glucose residues with a small amount of α-1,6-linked D-glucose as side chains. Amylopectin is a larger structure both with respect to the length of the backbone as well as the number of branches. Nearly all plants contain starch stored in granules, which are small chloroplast-like organelles inside the plant cell. Inulin is the most simple fructan found in plants (Ritsema and Smeekens, 2003) and consists a sucrose residue linked to β-(1-2)-linked D-fructose polymers.

3. Plant polysaccharide degradation by *Aspergillus*

Degradation of these polysaccharides by *Aspergillus* has been studied for many years due to the applications of polysaccharides degrading enzymes in biotechnology (De Vries and Visser, 2001; De Vries, 2003). Most of the studies have therefore been performed in industrially relevant species, such as *A. niger, A. oryzae, A. aculeatus* and *A. kawachii*. Analysis of the enzyme systems of these fungi during growth on plant derived substrates already indicated the broad range of enzymes they can produce. Nearly all enzyme activities indicated above have been described for *Aspergillus* (De Vries and Visser, 2001) although the enzymes responsible for some activities have not been identified to date.

On average, a larger number of isoenzymes can be found that act on parts of the polysaccharides that are either very abundant (e.g. endoglucanase – cellulose) or have highly different levels of substitutions (e.g. polygalacturonase – pectin). In contrast, highly specialised activities (e.g. α-glucuronidase) usually only occur as single enzymes in enzyme preparations. Some years ago an extensive overview of the available knowledge on *Aspergillus* plant cell wall polysaccharide degrading enzymes and their corresponding genes was published (De Vries and Visser, 2001), which contains detailed information on the enzymes with respect to MW, pI, and T and pH optimum. Although additional enzymes have been reported since then, this overview still covers most of the current knowledge on the enzymes involved in this process. In contrast, much more is known on the spectrum of genes encoding these enzymes since the genome sequences of several Aspergilli have become available (see below) (Galagan *et al.*, 2003; Machida *et al.*, 2005; Nierman *et al.*, 2005; Pel *et al.*, 2007). This effectively means that while the enzyme data used to outnumber the gene data, this has now reversed.

4. Presence of plant polysaccharide degrading functions in *Aspergillus* genomes

The availability of fungal genome sequences has allowed us to evaluate to a much larger degree the potential of fungi to degrade plant polysaccharides as evidenced by a recent study in *A. nidulans* (De Vries *et al.,* 2005b). Several approaches can be used to screen for the presence of novel genes in fungi. BLAST analysis against fungal genomes reveals homologous genes in the genome and the score and expect value provide some kind of value as to the likelihood that they encode the same enzymatic function. For the Aspergilli, this can currently be done most efficiently at the Broad Aspergillus genome server (http://www.broad.mit.edu/annotation/genome/aspergillus_group/MultiHome.html) where BLAST can be performed simultaneously against *A. niger, A. nidulans, A. fumigatus, A. oryzae, A. terreus, A. clavatus, A. flavus* and *N. fischeri*. An alternative is a keyword search in the annotation of these genomes (e.g. with endoxylanase) which will also provide a list of putative genes for this enzymatic function. However, in my opinion the most reliable method to compare the content of genomes with respect to plant polysaccharide degradation is through the CAZy web site (http://www.cazy.org/) (Coutinho and Henrissat, 1999). The Carbohydrate Active enzyme database allows for a division of the enzymatic functions over families based on amino acid signatures. Several fungal genomes have already been taken up in this database, including *A. niger, A. nidulans,* and *A. oryzae.* The algorithms used for this database will mine genomes for specific signatures and then assign these to the different CAZy families, providing a much more powerful and reliable tool for function prediction. For instance, a search for α-galactosidase at the Broad database resulted in 4, 6 and 4 candidate genes for *A. nidulans*, *A. niger* and *A. oryzae*, respectively. In the CAZy database, α-galactosidases are assigned to glycosyl hydrolase families 27 and 36, resulting in 7, 7 and 6 candidate genes for *A. nidulans*, *A. niger* and *A. oryzae*, respectively.

A global comparison of the plant polysaccharide degradation content of 4 Aspergilli, was published recently as part of the paper on the genome sequence of *A. niger* (Pel *et al.*, 2007). This study revealed large differences between *A. niger*, *A. nidulans*, *A. oryzae* and *A. fumigatus* with respect to the number of genes encoding specific enzymatic functions as well as the absence of specific functions in some genomes. For instance, endoinulinase is absent in *A. nidulans* and *A. oryzae*, while exoarabinanase is absent in *A. niger*. These data suggest that different Aspergilli employ different strategies to degrade plant matter or have specialised to different niches in the environment.

5. Fungal carbon catabolic pathways related to plant matter

Degradation of plant polysaccharides into monomeric components is only part of the strategy of fungi to utilise crude plant matter. These monomeric compounds (predominantly monosaccharides) are then taken up by the fungal cell and converted to required molecules using several catabolic pathways (David *et al.*, 2003). D-Glucose, D-fructose, and D-mannose all enter glycolysis after a small number of enzymatic steps. D-galactose can be converted using two pathways, the Leloir pathways and a recently identified alternative pathway that involves some enzymes of the pentose catabolic pathway (Fekete *et al.*, 2004). L-arabinose and D-xylose are both converted to D-xylulose-5-phosphate through the pentose catabolic pathway (Witteveen *et al.*, 1989). This compound then enters the pentose phosphate pathway that links to glycolysis. Little is known about the fungal pathway to catabolise L-rhamnose and D-glucuronic acid, but the D-galacturonic acid pathway has recently been identified in a different fungus, *Trichoderma reesei* (Richard, personal communication). This compound is converted through several enzymes to dihydroxyacetone and enters glycolysis at this point. All *Aspergillus* genomes contain orthologues for the genes from the central carbon catabolism, indicating that all have the potential to utilise the same carbon sources.

However, published data raises the question whether all fungi, and in particular all Aspergilli apply identical pathways to utilise these compounds and suggest that minor and/or major variations exist between the species. Growth tests of several Aspergilli on different carbon sources suggest significant differences in sugar uptake and/or catabolism. For instance, several Aspergilli (e.g. *A. nidulans*) can grow on D-galactose as a sole carbon source, while others (e.g. *A. niger*) can not.

A comparison of *A. niger*, *A. tubingensis*, *A. vadensis*, *A. foetidus* and *A. japonicus* demonstrated that all 5 species are able to use D-glucose, D-xylose, glycerol, xylitol, D-galacturonic acid, acetate and citric acid as sole carbon source (De Vries *et al.*, 2005a). However, growth on these carbon sources showed significant differences between the species. While *A. niger* grew well on all carbon sources, growth of *A. vadensis* was poor on glycerol, D-galacturonic acid and acetate. *A. foetidus* and *A. japonicus* grew poorly on xylitol and *A. tubingensis* on citric acid. A more recent study demonstrated a significant difference between *A. brasiliensis* and the other black Aspergilli in that this species was the only one of this group that is able to use D-galactose as the sole carbon source (Meijer, Samson and De Vries, unpublished results). These data demonstrate that carbon catabolism differs even in closely related fungal species, suggesting adaptation to their natural habitat.

6. Regulatory aspects of carbon source utilisation in *Aspergillus*

The regulation of consumption of natural carbon sources by fungi is a complex issue involving several transcriptional regulators. So far, only two positively acting regulators (AmyR, XlnR) and 1 negatively acting regulator (CreA) have been described for this genus.

6.1. The amylolytic regulator AmyR

AmyR is a transcriptional activator of the Zn_2Cys_6 family that regulates the degradation of starch (Petersen *et al.*, 1999; Tani *et al.*, 2001; Tsukagoshi *et al.*, 2001). In the presence of starch or maltose this regulator is activated resulting in expression of genes encoding starch degrading enzymes, such as glucoamylase, α-amylase and α-glucosidase. AmyR can bind to two specific sequences in the promoter of its target genes: $CGGN_8CGG$ and CGGAAATTTAA. AmyR has been described in detail for *A. oryzae* and *A. nidulans* but orthologues can be identified in all *Aspergillus* genomes (Table 1).

6.2. The xylanolytic regulator XlnR

XlnR is also a member of the Zn_2Cys_6 family and was originally described as the transcriptional activator controlling xylan degradation (Van Peij *et al.*, 1998b). Later studies demonstrated that XlnR not only controls xylan degradation, but also cellulose degradation and some steps of the intracellular pentose catabolic pathway in *Aspergillus* (Van Peij *et al.*, 1998a; De Vries *et al.*, 1999; Gielkens *et al.*, 1999; Hasper *et al.*, 2000, 2002; De Groot *et al.*, 2007). XlnR is activated in the presence of D-xylose and then binds to GGCTAR sequences in the promoter of its target genes

Table 1. Aspergillus *transcriptional regulators related to plant polysaccharide degradation (protein accession numbers).*

	AmyR	XlnR	CreA
A. clavatus	EAW10466	EAW14420	EAW14040
A. flavus[1]	1918.m01115	1866.m00247	2911.m00346
A. fumigatus	EAL89772	EAL93859	EAL93472
A. nidulans	BAA78564	EAA61796	AAR02858 (Dowzer and Kelly, 1991)
A. niger	CAK44691 (Pel *et al.*, 2007)	CAA05082 (Van Peij *et al.*, 1998)	Q05620 (Drysdale *et al.*, 1993)
A. oryzae	BAA25754 (Gomi *et al.*, 2000)	BAA95967 (Marui *et al.* 2002)	CAB89774
A. terreus	EAU39368	EAU37394	EAU37875
N. fisheri	EAW25101	EAW19127	EAW18746

[1]Identification numbers from the *A. flavus* genome website (http://www.aspergillusflavus.org/genomics/).

(De Vries *et al.*, 2002d). The most detailed studies have been performed with *A. niger*, but orthologues can be identified in all *Aspergillus* genomes (Table 1).

6.3. The carbon catabolite repressor protein CreA

Carbon catabolite repression is the system in which the expression of genes devoted to secondary carbon sources is repressed in the presence of sufficient amounts of a good primary carbon source. For example, addition of high concentrations of D-glucose (e.g. 3%) to media with xylan or starch will inhibit the expression of xylanolytic and amylolytic genes. In filamentous ascomycetes this regulatory effect is mediated through the repressor protein CreA (Dowzer and Kelly, 1991; Ruijter and Visser, 1997). CreA is a member of the C2H2 class of regulators and can bind to SYGGRG sequences in the promoter of its target genes (Kulmburg *et al.*, 1993). CreA not only affects polysaccharide degradation, but many gene systems including secondary metabolism, specific parts of primary metabolism, and extracellular functions. CreA has been studied in detail in *A. nidulans*, but is present in all *Aspergillus* genomes as well as all other ascomycetes studied so far (Table 1).

6.4. Other regulatory systems

Additional regulators involved in polysaccharide degradation has been identified recently using a bioinformatics approach. While these have been presented at conferences, no papers have been published dealing with these regulators.

AraR is the arabinanoloytic regulator that controls expression of genes encoding extracellular L-arabinose and D-galactose releasing enzymes as well as the L-arabinose catabolic pathway (Battaglia and De Vries, unpublished results). It is therefore important for the degradation of pectic side chains, arabinoxylan and probably xyloglucan. It was identified in *A. niger* as a homologue of XlnR and is also a member of the Zn_2Cys_6 family of regulators. It is present in all *Aspergillus* genomes. In the presence of L-arabinose or L-arabitol it activates its target genes by binding to an as yet unidentified sequence.

GalR is also a member of the Zn_2Cys_6 family of regulators and was identified in *A. nidulans* (Christensen *et al.*, 2005). In the presence of D-galactose it activates expression of genes encoding D-galactose releasing enzymes, indicating a role in pectin and galactomannan degradation and is possibly involved in D-galactose catabolism. Despite the inability of *A. niger* to use D-galactose as a carbon source, all *Aspergillus* genomes contain orthologues of this regulator including *A. niger*.

InuR controls inulin degradation and was identified in *A. niger* (A.F. Ram *et al.*, unpublished results). It is a homologue of AmyR. In the presence of sucrose it activates expression of genes encoding endo- and exoinulinases and invertases. All *Aspergillus* genomes contain an orthologue of InuR.

In addition to the regulators mentioned above, indications for other regulators have been reported. In the presence of D-galacturonic acid most genes encoding pectinolytic enzymes are expressed, while some of them also respond to the presence of L-rhamnose suggesting the presence of a D-galacturonic acid related and an L-rhamnose related activator (De Vries *et al.*, 2002a). Feruloyl esterase encoding genes and genes encoding aromatic catabolic enzymes are expressed in the presence of ferulic acid and/or other aromatic compounds suggesting the presence of multiple regulatory systems related to aromatic compounds (De Vries and Visser, 1999; De Vries *et al.*, 2002b). Most genes encoding galactomannan degrading enzymes are expressed in the presence of D-mannose, indicating a D-mannose specific activator (Ademark *et al.,* 2001).

7. Suitability of genes encoding plant polysaccharide degrading enzymes for phylogeny

Modern phylogenetic studies often use a combination of 2-6 genes to construct a phylogenetic tree. However, this strategy can be unreliable if the genes used are not present in all organisms or their rate of mutagenesis differs in different species. To test whether genes encoding plant polysaccharide degrading enzymes would be suitable for phylogeny studies, a number of genes from *A. niger* (Flipphi *et al.*, 1993; Gielkens *et al.*, 1999; Dan *et al.*, 2000; De Vries *et al.*, 2000, 2002c; Ademark *et al.*, 2001), *A. aculeatus* (Christgau *et al.*, 1995; Kauppinen *et al.*, 1995; Takada *et al.*, 1998; Manzanares *et al.*, 2001; Skjot *et al.*, 2001), *A. tubingensis* (Vlugt-Bergmans and Van Ooijen, 1999), and *A. kawachii* (Iwashita *et al.*, 1999) were used in a BLAST analysis to find the closest homologue in all 7 *Aspergillus* genomes. The amino acid sequences of these genes were then used in a ClustalW analysis to determine whether closely related species also cluster together based on these genes.

All 7 Aspergillus genome sequences contain an orthologue for the *A. aculeatus* endoarabinanase (Figure 1). The phylogeny of these genes follows the general distance between the different *Aspergillus* species. *A. fumigatus* and *N. fischeri* cluster together and are most closely related to *A. clavatus*. The black Aspergilli (*A. aculeatus* and *A.*

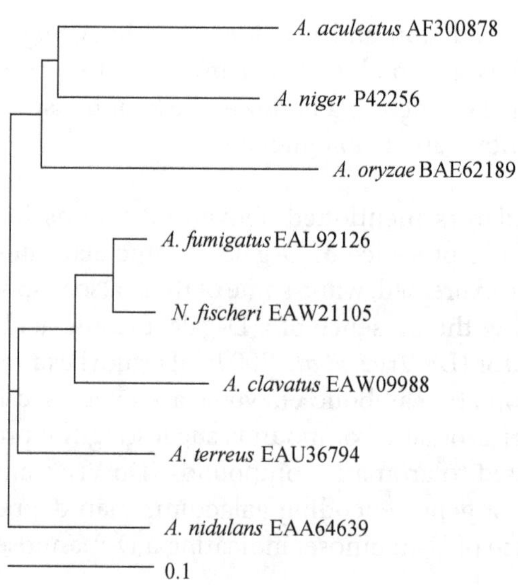

Figure 1. Phylogenetic tree of Aspergillus *endoarabinanases. The numbers are protein accession numbers.*

niger) cluster together. The same is true for the *A. aculeatus* β-glucosidase (Figure 2), β-mannosidase (Figure 3) and rhamnogalacturonan acetyl esterase (Figure 4). Although the phylogeny remains intact no orthologue was detected in *A. oryzae* for *A. aculeatus* cellobiohydrolase (Figure 5), while no orthologue was detected in *A. clavatus* for *A. tubingensis* endogalactanase (Figure 6). Two α-rhamnosidases were

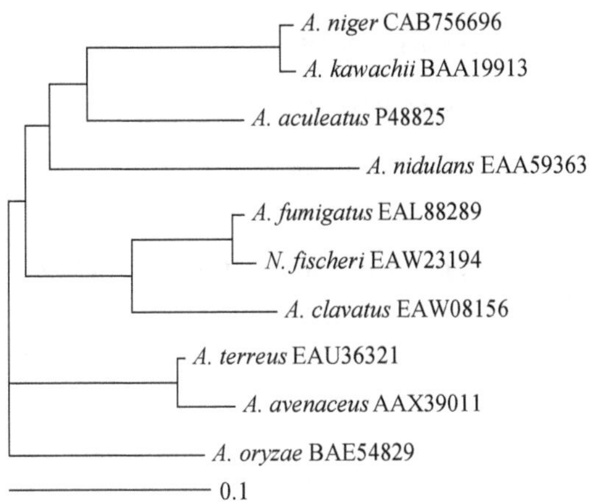

Figure 2. Phylogenetic tree of Aspergillus *β-glucosidases. The numbers are protein accession numbers.*

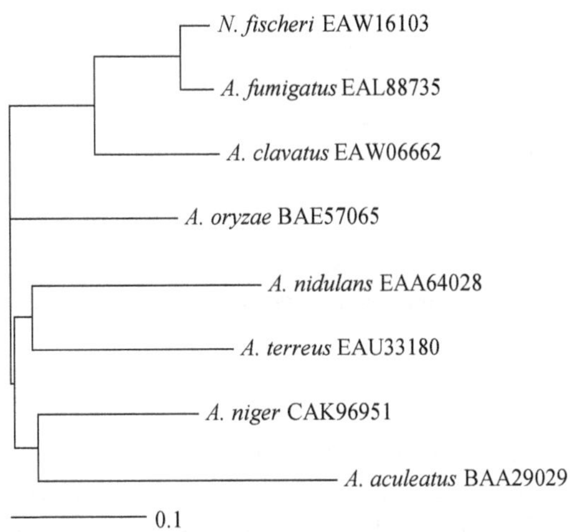

Figure 3. Phylogenetic tree of Aspergillus *β-mannosidases, The numbers are protein accession numbers.*

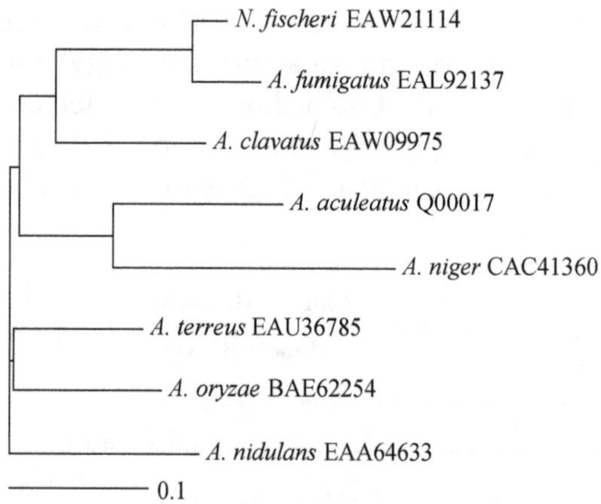

Figure 4. Phylogenetic tree of Aspergillus *rhamnogalacturonan acetyl esterases. The numbers are protein accession numbers.*

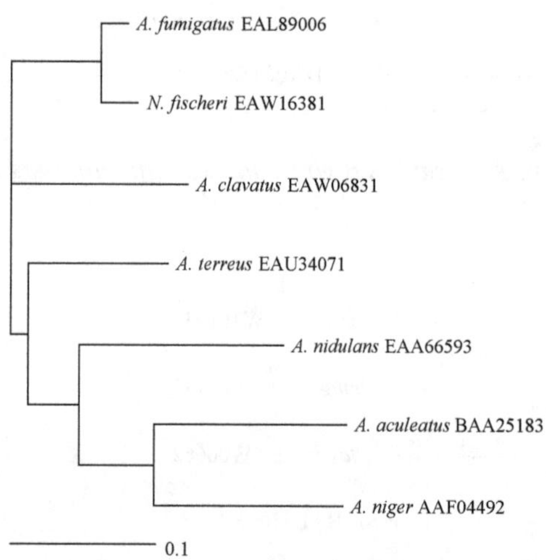

Figure 5. Phylogenetic tree of Aspergillus *cellobiohydrolases. The numbers are protein accession numbers.*

identified in *A. aculeatus* but only a single orthologue was identified in *A. niger, A. fumigatus, N. fischeri* and *A. oryzae* and no orthologue in *A. nidulans, A. terreus* and *A. clavatus* (Figure 7). *A. kawachii* bgl clusters with *A. niger* and *A. aculeatus* and *A. avenaceus* with *A. terreus* indicating a close relatedness between these species. In general, it can therefore be concluded that the genes related to plant polysaccharide

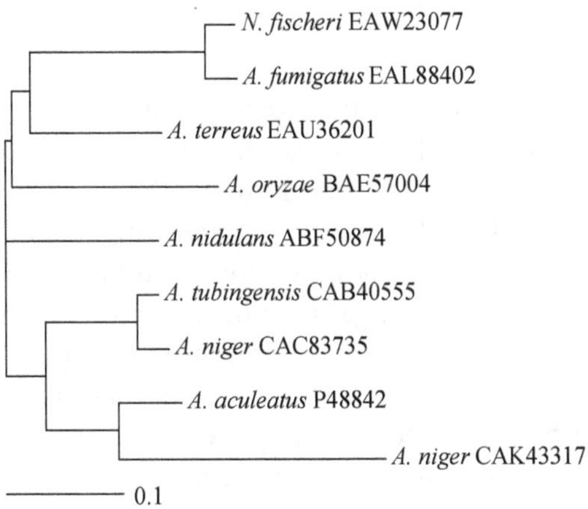

Figure 6. Phylogenetic tree of Aspergillus *endogalactanases. The numbers are protein accession numbers.*

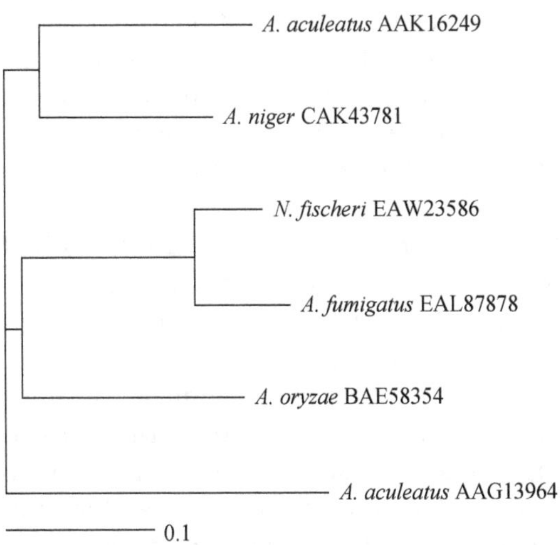

Figure 7. Phylogenetic tree of Aspergillus *α-rhamnosidases. The numbers are protein accession numbers.*

degradation follow the species phylogeny. However, as not all species contain orthologues of every gene, I would not recommend using these genes for species phylogeny (especially for bar coding).

8. Concluding remarks

The availability of genome sequences and post-genomic technologies (transcriptomics, proteomics, and metabolomics) allows research to be conducted in a more global manner studying gene systems rather than individual genes. In addition it allows a much broader analysis of the differences between fungal species which in time will result in a better understanding of evolution, adaptation to the environment and physiology. The rate at which fungal genomes become available is constantly increasing and the Aspergilli are the best represented group of filamentous fungi among the fungal genomes. As the *Aspergillus* research community is also one of the largest in the fungal field this offers a unique opportunity to perform genomic and post-genomic studies covering every aspect of fungal life as well as detailed phylogenetics.

The genome sequences have demonstrated that in the pre-genomics era we knew only a very small amount of the enzymatic potential of fungi and have enabled the identification of many novel enzymes and additional genes encoding a particular enzyme activity. Detailed analysis of the function of these enzymes will require more specific substrates based on natural compounds that are readily available and can be used in (semi-)high throughput screens. Development of these substrates should therefore be considered an important goal in post-genomic enzyme analysis.

Another step forward in the genomics of *Aspergillus* could be the availability of genome sequences of species with distinctly different lifestyles. Although some of the species sequenced so far are opportunistic pathogens, all species have a general saprophytic life style aimed at consuming plant matter. However, the genus *Aspergillus* also contains some species that are adapted to significantly different biotopes. One such species is *A. sydowii* which is a pathogen of Caribbean sea fan corals (Alker *et al.*, 2001). This biotope indicates that *A. sydowii* has adapted its physiology to withstand the high salt conditions, a marine life style and is able to degrade corral components to be used as a carbon source. Genomic and post-genomic comparisons of such species to the currently available genomes is likely to shed more light on fungal adaptation to its environment and evolution.

References

Ademark, P., R.P. de Vries, P. Hagglund, H. Stalbrand and J. Visser, 2001. Cloning and characterization of *Aspergillus niger* genes encoding an alpha-galactosidase and a beta-mannosidase involved in galactomannan degradation. European Journal of Biochemistry 268: 2982-2990.

Alker, A.P., G.W. Smith and K. Kim, 2001. Characterization of *Aspergillus sydowii* (Thom et Church), a fungal pathogen of Caribbean sea fan corals. Hydrobiology 460: 105-111.

Brillouet, J.-M. and J.P. Joseleau, 1987. Investigation of the structure of a heteroxylan from the outer pericarp (beeswing bran) of wheat kernel. Carbohydrate Research 159: 109-126.

Christensen, U., S. Hansen, H. Mulder, S. Madrid and I. Nikolaev, 2005. A novel transcription factor GalA, homologous to the transcriptional activator of xylanolytic and cellulolytic systems XlnR, regulates D-galactose metabolism in *Aspergillus nidulans.* Poster nr. 379 presented at the Fungal Genetics conference at Asilomar. Available at: http://www.fgsc.net/asil2005/GeneRegulation.htm

Christgau, S., T. Sandal, L.V. Kofod and H. Dalboge, 1995. Expression cloning, purification and characterization of a β-1,4-galactanase from *Aspergillus aculeatus.* Current Genetics 27: 135-141.

Coutinho, P.M. and B. Henrissat, 1999. Carbohydrate-active enzymes: An integrated database approach. In: Gilbert, H.J., Davies, G.J., Henrissat, B. and Svensson, B. (eds.), Recent Advances in Carbohydrate Bioengineering, The Royal Society of Chemistry, Cambridge, pp 3-14.

Dan, S., I. Marton, M. Dekel, B.A. Bravdo, S. He, S.G. Withers and O. Shoseyov, 2000. Cloning, expression, characterization, and nucleophile identification of family 3, *Aspergillus niger* beta-glucosidase. Journal of Biological Chemistry 275: 4973-4980.

David, H., M. Akesson and J. Nielsen, 2003. Reconstruction of the central carbon metabolism of *Aspergillus niger.* European Journal Biochemistry 270: 4243-4253.

De Groot, M.J.L., C. van den Dool, H.A.B. Wosten, M. Levisson, P.A. vanKuyk, G.J.G. Ruijter and R.P. de Vries, 2007. Regulation of pentose catabolic pathway genes of *Aspergillus niger.* Food Technology and Biotechnology 45: 134-138.

De Vries, R.P., 2003. Regulation of *Aspergillus* genes encoding plant cell wall polysaccharide degrading enzymes; relevance for industrial production. Applied Microbiology and Biotechnology 61: 10-20.

De Vries, R.P. and J. Visser, 1999. Regulation of the feruloyl esterase (*faeA*) gene from *Aspergillus niger.* Applied and Environmental Microbiology 65: 5500-5503.

De Vries, R.P. and J. Visser, J., 2001. *Aspergillus* enzymes involved in degradation of plant cell wall polysaccharides. Microbiology and Molecular Biology Reviews 65: 497-522.

De Vries, R.P., C. van Grieken, P.A. vanKuyk and H.A.B. Wösten, 2005b. The value of genome sequences in the rapid identification of novel genes encoding specific plant cell wall degrading enzymes. Current Genomics 6: 157-187.

De Vries, R.P., H.C.M. Kester, C.H. Poulsen, J.A.E. Benen and J. Visser, 2000. Synergy between accessory enzymes from *Aspergillus* in the degradation of plant cell wall polysaccharides. Carbohydrate Research 327: 401-410.

De Vries, R.P., H.C.M. Kester, P.A. vanKuyk and J. Visser, J., 2002b. The *Aspergillus niger faeB* gene encodes a second feruloyl esterase involved in pectin and xylan degradation, and is specifically induced on aromatic compounds. Biochemical Journal 363: 377-386.

De Vries, R.P., H.C. van den Broeck, E. Dekkers, P. Manzanares, L.H. de Graaff and J. Visser, 1999. Differential expression of three α-galactosidase genes and a single β-galactosidase gene from *Aspergillus niger*. Applied and Environmental Microbiology 65: 2453-2460.

De Vries, R.P., J. Jansen, G. Aguilar, L. Parenicová, J.A.E. Benen, V. Joosten, F. Wulfert and J. Visser, 2002a. Expression profiling of pectinolytic genes from *Aspergillus niger*. Febs Letters 530: 41-47.

De Vries, R.P., K. Burgers, P.J.I. van de Vondervoort, J.C. Frisvad, R.A. Samson and J. Visser, 2005a. *Aspergillus vadensis*, a new species of the group of black Aspergilli. Antonie van Leeuwenhoek 87: 195-203.

De Vries, R.P., L. Parenicová, S. Hinz, H. Kester, J.A.E. Benen, G. Beldman and J. Visser, J., 2002c. Endogalactanase a from *Aspergillus niger* is specifically induced on l-arabinose and galacturonic acid and acts synergistically with *Aspergillus* rhamnogalacturonases. European Journal Biochemistry 269: 4985-4993.

De Vries, R.P., P.J.I. van de Vondervoort, L. Hendriks, M. van de Belt and J. Visser, 2002d. Regulation of the α-glucuronidase encoding gene (*aguA*) from *Aspergillus niger.* Molecular and General Genetics 268: 96-102.

Dea, I.C.M. and A. Morrison, 1975. Chemistry and interactions of seed galactomannans. Advances in Carbohydrate Chemistry and Biochemistry 31: 241-312.

Dey, P.M., 1978. Biochemistry of plant galactomannans. Advances in Carbohydrate Chemistry and Biochemistry 35: 3411-3376.

Dowzer, C.E.A. and J.M. Kelly, 1991. Analysis of the *cre*a gene, a regulator of carbon catabolite repression in *Aspergillus nidulans*. Molecular and Cellular Biology 11: 5701-5709.

Drysdale, M.R., S.E. Kolze and J.M. Kelly, 1993. The *Aspergillus niger* carbon catabolite repressor gene, *creA*. Gene 130: 241-245.

Ebringerová, A. and T. Heinze, 2000. Xylan and xylan derivatives - biopolymers with valuable properties, 1 naturally occurring xylans structures, isolation procedures and properties. Macromolocules Rapid Communications. 21: 542-556.

Ebringerová, A., Z. Hromádková, E. Petráková and M. Hricovíni, 1990. Structural features of a water-soluble l-arabinoxylan from rye bran. Carbohydrate Research 198: 57-66.

Eda, S., A. Ohnishi and K. Kato, 1976. Xylan isolated from the stalk of *Nicotiana tabacum*. Agricultural and Biological Chemistry 40: 359-364.

Fekete, E., L. Karaffa, E. Sándor, I. Bányai, B. Seiboth, G. Gyémánt, A. Sepsi, A. Szentirmai and C.P. Kubicek, 2004. The alternative d-galactose degrading pathway of *Aspergillus nidulans* proceeds via l-sorbose. Archives Microbiology 181: 35-44.

Flipphi, M.J.A., H. Panneman, P. van der Veen, J. Visser and L.H. de Graaff, 1993. Molecular cloning, expression and structure of the endo-1,5-α-l-arabinase gene of *Aspergillus niger*. Applied Microbiology and Biotechnology 40: 318-326.

Galagan, J.E., S.E. Calvo, K.A Borkovich, E.U. Selker, N.D.Read, D.Jaffe, W. FitzHugh, L.-J. Ma, S. Smirnov, S. Purcell, B. Rehman, T. Elkins, R. Engels, S. Wang, C.B. Nielsen, J. Butler, M. Endrizzi, D. Qui, P. Ianakiev, D. Pedersen, M. Nelson, M. Washburne, C.P. Selitrennikoff, J.A. Kinsey, E.L. Braun, A. Zelter, U. Schulte, G.O. Kothe, G. Jedd, W. Mewes, C. Staben, E. Marcotte, D. Greenberg, A. Roy, K. Foley, J. Naylor, N. Thomann, R. Barrett, S. Gnerre, M. Kamal, M. Kamvysselis, E. Mauceli, C. Bielke, S. Rudd, D. Frishman, S. Krystofova, C. Rasmussen, R.L. Metzenberg, D.D. Perkins, S. Kroken, C. Cogoni, G. Macino, D. Catcheside, W. Li, R.J. Pratt, S.A. Osmani, C.C. DeSouza, L. Glass, M.J. Orbach, J, Berglund, R. Voelker, O. Yarden, M. Plamann, S. Seiler, J. Dunlap, A. Radford, R. Aramayo, D.O. Natvig, L.A. Alex, G. Mannhaupt, D.J. Ebbole, M. Freitag, I. Paulsen, M.S. Sachs, E.S. Lander, C. Nusbaum and B. Birren, 2003. The genome sequence of the filamentous fungus *Neurospora crassa*. Nature 422: 859-868.

Gielkens, M.M.C., E. Dekkers, J. Visser and L.H. de Graaff, 1999. Two cellobiohydrolase-encoding genes from *Aspergillus niger* require d-xylose and the xylanolytic transcriptional activator *xlnR* for their expression. Applied and Environmental Microbiology 65: 4340-4345.

Gomi, K., T. Akeno, T. Minetoki, K. Ozeki, C. Kumagai, N. Okazaki and Y. Iimura, 2000. Molecular cloning and characterization of a transcriptional activator gene, *amyR*, involved in the amylolytic gene expression in *Aspergillus oryzae*. Bioscience, Biotechnology and Biochemistry 64: 816-827.

Hasper, A.A., E. Dekkers, M. van Mil, P.J.I. van de Vondervoort, and L.H. de Graaff, 2002. EglC, a new endoglucanase from *Aspergillus niger* with major activity towards xyloglucan. Applied and Environmental Microbiology 68: 1556-1560.

Hasper, A.A., J. Visser and L.H. de Graaff, 2000. The *Aspergillus niger* transcriptional activator XlnR, which is involved in the degradation of the polysaccharides xylan and cellulose, also regulates d-xylose reductase gene expression. Molecular Microbiology 36: 193-200.

Hoover, R. and F.W. Sosulski, 1991. Composition, structure, functionality, and chemical modification of legume starches: A review. Canadian Journal of Physiology and Pharmacology 69: 79-92.

Huisman, M.M.H., H.A. Schols and A.G.J. Voragen, 2000. Glucuronoarabinoxylans from maize kernel cell walls are more complex than those from sorghum kernel cell walls. Carbohydrate Polymers 43: 269-279.

Iwashita, K., T. Nagahara, H. Kimura, M. Takano, H. Shimoi and K. Ito, 1999. The *bglA* gene from *Aspergillus kawachii* encodes both extracellular and cell wall bound β-glucosidases. Applied and Environmental Microbiology 65: 5546-5553.

Kauppinen, S., S. Christgau, L.V. Kofod, T. Halkier, K. Dorreich and H. Dalboge, 1995. Molecular cloning and characterization of a rhamnogalacturonan acetylesterase from *aspergillus aculeatus*. Synergism between rhamnogalacturonan degrading enzymes. Journal of Biological Chemistry 270: 27172-27178.

Kolpak, F.J. and J. Blackwell, 1976. Determination of the structure of cellulose ii. Macromolecules 9: 273-278.

Kulmburg, P., M. Mathieu, C. Dowzer, J. Kelly and B. Felenbok, 1993. Specific binding sites in the *alcR* and *alcA* promoters of the ethanol regulon for the CreA repressor mediating carbon catabolite repression in *Aspergillus nidulans*. Molecular Microbiology 7: 847-857.

Machida, M., K. Asai, M. Sano, T. Tanaka, T. Kumagai, G. Terai, K. Kusumoto, T. Arima, O. Akita, Y. Kashiwagi, K. Abe, K. Gomi, H. Horiuchi, K. Kitamoto, T. Kobayashi, M. Takeuchi, D.W. Denning, J.E. Galagan, W.C. Nierman, J. Yu, D.B. Archer, J.W. Bennett, D. Bhatnagar, T.E. Cleveland, N.D. Fedorova, O. Gotoh, H. Horikawa, A. Hosoyama, M. Ichinomiya, R. Igarashi, K. Iwashita, P.R. Juvvadi, M. Kato, Y. Kato, T. Kin, A. Kokubun, H. Maeda, N. Maeyama, J. Maruyama, H. Nagasaki, T. Nakajima, K. Oda, K. Okada, I. Paulsen, K. Sakamoto, T. Sawano, M. Takahashi, K. Takase, Y. Terabayashi, J.R. Wortman, O. Yamada, Y. Yamagata, H. Anazawa, Y. Hata, Y. Koide, T. Komori, Y. Koyama, T. Minetoki, S. Suharnan, A. Tanaka, K. Isono, S. Kuhara, N. Ogasawara and H. Kikuchi, 2005. Genome sequencing and analysis of *Aspergillus oryzae*. Nature 438: 1157-1161.

Manzanares, P., H.C. van den Broeck, L.H. de Graaff and J. Visser, 2001. Purification and characterization of two different alpha-l- rhamnosidases, RhaA and RhaB, from *Aspergillus aculeatus*. Applied and Environmental Microbiology 67: 2230-2234.

Marui, J., A. Tanaka, S. Mimura, L.H. de Graaff, J. Visser, N. Kitamoto, M. Kato, T. Kobayashi and N. Tsukagoshi, 2002. A transcriptional activator, AoxlnR, controls the expression of genes encoding xylanolytic enzymes in *Aspergillus oryzae*. Fungal Genetics and Biology 35: 157-169.

Mazeau, K. and S. Perez, 1998. The preferred conformations of the four oligomeric fragments of rhamnogalacturonan ii. Carbohydrate Research 311: 203-217.

McNeill, M., A.G. Darvill, S.C. Fry and P. Albersheim, 1984. Structure and function of the primary cell walls of plants. Annual Review Biochemistry 53: 625-663.

Montgomery, R., F. Smith and H. Srivastava, 1956. Structure of corn hull hemicellulose. I. Partial hydrolysis and identification of 2-o-(α-glucopyranosuluronic acid)-d-xylopyranose. Journal of the American Chemical Society 78: 2837-2839.

Nierman, W.C., A. Pain, M.J. Anderson, J.R. Wortman, H.S. Kim, J. Arroyo, M. Berriman, K. Abe, D.B. Archer, C. Bermejo, J. Bennett, P. Bowyer, D. Chen, M. Collins, R. Coulsen, R. Davies, P.S. Dyer, M. Farman, N. Fedorova, N. Fedorova, T.V. Feldblyum, R. Fischer, N. Fosker, A. Fraser, J.L. Garcia, M.J. Garcia, A. Goble, G.H. Goldman, K. Gomi, S. Griffith-Jones, R. Gwilliam, B. Haas, H. Haas, D. Harris, H. Horiuchi, J. Huang, S. Humphray, J. Jimenez, N. Keller, H. Khouri, K. Kitamoto, T. Kobayashi, S. Konzack, R. Kulkarni, T. Kumagai, A. Lafon, J.P. Latge, W. Li, A. Lord, C. Lu, W.H. Majoros, G.S. May, B.L. Miller, Y. Mohamoud, M. Molina, M. Monod, I. Mouyna, S. Mulligan, L. Murphy, S. O'Neil, I. Paulsen, M.A. Penalva, M. Pertea, C. Price, B.L. Pritchard, M.A. Quail, E. Rabbinowitsch, N. Rawlins, M.A. Rajandream, U. Reichard, H. Renauld, G.D. Robson, S. Rodriguez de Cordoba, J.M. Rodriguez-Pena, C.M. Ronning, S. Rutter, S.L. Salzberg, M. Sanchez, J.C. Sanchez-Ferrero, D. Saunders, K. Seeger, R. Squares, S. Squares, M. Takeuchi, F. Tekaia, G. Turner, C.R. Vazquez de Aldana, J. Weidman, O. White, J. Woodward, J.H. Yu, C. Fraser, J.E. Galagan, K. Asai, M. Machida, N. Hall, B. Barrell and D.W. Denning, 2005. Genomic sequence of the pathogenic and allergenic filamentous fungus *Aspergillus fumigatus*. Nature 438: 1151-1156.

Pel, H.J., J.H. de Winde, D.B. Archer, P.S. Dyer, G. Hofmann, P.J. Schaap, G. Turner, R.P. de Vries, R. Albang, K. Albermann, M.R. Andersen, J.D. Bendtsen, J.A. Benen, M. van den Berg, S. Breestraat, M.X. Caddick, R. Contreras, M. Cornell, P.M. Coutinho, E.G. Danchin, A.J. Debets, P. Dekker, P.W. van Dijck, A. van Dijk, L. Dijkhuizen, A.J. Driessen, C. d'Enfert, S. Geysens, C. Goosen, G.S. Groot,

P.W. de Groot, T. Guillemette, B. Henrissat, M. Herweijer, J.P. van den Hombergh, C.A. van den Hondel, R.T. van der Heijden, R.M. van der Kaaij, F.M. Klis, H.J. Kools, C.P. Kubicek, P.A. van Kuyk, J. Lauber, X. Lu, M.J. van der Maarel, R. Meulenberg, H. Menke, M.A. Mortimer, J. Nielsen, S.G. Oliver, M. Olsthoorn, K. Pal, N.N. van Peij, A.F. Ram, U. Rinas, J.A. Roubos, C.M. Sagt, M. Schmoll, J. Sun, D. Ussery, J. Varga, W. Vervecken, P.J. van de Vondervoort, H. Wedler, H.A. Wosten, A.P. Zeng, A.J. van Ooyen, J. Visser and H. Stam, 2007. Genome sequencing and analysis of the versatile cell factory *Aspergillus niger* CBS 513.88. Nature Biotechnology 25: 221-231.

Perez, S., K. Mazeau and C. Herve du Penhoat, 2000. The three-dimensional structures of the pectic polysaccharides. Plant Physiology and Biochemistry 38: 37-55.

Petersen, K.L., J. Lehmbeck and T. Christensen, 1999. A new transcriptional activator for amylase genes in aspergillus. Molecular and General Genetics 262: 668-676.

Ritsema, T. and S. Smeekens, 2003. Fructans: Beneficial for plants and humans. Current Opinions in Plant Biology 6: 223-230.

Ruijter, G.J.G. and J. Visser, 1997. Carbon repression in Aspergilli. Fems Microbiology Letters 151: 103-114.

Saulnier, L., J. Vigouroux and J.-F. Thibault, 1995. Isolation and partial characterization of feruloylated oligosaccharides from maize bran. Carbohydrate Research 272: 241-253.

Schooneveld-Bergmans, M.E.F., A.M.C.P. Hopman, G. Beldman and A.G.J. Voragen, 1998. Extraction and partial characterization of feruloylated glucuronoarabinoxylans from wheat bran. Carbohydrate Polymers 35: 39-47.

Skjot, M., S. Kauppinen, L.V. Kofod, C. Fuglsang, M. Pauly, H. Dalboge and L.N. Andersen, 2001. Functional cloning of an endo-arabinanase from *Aspergillus aculeatus* and its heterologous expression in *A. oryzae* and tobacco. Molecular Genetics and Genomics 265: 913-921.

Smith, M.M. and R.D. Hartley, 1983. Occurrence and nature of ferulic acid substitution of cell wall polysaccharides in graminaceous plants. Carbohydrate Research 118: 65-80.

Takada, G., T. Kawaguchi, J.-I. Sumitani and M. Arai, 1998. Cloning, nucleotide sequence, and transcriptional analysis of *Aspergillus aculeatus* no.F-50 cellobiohydrolase I (*cbh*I) gene. Journal of Fermention and Bioengineering 85: 1-9.

Tani, S., Y. Katsuyama, T. Hayashi, H. Suzuki, M. Kato, K. Gomi, T. Kobayashi and N. Tsukagoshi, 2001. Characterisation of the *amyR* gene encoding a transcriptional activator for the amylase genes in *Aspergillus nidulans*. Current Genetics 39: 10-15.

Timell, T.E., 1967. Recent progress in the chemistry of wood hemicelluloses. Wood Science and Technology 1: 45-70.

Tsukagoshi, N., T. Kobayashi and M. Kato, 2001. Regulation of the amylolytic and (hemi-) cellulolytic genes in aspergilli. Journal of General and Applied Microbiology 47: 1-19.

Van Peij, N., M.M.C. Gielkens, R.P. de Vries, J. Visser and L.H. de Graaff, 1998a. The transcriptional activator XlnR regulates both xylanolytic and endoglucanase gene expression in *Aspergillus niger*. Applied and Environmental Microbiology 64: 3615-3619.

Van Peij, N.N.M.E., J. Visser and L.H. de Graaff, 1998b. Isolation and analysis of *xlnR*, encoding a transcriptional activator coordinating xylanolytic expression in *Aspergillus niger*. Molecular Microbiology 27: 131-142.

Vincken, J.-P., W.S. York, G. Beldman and A.G.J. Voragen, 1997. Two general branching patterns of xyloglucan, XXXG and XXGG. Plant Physiology 114: 9-13.

Vlugt-Bergmans, C.J.B. and A.J.J. van Ooijen, 1999. Expression cloning in *Kluyveromyces lactis*. Biotechniques 13: 87-92.

Wende, G. and S.C. Fry, 1997. O-feruloylated, o-acetylated oligosaccharides as side-chains of grass xylans. Phytochemistry 44: 1011-1018.

Wilkie, K.C.B., 1979. The hemicelluloses of grasses and cereals. Advances in Carbohydrate Chemistry and Biochemistry 36: 215-264.

Witteveen, C.F.B., R. Busink, P. van de Vondervoort, C. Dijkema, K. Swart and J. Visser, 1989. L-arabinose and D-xylose catabolism in *Aspergillus niger*. Journal of General Microbiology 135: 2163-2171.

Sexual genes in the asexual filamentous fungus *Aspergillus niger* and related Aspergilli

Károly Pál[1,2], Anne D. van Diepeningen[1], János Varga[2,3], Alfons J.M. Debets[1] and Rolf F. Hoekstra[1]
[1]Laboratory of Genetics, Plant Sciences, Wageningen University, Wageningen, The Netherlands
[2]University of Szeged, Faculty of Science, Dept. of Microbiology, Szeged, H-6701 Hungary
[3]CBS Fungal Biodiversity Centre, Uppsalalaan 8, 3584 CT Utrecht, The Netherlands

Abstract

Sexual reproduction is a general feature in higher eukaryotes, although the genetic machinery to produce sexually derived offspring can be very different. Contrary to the megabase-sized sexual chromosomes of animals, fungal genomes harbour much shorter sequences located at the *MAT* locus which governs the mating processes. For heterothallic fungi, the two mating-types are characterised by the presence of either of the two idiomorphs of the *MAT* locus, *MAT-1* or *MAT-2*, while homothallic species may contain both *MAT* idiomorphs. The black mould *Aspergillus niger* is known as an asexual species, but some related presumed asexual species have recently been shown to have a cryptic sexual cycle. The scope of this work was to screen for and compare mating and meiosis related genes in the two sequenced *A. niger* genomes and to compare them to those in related Aspergilli. The *MAT-1* mating type gene was found in both sequenced *A. niger* strains, a *MAT-2* homologue was not detected in either, indicating that this species is either heterothallic or truly asexual. In comparison, the sequenced homothallic *A. nidulans* genome contains both MAT genes on different chromosomes, whereas the *A. fumigatus, A. oryzae,* and *A. terreus* genomes contain one MAT gene per sequenced genome. Many of the proteins known from *Saccharomyces cerevisiae* to be involved in meiosis proved well conserved in all five tested sexual and presumed asexual *Aspergillus* species. The Aspergilli also proved to have very similar sets of pheromone processing and response pathway proteins. Comparison of the two *A. niger* genomes for their gene sets show that many of these proteins may have essential pleiotropic functions next to mating and meiosis.

Keywords: asexuality, mating type genes, meiosis related genes

1. Introduction

Sexual reproduction involves the regular alternation in the life cycle of an organism of karyogamy and meiosis. Its main biological function lies in the fact that it generates genetic variation by recombination to promote the creation and spread of advantageous traits and to purge the genome of deleterious mutations. However, there are still continuing debates about the origin and advantages of sex (Hurst and Peck, 1996; Barton and Charlesworth, 1998; Cavalier-Smith, 2002; De Visser and Elena, 2007).

In fungi mating processes are governed by mating type genes. Most heterothallic filamentous ascomycetes have a mating system with two alleles (called idiomorphs, as they do not share any significant sequence similarity) located at a single locus (Turgeon and Yoder, 2000). One idiomorph (*MAT-1*) contains an ORF encoding a protein with a motif called an alpha box, which is also present in the MATα1 protein of the yeast *Saccharomyces cerevisiae* (Turgeon and Yoder, 2000). The other idiomorph (*MAT-2*) contains a single ORF encoding a regulatory protein with a DNA-binding domain of the high mobility group (HMG) type. In homothallic species both mating type genes are present in one genome making the strains self fertile, though often also capable of outcrossing. Mating type genes regulate plasmogamy and the production of ascogenous hyphae in ascomycetes, but they can also have other functions in some species, for example they can affect vegetative incompatibility in *Neurospora crassa*. Mating type genes have been identified in a number of filamentous ascomycetes belonging to fungal groups that are widely separated in evolutionary terms (e.g. in species belonging to the Pyrenomycetes, Discomycetes and Loculoascomycetes). Recently, mating type genes have been identified in *A. fumigatus*, *A. nidulans* and *A. oryzae* (Pöggeler, 2002; Varga, 2003; Dyer *et al.*, 2003; Paoletti *et al.*, 2005).

Knowledge about the reproductive strategy of fungi is not only of fundamental interest but is also crucial for applied research and e.g. disease management (Milgroom and Fry, 1997). Observation of a high genetic diversity in a supposedly asexual fungus might be explained by meiotic exchanges in the near past, by mitotic recombination via the so-called parasexual cycle (Pontecorvo *et al.*, 1953; Pontecorvo, 1956) or the existence of a cryptic sexual state. Recently, in different isolates of the opportunistic human pathogen *A. fumigatus* (Latgé, 1999), which was considered an asexual species, both mating-type genes and meiosis-related genes have been identified (Pöggeler, 2002; Varga, 2003; Dyer and Paoletti, 2005; Paoletti *et al.*, 2005).

The ascomycete *Aspergillus* genus includes many different fungi which are mostly saprobic, but human and plant pathogens are also common. So far 114 out of 186

Aspergillus species were found to produce only asexual mitotic conidiospores. The other 72 strains produce both asexual conidia and sexual meiotic ascospores, while two species are known to propagate only via sexual spores. Phylogenetic analysis of 15 species revealed that sexual and asexual Aspergilli are closely related, and that the ability to undergo a sexual cycle has been lost several times independently during the evolution of the genus (Geiser *et al.,* 1996).

The black mould *A. niger* is an industrially important fungus and occasionally an opportunistic human pathogen. *Aspergillus* section *Nigri* consists of several presumably asexual species next to *A. niger,* including e.g. *A. tubingensis, A. brasiliensis, A. japonicus, A. carbonarius* and several others (Samson *et al.,* 2004). These species have minor morphological differences, but can be distinguished by mitochondrial DNA analysis (Varga *et al.,* 1993, 1994; Kevei *et al.,* 1996; Hamari *et al.,* 1997), restriction analysis of the ITS (intergenic spacer) region (Accensi *et al.,* 1999), and by sequence analysis (Samson *et al.,* 2004). In general, asexual fungi can recombine via the parasexual cycle, but this process seems to be limited in the natural isolates of the black Aspergilli as they have a high degree of heterokaryon incompatibility (Van Diepeningen *et al.,* 1997). Also looking at recombination between molecular phylogenies of different nuclear genes no evidence could be found for recombination between them (Van Diepeningen, 1999).

In this study, we examined five, recently sequenced *Aspergillus* genomes for their *MAT* genes and other genes involved in meiotic processes. Our specific focus is on the two sequenced genomes of presumed asexual *A. niger,* but four closely related sexual and asexual Aspergilli were also examined. Our aim was to compare the genetic architecture of the mating related genes in sexual and asexual Aspergilli with those of the yeast *Saccharomyces cerevisiae,* a model organism for sexual processes.

2. Results

2.1. Identification of MAT homologous sequences in Aspergillus niger

Mating type genes govern the mating processes in fungi. Recently mating type genes have been identified in *A. nidulans, A. oryzae* and even in *A. fumigatus* which has been presumed for a long-time to be able to reproduce only asexually (Pöggeler, 2002; Varga, 2003; Dyer *et al.,* 2003; Dyer and Paoletti, 2005; Paoletti *et al.,* 2005). Homothallic species are both self-fertile and capable of outbreeding. They have a combination of both mating type gene idiomorphs in their genome. Heterothallic ascomycetes have two alternate mating idiomorphs located at a single locus.

Therefore, a heterothallic strain contains only one of the idiomorphs and needs a strain of the opposite mating type for sexual reproduction.

We selected five *Aspergillus* species of which the total genomes have been sequenced for our study of mating and meiosis related genes, and first examined them for their mating type genes and organisation. The *A. fumigatus* Af 293 sequence data was obtained from The Institute for Genomic Research website at http://www.tigr.org. *A. nidulans* FGSC A4 and *A. terreus* NIH 2624 sequence data were obtained from the *Aspergillus nidulans* and *Aspergillus terreus* Sequencing Project, Broad Institute of MIT and Harvard (http://www.broad.mit.edu). *A. oryzae* RIB40 sequences were available on the server of National Institute of Technology and Evaluation (NITE). Finally, we searched two *A. niger* databases: the genome of the culture collection strain CBS 513.88 from DSM (prepared by shotgun sequencing, 7.5 times coverage; http://www.dsm.com/en_US/html/dfs/genomics_aniger.htm) and the genome of the culture collection strain ATCC 1015 from DOE Joint Genome Institute (assembly release version 1.0 of whole genome shotgun reads was constructed with the JGI assembler, Jazz, using paired end sequencing reads at a coverage of 8.9X; http://www.jgi-psf.org/cgi-bin/runAlignment?db=Aspni1&advanced=1). These two *A. niger* strains were shown in our lab to be heterokaryon incompatible with one another and as all *A. niger* strains they are presumably asexual.

In the genome of homothallic *A. nidulans* indeed both mating type idiomorphs can be identified at two distinct loci. In the genomes of *A. fumigatus*, *A. oryzae* and *A. terreus* only the *MAT*-1 gene could be detected, hinting on either heterothallism or asexuality if the *MAT*-2 gene would be completely absent from the population. The *MAT*-1 gene could be detected in the genomes of both *A. niger* strains, while no *MAT*-2 homologue was found in any of them (the used criterium for homologues is an E-value of $< e^{-10}$). Homologues of the *MAT-1* idiomorph have also been identified in the genomes of two other *Aspergillus* species, *A. clavatus* NRRL 1 and *Neosartorya fischeri* NRRL 181. Alignment of the *MAT-1* proteins of Aspergilli is presented in Figure 1. Note the extensive homology between different *Aspergillus* species. A phylogenetic tree showing the relationships of *MAT-1* proteins of Aspergilli and other species belonging to the Eurotiales order to those identified in other filamentous fungi is presented in Figure 2. Species belonging to the Eurotiales order (*Aspergillus* and *Penicillium marneffei*; Woo *et al.*, 2006) form a well-defined clade closely related to species of the Onygenales order (*Ajellomyces*, *Coccidioides* spp.; Fraser *et al.*, 2007), in accordance with their taxonomic placement (Geiser *et al.*, 2006).

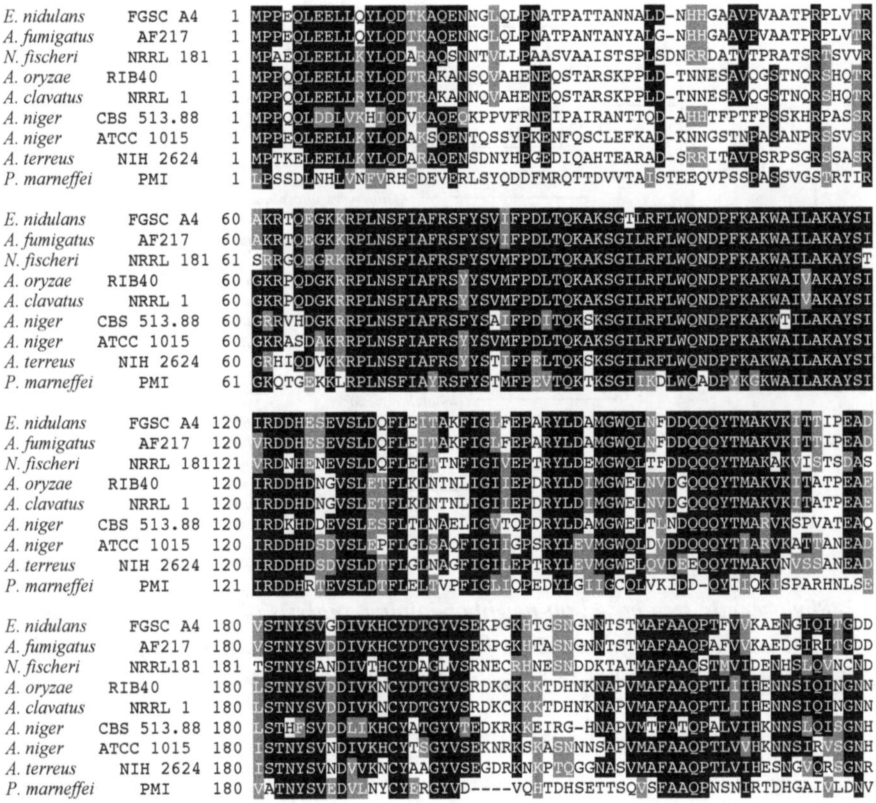

Figure 1. Alignment of partial sequences of the MAT-1 *idiomorph of* Aspergillus *species with that of* Penicillium marneffei. *The alignment was created by Boxshade 3.21 (http://www.ch.embnet.org/software/BOX_form.html)*

2.2. Comparison of mating and meiosis related genes between S. cerevisiae and Aspergilli

Mating type genes regulate plasmogamy and the production of ascogenous hyphae in ascomycetes, but of course they are not the only genes involved in the regulation of mating and meiosis. In *S. cerevisiae* three different groups of genes have been identified that are involved in mating and meiosis. These groups are the meiotic proteins that are involved in meiosis itself, the pheromone processing proteins involved in signalling to the other mating partner and the pheromone response proteins involved in the response to mating pheromones. A list of these *S. cerevisiae* genes based on the article of Woo *et al.* (2006) is given in Table 1.

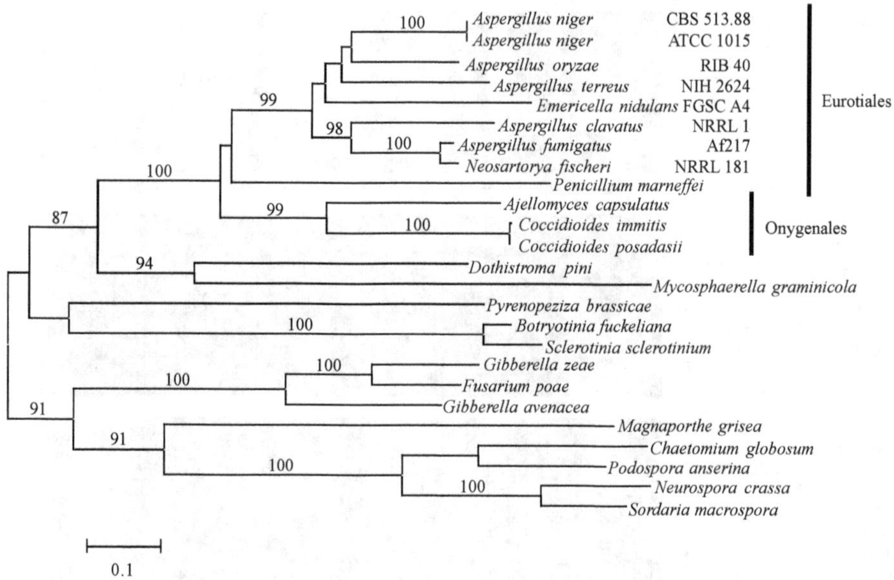

Figure 2. Phylogenetic tree based on neighbor-joining analysis of MAT-1 *gene sequences of Aspergilli and various other filamentous fungi. Phylogenetic analysis was performed by the MEGA 3.1 software (Kumar* et al., *2004). Numbers above branches are bootstrap values.*

Table 1. Mating related proteins in S. cerevisiae *and their functions. The genes fall into three functional classes: the meiotic proteins, the pheromone processing proteins and the pheromone response proteins (Woo* et al., *2006).*

Protein name (ID)	Function
Meiotic proteins	
DMC1 (YER179W)	meiosis-specific protein, repair of double-strand breaks
HOP1 (YIL072W)	meiosis-specific DNA binding protein
HOP2 (YGL033W)	meiosis-specific protein, prevents synapsis between nonhomologous chromosomes
MLH1 (YMR167W)	required for mismatch repair in mitosis and meiosis
MLH2 (YLR035C)	required for DNA mismatch repair in mitosis and meiosis
MLH3 (YPL164C)	involved in DNA mismatch repair
MND1 (YGL183C)	required for recombination and meiotic nuclear division
MRE11 (YMR224C)	subunit of a complex that functions in repair of DNA double-strand breaks and in telomere stability
MSH2 (YOL090W)	forms heterodimers with Msh3p and Msh6p that bind to DNA mismatches to initiate the mismatch repair process

Table 1. Continued.

Protein name (ID)	Function
Meiotic proteins (Continued)	
MSH4 (YFL003C)	involved in meiotic recombination
MSH5 (YDL154W)	forms a dimer with Msh4p that facilitates crossovers between homologues during meiosis
MSH6 (YDR097C)	required for mismatch repair in mitosis and meiosis
PMS1 (YNL082W)	ATP-binding protein required for mismatch repair in mitosis and meiosis
RAD50 (YNL250W)	subunit of MRX complex, involved in processing double-strand DNA breaks in vegetative cells
RAD51 (YER095W)	DNA strand exchange protein
RAD52 (YML032C)	stimulates DNA strand exchange
SPO11 (YHL022C)	meiosis-specific protein, initiates meiotic recombination
Pheromone processing proteins	
KEX1 (YGL203C)	protease involved in the processing of killer toxin and alpha factor precursor
KEX2 (YNL238W)	subtilisin-like protease (proprotein convertase)
RAM1 (YDL090C)	beta subunit of the CAAX farnesyltransferase (FTase)
RAM2 (YKL019W)	alpha subunit of both the farnesyltransferase and type I geranylgeranyltransferase
RCE1 (YMR274C)	type II CAAX prenyl protease involved in the proteolysis and maturation of Ras and the a-factor mating pheromone
STE6 (YKL209C)	ATP-binding cassette (ABC) transporter required for the export of a-factor
STE13 (YOR219C)	dipeptidyl aminopeptidase, required for maturation of alpha factor
STE14 (YDR410C)	farnesyl cysteine-carboxyl methyltransferase
STE23 (YLR389C)	metalloprotease involved in N-terminal processing of pro-a-factor to the mature form
STE24 (YJR117W)	zinc metalloprotease that functions in two steps of a-factor maturation
Pheromone response proteins	
BEM1 (YBR200W)	protein containing SH3-domains, involved in establishing cell polarity and morphogenesis
CDC24 (YAL041W)	guanine nucleotide exchange factor, required for polarity establishment and maintenance
CDC42 (YLR229C)	small rho-like GTPase, essential for establishment and maintenance of cell polarity
DIG1 (YPL049C)	regulatory protein of unknown function
DIG2 (YDR480W)	regulatory protein of unknown function
FAR1 (YJL157C)	cyclin-dependent kinase inhibitor that mediates cell cycle arrest in response to pheromone

Table 1. Continued.

Protein name (ID)	Function
Pheromone response proteins (Continued)	
FUS3 (YBL016W)	mitogen-activated protein kinase involved in mating pheromone response
GPA2 (YER020W)	nucleotide binding alpha subunit of the heterotrimeric G protein that interacts with the receptor Gpr1p
KSS1 (YGR040W)	mitogen-activated protein kinase (MAPK) involved in signal transduction pathways that control filamentous growth and pheromone response
STE2 (YFL026W)	receptor for alpha-factor pheromone
STE3 (YKL178C)	receptor for a-factor lipoprotein
STE4 (YOR212W)	G protein beta subunit, forms a dimer with Ste18p to activate the mating signaling pathway
STE5 (YDR103W)	scaffold protein that, in response to pheromone, shuttles from the nucleus to the plasma membrane
STE7 (YDL159W)	signal transducing MAP kinase kinase involved in pheromone response
STE11 (YLR362W)	signal transducing MEK kinase involved in pheromone response
STE12 (YHR084W)	transcription factor, activates genes involved in mating or pseudohyphal/ invasive growth pathways
STE18 (YJR086W)	G protein gamma subunit, forms a dimer with Ste4p to activate the mating signaling pathway
STE20 (YHL007C)	signal transducing kinase of the PAK (p21-activated kinase) family, involved in pheromone response
STE50 (YCL032W)	protein involved in mating response

The *S. cerevisiae* mating and meiosis related proteins were used to blastp search (Altschul *et al.*, 1990) the genomes of *A. fumigatus*, *A. nidulans*, *A. oryzae*, *A. terreus*, and *A. niger* for homologues. Bi-directional best hit analyses were performed with as criterion for homology an accepted E-value of $<e^{-10}$. Homologues could be found in the genomes of the filamentous fungi for the majority of *S. cerevisiae* mating related genes (Table 2). However, 9 out of 46 (20%) of the *S. cerevisiae* mating related genes could not be detected in any of the Aspergilli. The number of *S. cerevisiae* mating related genes that were missing from the individual genomes of different *Aspergillus* species was between 11 and 13.

Some of the mating and meiosis related proteins were highly conserved in all genomes examined even in those of the presumably asexual fungi (with E-values

Table 2. S. cerevisiae, N. crassa *and* A. nidulans *mating related proteins and their homologues (threshold E-value <e^{-10}) in different* Aspergillus *species.*

			Strain					
			A. fumigatus (Af 293)	*A. nidulans* (FGSC A4)	*A. niger* (CBS 513.88)	*A. niger* (ATCC 1015)	*A. oryzae* (RIB40)	*A. terreus* (NIH 2624)
S. cerevisiae	Meiotic proteins							
		DMC1	•	•	•	•	•	•
		HOP1	•	•	•	•		•
		HOP2						
		MLH1	•	•	•	•	•	•
		MLH2						
		MLH3	•	•	•	•	•	•
		MND1		•			•	
		MRE11	•	•	•	•	•	•
		MSH2	•	•	•	•	•	•
		MSH4	•		•		•	•
		MSH5	•		•	•		•
		MSH6	•	•		•	•	•
		PMS1	•	•	•	•	•	•
		RAD50	•	•	•	•	•	•
		RAD51	•	•	•	•	•	•
		RAD52	•	•	•	•	•	•
		SPO11		•				
	Pheromone processing proteins							
		KEX1	•	•	•	•	•	•
		KEX2	•	•	•	•	•	•
		RAM1	•	•	•	•	•	•
		RAM2	•	•	•	•	•	•
		RCE1	•	•	•	•	•	•
		STE6	•	•	•	•	•	•
		STE13						
		STE14	•	•	•	•	•	•
		STE23	•	•	•	•	•	•
		STE24	•	•	•	•	•	•

Table 2. Continued.

		Strain					
		A. fumigatus (Af 293)	*A. nidulans* (FGSC A4)	*A. niger* (CBS 513.88)	*A. niger* (ATCC 1015)	*A. oryzae* (RIB40)	*A. terreus* (NIH 2624)
S. cerevisiae	Pheromone response pathway proteins						
	BEM1	•	•	•	•	•	•
	CDC24	•	•	•	•	•	•
	CDC42	•	•	•	•	•	•
	DIG1						
	DIG2						
	FAR1						
	FUS3	•	•	•	•	•	•
	GPA2	•	•	•	•	•	•
	KSS1						
	STE2	•	•	•		•	•
	STE3	•	•	•	•	•	•
	STE4	•	•	•	•	•	•
	STE5						
	STE7	•	•	•	•	•	•
	STE11	•	•	•	•	•	•
	STE12	•	•	•	•	•	•
	STE18						
	STE20	•	•	•	•	•	•
	STE50	•	•	•	•	•	•
N. crassa							
	MAT-A1		•	•	•	•	•
	MAT-A2						
	MAT-A3						
A. nidulans[1]							
	MAT-1	[2]	•	•	•	•	•
	MAT-2	•	•			[3]	

[1]*A. nidulans* MAT-1 and MAT-2 proteins have identical function as the *N. crassa* MAT-A1 and MAT-A2, respectively.
[2]The MAT-1 idiomorph has been found in other isolates of *A. fumigatus* (Paoletti *et al.*, 2005).
[3]The MAT-2 idiomorph has been found in other isolates of *A. oryzae* (Paoletti *et al.*, 2005).

$<e^{-100}$; Table 3), and these may have an additional role next to their function in mating processes (i.e. DNA repair). In other cases the E-values of the proteins of the Aspergilli were just above the threshold when we did the blast search with the *S. cerevisiae* proteins, but comparing the homologues in the different Aspergilli these proteins showed a much higher similarity to one another.

The number of protein homologues was generally not different between presumed sexual and asexual lines, except for two proteins: the protease KEX1 and the ABC transporter STE6 (Table 3). Both of these genes with a more general function have multiple homologues. KEX1 has 11-14 homologues in the asexual Aspergilli (including *A. fumigatus* that may have a sexual cycle) and only 5 in the sexual *A. nidulans*. We identified 24 to 45 homologues of STE6 in the Aspergilli. In contrast, the SPO11 protein, involved in the initiation of meiotic recombination, had only a homologue in *A. nidulans,* with a rather low homology, and none in any of the other Aspergilli (Table 3).

2.3. Comparison of mating and meiosis related genes between two **A. niger** *genomes*

Thus, our findings so far show that *A. niger* does contain one of the mating type idiomorphs, and many of the mating and meiosis related genes have homologues in the species as well as in related Aspergilli. As two distinct *A. niger* genomes have been completely sequenced by the DOE Joint Genome Institute and DSM respectively, we can compare the sets of mating and meiosis related genes between them.

Three of the tested *S. cerevisiae* proteins were found only in one of the two *A. niger* strains (Table 3): the meiotic recombination protein MSH4 (no hit) and alpha factor pheromone receptor STE2 (E-value below threshold) were not present in strain ATCC 1015 and mismatch repair protein MSH6 (counterblasting with the *A. niger* protein resulted SRP40 protein as best hit in the *S. cerevisiae* database) was not present in CBS 513.88.

For the rest, the two *A. niger* strains seemed to possess a very similar set of mating-related genes. These seemingly homologous polypeptides were aligned with ClustalX (Thompson *et al.,* 1997) and edited with Boxshade 3.21 (written by Kay Hofmann and Michael D. Baron and available at http://www.ch.embnet.org/software/BOX_form.html). The degree of identity of proteins was described with blasting the proteins against each other by BLOSUM62 matrix (Tatusova and Madden, 1999, at http://www.ncbi.nlm.nih.gov/blast/bl2seq/wblast2.cgi,). When proteins from the two *A. niger* databases were compared, we found many, smaller or bigger annotation errors

Table 3. Homologues of S. cerevisiae *proteins involved in mating and pheromone processing in Aspergilli. Only proteins with an E-value of <e^{-10} are counted. First the number of homologues is given, than the ID of the most homologous protein is shown with its E-value between brackets. E-values below e^{-100} are shown as 0.0.*

	Number of homologues in Aspergilli and ID of most homologous protein with E-value					
	A. fumigatus (Af 293)	*A. nidulans* (FGSC A4)	*A. niger* (CBS 513.88)	*A. niger* (ATCC 1015)	*A. oryzae* (RIB40)	*A. terreus* (NIH 2624)
Meiotic proteins						
DMC1	3 Afu7g02200 (0.0)	3 AN9092.3 (0.0)	3 An12g00460 (0.0)	?[a] 43615 (0.0)	3 AO090038000596 ($7e^{-99}$)	3 ATEG_01853.1 (0.0)
HOP1	1 Afu6g13050 ($3e^{-12}$)	1 AN5516.3 ($3e^{-14}$)	1 An08g10440 ($8e^{-19}$)	1 175768 ($2e^{-11}$)	0	1 ATEG_03335.1 ($2e^{-14}$)
HOP2	0	0	0	0	0	0
MLH1	3 Afu5g11700 (0.0)	3 AN0126.3 (0.0)	3 An18g03030 (0.0)	3 42780 (0.0)	3 AO090120000288 (0.0)	3 ATEG_02100.1 (0.0)
MLH2	0	0	0	0	0	0
MLH3	4 Afu4g06490 ($1e^{-34}$)	3 AN4365.3 ($6e^{-28}$)	3 An04g00870 ($8e^{-29}$)	3 54646 ($2e^{-34}$)	3 AO090023000933 ($1e^{-30}$)	3 ATEG_05571.1 ($3e^{-32}$)
MND1	0	1 AN1843.3 ($6e^{-12}$)	0	0	1 AO090003001276 ($8e^{-15}$)	0
MRE11	1 Afu6g11410 (0.0)	1 AN0556.3 (0.0)	1 An08g07330 (0.0)	1 175727 (0.0)	1 AO090023000471 (0.0)	1 ATEG_05022.1 (0.0)
MSH2	6 Afu3g09850 (0.0)	4 AN10621.3 (0.0)	6 An16g03520 (0.0)	5 183201 (0.0)	5 AO090005001261 (0.0)	6 ATEG_04508.1 (0.0)

	Number of homologues in Aspergilli and ID of most homologous protein with E-value					
	A. fumigatus (Af 293)	*A. nidulans* (FGSC A4)	*A. niger* (CBS 513.88)	*A. niger* (ATCC 1015)	*A. oryzae* (RIB40)	*A. terreus* (NIH 2624)
MSH4	5 Afu1g02000 ($5e^{-66}$)	0	6 An01g06260 (0.0)	0	5 AO090005000895 ($5e^{-98}$)	6 ATEG_04890.1 (0.0)
MSH5	6 Afu1g11170 ($2e^{-45}$)	0	6 An08g03470 ($7e^{-73}$)	5 122070 ($1e^{-43}$)	0	6 ATEG_00319.1 (0.0)
MSH6	1 Afu4g08300 (0.0)	4 AN1708.3 (0.0)	0	4 50490 (0.0)	5 AO090023000729 (0.0)	6 ATEG_05365.1 (0.0)
PMS1	1 Afu2g13410 (0.0)	3 AN6316.3 (0.0)	3 An02g02280 (0.0)	4 36742 (0.0)	3 AO090026000318 ($9e^{-79}$)	3 ATEG_01317.1 (0.0)
RAD50	7 Afu4g12680 (0.0)	8 AN3619.3 (0.0)	15 An01g08180 (0.0)	?[a] 138700 (0.0)	6 AO090009000296 (0.0)	6 ATEG_03151.1 (0.0)
RAD51	2 Afu1g10410 (0.0)	2 AN1237.3 (0.0)	3 An08g02350 (0.0)	?[a] 47657 (0.0)	2 AO090038000386 (0.0)	3 ATEG_00230.1 (0.0)
RAD52	1 Afu4g06970 ($4e^{-53}$)	1 AN4407.3 (0.0)	1 An04g01290 ($4e^{-47}$)	1 44271 ($3e^{-32}$)	1 AO090023000890 ($1e^{-47}$)	1 ATEG_05536.1 (0.0)
SPO11	0	1 AN8259.3 ($1e^{-12}$)	0	0	0	0

Table 3. Continued.

	Number of homologues in Aspergilli and ID of most homologous protein with E-value					
	A. fumigatus (Af 293)	*A. nidulans* (FGSC A4)	*A. niger* (CBS 513.88)	*A. niger* (ATCC 1015)	*A. oryzae* (RIB40)	*A. terreus* (NIH 2624)
Proteins involved in pheromone processing						
KEX1	11	5	14	12	12	11
	Afu1g08940	AN10184.3	An08g00430	208486	AO090005001632	ATEG_08515.1
	($1e^{-83}$)	(0.0)	($2e^{-98}$)	($2e^{-94}$)	($6e^{-92}$)	(0.0)
KEX2	1	1	1	1	1	1
	Afu4g12970	AN3583.3	An01g08530	55344	AO090009000291	ATEG_03179.1
	(0.0)	(0.0)	(0.0)	(0.0)	(0.0)	(0.0)
RAM1	2	2	3	2	2	2
	Afu4g10330	AN2002.3	An04g06620	133160	AO090003001188	ATEG_00696.1
	($4e^{-47}$)	(0.0)	($2e^{-59}$)	($3e^{-47}$)	($5e^{-37}$)	($3e^{-40}$)
RAM2	1	2	2	2	2	1
	Afu4g07800	AN3867.3	An04g02210	190580	AO090023000791	ATEG_05421.1
	($9e^{-31}$)	($1e^{-45}$)	($4e^{-43}$)	($4e^{-39}$)	($1e^{-41}$)	($4e^{-43}$)
RCE1	1	1	1	1	1	1
	Afu6g04890	AN6528.3	An14g03420	211108	AO090701000050	ATEG_06959.1
	($1e^{-18}$)	($1e^{-19}$)	($2e^{-17}$)	($1e^{-12}$)	($5e^{-20}$)	($7e^{-16}$)
STE6	29	24	25	35	45	25
	Afu4g08800	AN2300.3	An04g03690	190859	AO090023000664	ATEG_09424.1
	(0.0)	(0.0)	(0.0)	(0.0)	(0.0)	(0.0)
STE13	0	0	0	0	0	0
STE14	1	1	1	1	1	1
	Afu2g08420	AN6162.3	An12g03660	186640	AO090011000860	ATEG_09679.1
	($1e^{-43}$)	(0.0)	($5e^{-44}$)	($3e^{-48}$)	($2e^{-37}$)	(0.0)
STE23	1	1	2	2	1	1
	Afu5g02010	AN8044.3	An16g01860	41311	AO090003001317	ATEG_09820.1
	(0.0)	(0.0)	(0.0)	(0.0)	(0.0)	(0.0)

	Number of homologues in Aspergilli and ID of most homologous protein with E-value					
	A. fumigatus (Af 293)	*A. nidulans* (FGSC A4)	*A. niger* (CBS 513.88)	*A. niger* (ATCC 1015)	*A. oryzae* (RIB40)	*A. terreus* (NIH 2624)
STE24	1	1	1	1	1	1
	Afu4g07590	AN11231.3	An04g01950	50547	AO090023000816	ATEG_05444.1
	(0.0)	(0.0)	(0.0)	(0.0)	(0.0)	(0.0)
Proteins involved in pheromone response						
BEM1	1	1	1	1	1	1
	Afu4g04120	AN7030.3	An14g00710	53737	AO090206000084	ATEG_10121.1
	($1e^{-48}$)	(0.0)	($9e^{-67}$)	($3e^{-53}$)	($7e^{-55}$)	(0.0)
CDC24	1	1	2	1	1	2
	Afu4g11450	AN5592.3	An04g05150	129585	AO090003001078	ATEG_03936.1
	($1e^{-46}$)	($3e^{-38}$)	($4e^{-41}$)	($1e^{-44}$)	($1e^{-39}$)	($7e^{-39}$)
CDC42	17	15	17	9	14	11
	Afu2g05740	AN7487.3	An02g14200	52477	AO090001000693	ATEG_06763.1
	($1e^{-85}$)	(0.0)	($7e^{-90}$)	(0.0)	($4e^{-90}$)	(0.0)
DIG1	0	0	0	0	0	0
DIG2	0	0	0	0	0	0
FAR1	0	0	0	0	0	0
FUS3	63	66	25[b]	51	64	63
	Afu6g12820	AN3719.3	An08g10670	207710	AO090003000402	ATEG_03316.1
	(0.0)	(0.0)	(0.0)	(0.0)	(0.0)	(0.0)
GPA2	3	3	4	4	4	3
	Afu1g12930	AN1016.3	An08g05820	55775	AO090012000600	ATEG_00488.1
	(0.0)	(0.0)	(0.0)	(0.0)	(0.0)	(0.0)
KSS1	0	0	0	0	0	0
STE2	1	1	1	0	1	1
	Afu3g14330	AN2520.3	An09g04180		AO090701000605	ATEG_03500.1
	($3e^{-17}$)	($1e^{-18}$)	($5e^{-17}$)		($4e^{-16}$)	($5e^{-15}$)

Table 3. Continued.

	Number of homologues in Aspergilli and ID of most homologous protein with E-value					
	A. fumigatus (Af 293)	*A. nidulans* (FGSC A4)	*A. niger* (CBS 513.88)	*A. niger* (ATCC 1015)	*A. oryzae* (RIB40)	*A. terreus* (NIH 2624)
Proteins involved in pheromone response						
STE3	1	1	1	1	1	1
	Afu5g07880	AN7743.3	An03g03890	44420	AO090701000699	ATEG_08338.1
	$(7e^{-38})$	$(3e^{-15})$	$(2e^{-37})$	$(3e^{-19})$	$(4e^{-38})$	$(8e^{-36})$
STE4	11	14	13	8	15	11
	Afu5g12210	AN0081.3	An18g02090	54102	AO090120000339	ATEG_02052.1
	$(2e^{-55})$	(0.0)	$(7e^{-78})$	$(1e^{-76})$	$(4e^{-77})$	(0.0)
STE5	0	0	0	0	0	0
STE7	39	50	25[b]	24	49	49
	Afu3g05900	AN3422.3	An11g10690	209137	AO090020000060	ATEG_08950.1
	$(2e^{-57})$	(0.0)	$(3e^{-63})$	$(4e^{-66})$	$(4e^{-64})$	(0.0)
STE11	58	61	25[b]	55	61	59
	Afu5g06420	AN2269.3	An17g01280	214017	AO090009000610	ATEG_09389.1
	$(4e^{-91})$	(0.0)	$(4e^{-89})$	$(4e^{-99})$	$(1e^{-79})$	(0.0)
STE12	1	1	1	1	1	1
	Afu5g06190	AN2290.3	An17g01580	57283	AO090009000638	ATEG_09411.1
	$(2e^{-60})$	(0.0)	$(2e^{-60})$	$(3e^{-66})$	$(3e^{-62})$	(0.0)
STE18	0	0	0	0	0	0
STE20	60	58	25[b]	43	59	55
	Afu2g04680	AN2067.3	An11g04320	208780	AO090003000267	ATEG_06035.1
	(0.0)	(0.0)	(0.0)	(0.0)	(0.0)	(0.0)
STE50	1	1	1	1	1	1
	Afu2g17130	AN7252.3	An04g09220	51404	AO090102000104	ATEG_10030.1
	$(4e^{-19})$	$(2e^{-16})$	$(2e^{-18})$	$(3e^{-17})$	$(7e^{-17})$	$(5e^{-16})$

[a]Exact number of homologues is not known, since only blasting in the non-filtered database resulted hits, which were redundant.
[b]Maximum number of hits was limited in 25 hits per blast search.

in both genomes. As a solution to this problem we aligned the DNA sequences from both *A. niger* genomes and changed the ORFs of the shorter protein according to the longer protein. Thus compared, the investigated protein homologues in the *A. niger* strains showed no or little polymorphism between each other, only a few amino acid substitutions or single missing amino acids generate slight polymorphisms (Table 4).

Table 4. Differences between mating related proteins of the two A. niger *(CBS 513.88 and ATCC 1015) strains. If two proteins differ in size, the longer one is the basis for counting percentage of identities, similarities and gaps. Gaps are counted only in the homologous region.*

Protein	Type and size of difference	Identical	Gaps
Meiotic proteins			
DMC1 (YER179W)	no difference	344/344 (100%)	(0%)
HOP1 (YIL072W)	no difference	783/783 (100%)	(0%)
MLH1 (YMR167W)	substitutions: 251, 376, 526, 537, 548, 665, 726, 731-732, 744, 760-761, 765	759-772 (98%)	(0%)
MLH3 (YPL164C)	no difference	943/943 (100%)	(0%)
MRE11 (YMR224C)	indel region: 644 substitutions: 506, 584, 732, 742	737/742 (99%)	1/742 (0%)
MSH2 (YOL090W)	substitution: 511	944/945 (99%)	(0%)
MSH4 (YFL003C)	only present in CBS513.88	–	–
MSH5 (YDL154W)	substitutions: 8, 141, 487, 664	892/896 (99%)	(0%)
MSH6 (YDR097C)	only present in ATCC1015	–	–
PMS1 (YNL082W)	substitutions: 6, 395, 490, 518, 576, 591, 610, 656, 771, 1029, 1050	1044/1055 (99%)	(0%)
RAD50 (YNL250W)	substitutions: 425, 479, 482, 512, 553, 565, 597, 729, 775, 832, 929, 937, 942, 991,	1280/1294 (99%)	(0%)
RAD51 (YER095W)	no difference	–	–
RAD52 (YML032C)	substitutions: 234, 291, 395	595/598 (99%)	(0%)
Pheromone processing proteins			
KEX1 (YGL203C)	substitutions: 673, 721, 811	820/823 (99%)	(0%)
KEX2 (YNL238W)	no difference	–	–
RAM1 (YDL090C)		523/523 (100%)	(0%)
RAM2 (YKL019W)	no difference	–	–
RCE1 (YMR274C)	substitutions: 105, 113, 215	329/332 99%)0	(0%)

Table 4. Continued.

Protein	Type and size of difference	Identical	Gaps
Pheromone processing proteins (Continued)			
STE6 (YKL209C)	substitutions: 7, 125, 148, 208, 243, 247-248, 262, 452, 454, 468, 594, 634, 641, 658, 672, 740, 819, 912, 981-982, 985, 990, 1008, 1052, 1123	1306/1332 (98%)	(0%)
STE14 (YDR410C)	no difference	–	–
STE23 (YLR389C)	1st substitution: 774	1166/1167 (99%)	(0%)
	2nd pair substitutions: 269, 1019	1035/1037 (99%)	(0%)
	The two allele pairs differ from each other.		
STE24 (YJR117W)	no difference	–	–
Pheromone response proteins			
BEM1 (YBR200W)	substitutions: 96	600/601 (99%)	(0%)
CDC24 (YAL041W)	no difference	–	–
CDC42 (YLR229C)	1st pair: no difference	250/250 (100%) and	(0%) and
	2nd pair: substitution: 48	198/199 (99%)	(0%)
FUS3 (YBL016W)	no difference	–	–
GPA2 (YER020W)	no difference	–	–
STE2 (YFL026W)	only present in CBS513.88	–	–
STE3 (YKL178C)	substitutions: 134, 307	521/523 (99%)	(0%)
STE4 (YOR212W)	no difference	–	–
STE7 (YDL159W)	no difference	–	–
STE11 (YLR362W)	indel region: 87-88	903/905 (99%)	2/905 (0%)
STE12 (YHR084W)	no difference	–	–
STE20 (YHL007C)	substitution: 92	837/838 (99%)	0/838 (0%)
STE50 (YCL032W)	no difference	–	–

3. Discussion

Sexual and asexual propagation are two alternative ways of fungal reproduction. Whereas certain strains are able to undergo both, others reproduce exclusively sexually or asexually. The black mould *A. niger* and several closely related species constitute the *A. niger* aggregate and are generally considered to be strictly asexual fungi as no sexual stage has ever been observed. These asexual *Aspergillus* species are of high importance as human pathogens and as industrial microorganisms. It is crucial to know more about their reproductive cycles for better disease control or for applications in industry. Recently sequenced genomes may be adequate tools to compare fungi with different life cycles (i.e. sexual/asexual) and infer conclusions about the way of reproduction of related species.

Idenification of mating type genes may give a first clue about the sexuality of a species. Homothallic sexual fungi combine both mating type idiomorphs in their genome, enabling the species to reproduce sexually without mating partner. Heterothallic species depend on a partner of the opposite mating type for successful reproduction. The majority of species in the Trichocomaceae family are homothallic, with eight confirmed exceptions. Among those having *Aspergillus* anamorphs, six belong to the genus *Neosartorya*: *N. fennelliae, N.spathulata, N. nishimurae, N. udagawae, N. indohii* and *N. tsurutae* (Kwon-Chung and Kim, 1974; Takada and Udagawa, 1985; Takada *et al.,* 2001; Horie *et al.*, 1995; 2003), one to *Emericella*: *E. heterothallica* (Raper and Fennell, 1965; Malloch and Cain, 1972), while *Talaromyces derxii* belongs to *Penicillium* subgenus *Biverticillium* (Takada and Udagawa, 1988). Asexual species do not depend on mating type genes, as a relic from a sexual ancestor they may contain one or more mating type genes or these genes may have been lost.

Recently, mating type genes have been identified in several *Aspergillus* species making it possible to search for homologous sequences in related species (Pöggeler, 2002; Varga, 2003; Dyer *et al.,* 2003; Paoletti *et al.,* 2005). In the homothallic *A. nidulans* both mating idiomorphs have been identified within one genome. *A. nidulans* is a cosmopolitan species that occurs both in tropical and temperate regions. It is capable of reproducing both asexually and sexually, outcrossing often enough for recombination to be detectable (Geiser *et al.,* 1994). In the sequenced genomes of *A. fumigatus, A. oryzae* and *A. terreus* only one of the mating idiomorphs, either *MAT-1* or *MAT-2* could be identified. *A. fumigatus,* for a long time considered to be an asexual species, has been shown to carry the mating type genes in a heterothallic manner, each isolate carrying one mating type gene at a time. In *A. fumigatus* both mating type genes are present in the population at approximately equal ratios (Dyer *et al.,* 2003; Paoletti *et al.,* 2005).

Both *A. niger* strains examined were found to carry a *MAT-1* idiomorph, while a *MAT-2* homologue was not found. Since homothallic filamentous ascomycetes, including Fusaria, *Cochliobolus* species, *A. nidulans* and *N. fischeri* usually carry both mating type idiomorphs (Yun *et al.,* 1999, 2000; Dyer *et al.,* 2003; Rydholm *et al.,* 2007), we presume that the examined *A. niger* isolates represent one mating type of a heterothallic species. The other putative mating type either became extinct, or might exist in small endemic populations. Such extinction processes are also happening today in *Ophiostoma novo-ulmi* and *Cryptococcus neoformans* populations (Mitchell and Brasier, 1994; Yan *et al.*, 2002). If truly asexual, we presume that *A. niger* also evolved from a heterothallic ancestor. Heterothallism was proposed to be ancient in the genus *Cochliobolus* (Yun *et al.*, 1999), and the putative heterothallic nature of asexual *Fusarium* species has also been proposed earlier (Kerenyi *et al.*, 2004). Geiser *et al.* (1996) hypothesised that asexual Aspergilli derived from homothallic species. In *Aspergillus* section *Fumigati* in particular, homothallism was suggested to be an ancient character based on phylogenetic analysis of β-tubulin and hydrophobin sequences (Geiser *et al.*, 1998). However, more recently, Rydholm *et al.* (2007) proposed that *A. fumigatus* and other asexual Aspergilli derived from heterothallic ancestors based on comparison of the organisation of the mating type loci of homothallic and heterothallic Aspergilli (Figure 3). Loss of one mating type in heterothallic fungi could be a major cause of asexuality in filamentous ascomycetes.

Many more genes are involved in a successful sexual reproduction than just the mating type genes. In model organisms *S. cerevisiae* sets of genes have been identified coding for proteins involved in meiosis, pheromone production or pheromone response pathways. In this study we compared mating related genes of *S. cerevisiae* with those found in the sequenced genomes of five *Aspergillus* species, including the two sequenced genomes of *A. niger*.

The presence of homologues of the majority of *S. cerevisiae* genes in the filamentous fungi strongly supports the hypothesis that sexual reproduction arose early in a common ancestor of all living eukaryotes (Ramesh *et al.*, 2005). Comparing the five *Aspergillus* species, we can see that they share a similar set of mating related genes. The majority of the genes present in the homothallic *A. nidulans* are also present in asexual *Aspergillus* species, such as the two *A. niger* isolates.

The main difference between the two sequenced *A. niger* strains lies in the observation that three mating related genes (meiotic recombination protein MSH4, alpha factor pheromone receptor STE2, and mismatch repair protein MSH6) are unequally distributed between the two strains. For the rest we see only minor differences between the examined genes between the two. This leads to the question: why are meiosis-related genes preserved in an asexual species? This could either suggest a possible sexual cycle, a recently lost sexual cycle, or pleiotropy of many of the mating related proteins. Of course, whether these mating related genes are truly expressed in *A. niger* is not known. The detection of MAT-1 but not MAT-2 in the two sequenced strains is not conclusive, analysis of the MAT sequences of more natural isolates is needed to get a better insight into possible sexual reproduction of *A. niger*.

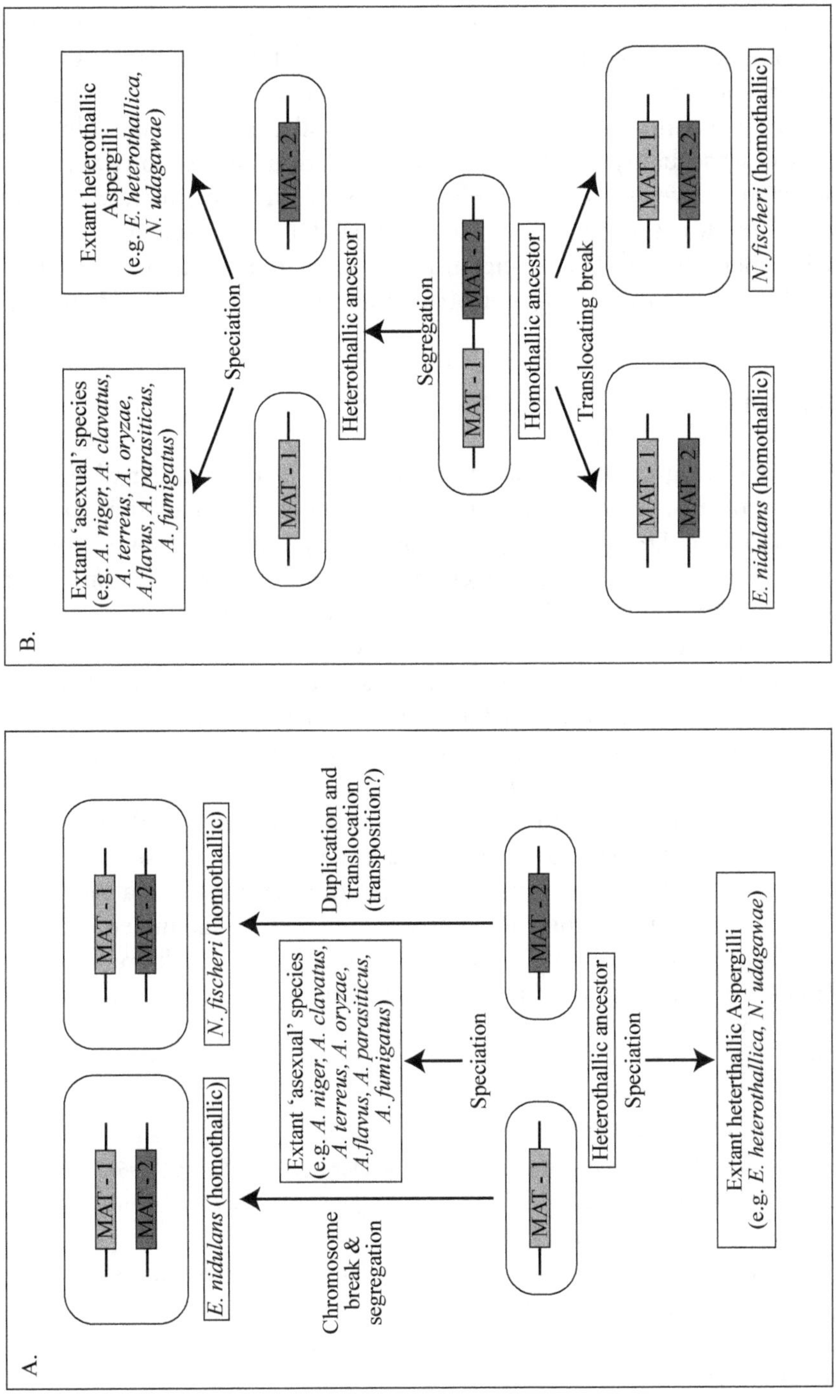

Figure 3. Schemes depicting two views of the evolution of sex in the Aspergillus *genus. (A) The ancestor was heterothallic (Rydholm* et al., *2007); (B) The ancestor was homothallic (Geiser* et al., *1996, 1998).*

There is one more intriguing observation in *A. niger*: the high genetic diversity and vegetative incompatibility observed amongst natural isolates of the black *Aspergillus* aggregate (Van Diepeningen *et al.,* 1997). Vegetative incompatibility effectively blocks parasexual recombination, but a cryptic sexual cycle or - in evolutionary terms - recently lost sexuality could explain the observed variation. However, looking at recombination between molecular phylogenies of different nuclear genes no evidence was found for recombination, whether sexual or parasexual in *A. niger* and related black Aspergilli (Van Diepeningen, 1999). Thus, our findings together with previous results indicate that *A. niger* is a truly asexual species unable to undergo a sexual cycle.

Acknowledgments

We thank the following institutes and companies and individuals involved in the genome projects of the species used in this study: genome of the culture collection strain CBS 513.88 from DSM (prepared by shotgun sequencing, 7.5 times coverage; http://www.dsm.com/en_US/html/dfs/genomics_aniger.htm) and genome of the culture collection strain ATCC 1015 from DOE Joint Genome Institute (assembly release version 1.0 of whole genome shotgun reads was constructed with the JGI assembler, Jazz, using paired end sequencing reads at a coverage of 8.9X; http://www.jgi-psf.org/cgi-bin/runAlignment?db=Aspni1&advanced=1). The *A. fumigatus* preliminary sequence data was obtained from The Institute for Genomic Research website at http://www.tigr.org. *A. nidulans* and *A. terreus* sequence data were from the *Aspergillus nidulans* and *Aspergillus terreus* Sequencing Project, Broad Institute of MIT and Harvard (http://www.broad.mit.edu). *A. oryzae* sequences were available on the server of National Institute of Technology and Evaluation (NITE). We thank the lab assistance in the Laboratory of Genetics, Wageningen University and Department of Microbiology, University of Szeged. We thank Paul Dyer and Mathieu Paoletti for comments and recommendations.

References

Accensi, F., J. Cano, L. Figuera, M.L. Abarca and F.J. Cabañes, 1999. New PCR method to differentiate species in the *Aspergillus niger* aggregate. FEMS Microbiology Letters 180: 191-196.

Altschul, S.F., W. Gish, W. Miller, E.W. Myers and D.J. Lipman, 1990. Basic local alignment search tool. Journal of Molecular Biology 215: 403-410.

Barton, N.H. and B. Charlesworth, 1998. Why sex and recombination? Science 281: 1986-1990.

Cavalier-Smith, T., 2002. Origins of the machinery of recombination and sex. Heredity 88: 125-141.

De Visser, J.A.G.M. and S.F. Elena, 2007. The evolution of sex: empirical insights into the roles of epistasis and drift. Nature Reviews Genetics 8: 139-149.

Dyer, P.S. and M. Paoletti, 2005. Reproduction in *Aspergillus fumigatus*: sexuality in a supposedly asexual species? Medical Mycology 43 (Suppl. 1): S7-S14.

Dyer, P.S., M. Paoletti and D.B. Archer, 2003. Genomics reveals sexual secrets of *Aspergillus*. Microbiology 149: 2301-2303.

Fraser, J.A., J.E. Stajich, E.J. Tarcha, G.T. Cole, D.O. Inglis, A. Sil and J. Heitman, 2007. Evolution of the mating type locus: insights gained from the dimorphic primary fungal pathogens *Histoplasma capsulatum, Coccidioides immitis,* and *Coccidioides posadasii.* Eukaryotic Cell 6: 622-629.

Geiser, D.M., C. Gueidan, J. Miadlikowska, F. Lutzoni, F. Kauff, V. Hofstetter, E. Fraker, C.L. Schoch, L. Tibell, W.A. Untereiner and A. Aptroot, 2006. Eurotiomycetes: Eurotiomycetidae and Chaetothyriomycetidae. Mycologia 98: 1054-1065.

Geiser, D.M., J.C. Frisvad and J.W. Taylor, 1998. Evolutionary relationships in *Aspergillus* section *Fumigati* inferred from partial β-tubulin and hydrophobin DNA sequences. Mycologia 90: 831-845.

Geiser, D.M., M.L. Arnold and W.E. Timberlake, 1994. Sexual origins of British Aspergillus nidulans isolates. Proceedings of the National Academy of Sciences of the United States of America 91: 2349-2352.

Geiser, D.M., W.E. Timberlake and M.L. Arnold, 1996. Loss of meiosis in *Aspergillus.* Molecular Biology and Evolution 13: 809-817.

Hamari, Z., F. Kevei, É. Kovács, J. Varga, Z. Kozakiewicz and J.H. Croft, 1997. Molecular and phenotypic characterization of *Aspergillus japonicus* strains with special regard to their mitochondrial DNA polymorphisms. Antonie van Leeuwenhoek 72: 337-347.

Horie, Y., M. Miyaji, K. Nishimura, M.F. Franco and K.I.R. Coelho, 1995. New and interesting species of *Neosartorya* from Brazilian soil. Mycoscience 36: 199-204.

Horie, Y., P. Abliz, K. Fukushima, K. Okada and G.M. Campos-Takagi, 2003. Two new species of *Neosartorya* from Amazonian soil, Brazil. Mycoscience 44: 397-402.

Hurst, L.D. and J.R. Peck, 1996. Recent advances in understanding of the evolution and maintenance of sex. Trends in Ecology and Evolution 11: 46-52.

Kerenyi, Z., A. Moretti, C. Waalwijk, B. Olah and L. Hornok, 2004. Mating type sequences in asexually reproducing *Fusarium* species. Applied and Environmental Microbiology 70: 4419-4423.

Kevei, F., Z. Hamari, J. Varga, Z. Kozakiewicz and J.H. Croft, 1996. Molecular polymorphism and phenotypic variation in *Aspergillus carbonarius.* Antonie van Leeuwenhoek 70: 59-66.

Kumar, S., K. Tamura and M. Nei, 2004. MEGA3: Integrated Software for Molecular Evolutionary Genetics Analysis and Sequence Alignment. Briefings in Bioinformatics 5: 150-163.

Kwon-Chung, K.J. and S.J. Kim, 1974. A second heterothallic *Aspergillus.* Mycologia 66: 628-638.

Latgé, J.P., 1999. *Aspergillus fumigatus* and aspergillosis. Clinical Microbiology Reviews 12: 310-350.

Malloch, D. and Cain, R. F. 1972. New species and combinations of cleistothecial ascomycetes. Canadian Journal of Botany 50: 61-72.

Milgroom, M.G. and W.E. Fry, 1997. Contributions of population genetics to plant disease epidemiology and management. Advances in Botanical Research 24: 1-30.

Mitchell, A.G. and C.M. Brasier, 1994. Contrasting structure of European and North American populations of *Ophiostoma ulmi.* Mycological Research 88: 576-582.

Paoletti, M., C. Rydholm, E.U. Schwier, M.J. Anderson, G. Szakacs, F. Lutzoni, J.P. Debeaupuis, J.P. Latgé, D.W. Denning and P.S. Dyer, 2005. Evidence for sexuality in the opportunistic fungal pathogen *Aspergillus fumigatus*. Current Biology 15: 1242-1248.

Pöggeler, S., 2002. Genomic evidence for mating abilities in the asexual pathogen *Aspergillus fumigatus*. Current Genetics 42: 153-160.

Pontecorvo, G., 1956. The parasexual cycle in fungi. Annual Review of Microbiology 10: 393-400.

Pontecorvo, G., J.A. Roper and E. Forbes, 1953. Genetic recombination without sexual reproduction in *Aspergillus niger*. Journal of General Microbiology 8: 198-210.

Ramesh, M.A., S.B. Malik and J.M. Logsdon Jr., 2005. A phylogenomic inventory of meiotic genes: evidence for sex in *Giardia* and an early eukaryotic origin of meiosis. Current Biology 15: 185-191.

Raper, K.B. and D.I. Fennell, 1965. The genus *Aspergillus*. Williams and Wilkins, Baltimore.

Rydholm, C., P.S. Dyer and F. Lutzoni, 2007. DNA sequence characterization and molecular evolution of MAT1 and MAT2 mating-type loci of the self-compatible ascomycete mold *Neosartorya fischeri*. Eukaryotic Cell 6: 868-874.

Samson, R.A., J.A.M.P. Houbraken, A.F.A. Kuijpers, J.M. Frank and J.C. Frisvad, 2004. New ochratoxin or sclerotium producing species in *Aspergillus* section *Nigri*. Studies in Mycology 50: 45-61.

Takada, M. and S. Udagawa, 1988. A new species of heterothallic *Talaromyces*. Mycotaxon 31: 417-425.

Takada, M., Y. Horie and P. Abliz, 2001. Two new heterothallic *Neosartorya* from African soil. Mycoscience 42: 361-367.

Tatusova, T.A. and T.L. Madden, 1999. Blast 2 sequences - a new tool for comparing protein and nucleotide sequences. FEMS Microbiology Letters. 174: 247-250.

Thompson, J.D., T.J. Gibson, F. Plewniak, F. Jeanmougin and D.G.Higgins, 1997. The CLUSTAL_X windows interface: flexible strategies for multiple sequence alignment aided by quality analysis tools. Nucleic Acids Research 25: 4876-4882.

Turgeon, B.G. and O.C. Yoder, 2000. Proposed nomenclature for mating type genes of filamentous ascomycetes. Fungal Genetics and Biology 31: 1-5.

Van Diepeningen, A.D., 1999. Horizontal transfer of genetic elements in the black Aspergilli. PhD Thesis, Wageningen University, The Netherlands.

Van Diepeningen, A.D., A.J.M. Debets and R.F. Hoekstra, 1997. Heterokaryon incompatibility blocks virus transfer among natural isolates of black *Aspergilli*. Current Genetics 32: 209-217.

Varga, J., 2003. Mating type gene homologs in *Aspergillus fumigatus*. Microbiology 149: 816-819.

Varga, J., F. Kevei, A. Vriesema, F. Debets, Z. Kozakiewicz and J.H. Croft, 1994. Mitochondrial DNA restriction fragment length polymorphisms in field isolates of the *Aspergillus niger* aggregate. Canadian Journal of Microbiology 40: 612-621.

Varga, J., F. Kevei, C. Fekete, A. Coenen, Z. Kozakiewicz and J.H. Croft, 1993. Restriction fragment length polymorphisms in the mitochondrial DNAs of the *Aspergillus niger* aggregate. Mycological Research 97: 1207-1212.

Woo, P.C.Y., K.T.K. Chong, H. Tse, J.J. Cai, C.C.Y. Lau, A.C. Zhou, S.K.P. Lau and K.Y. Yuen, 2006. Genomic and experimental evidence for a potential sexual cycle in the pathogenic thermal dimorphic fungus *Penicillium marneffei*. FEBS Letters 580: 3409-3416.

Yan, Z., X. Li and J. Xu, 2002. Geographic distribution of mating type alleles of *Cryptococcus neoformans* in four areas of the United States. Journal of Clinical Microbiology 40: 965-972.

Yun, S.H., M.L. Berbee, O.C. Yoder and B.G. Turgeon, 1999. Evolution of the fungal self-fertile reproductive life style from self-sterile ancestors. Proceedings of the National Academy of Sciences USA 96: 5592-5597.

Yun, S.H., T. Arie, I. Kaneko, O.C. Yoder and B.G. Turgeon, 2000. Molecular organization of mating type loci in heterothallic, homothallic and asexual *Gibberella/Fusarium* species. Fungal Genetics and Biology 31: 7-20.

Mycoviruses in the Aspergilli

Anne D. van Diepeningen[1], János Varga[2,3], Rolf F. Hoekstra[1] and Alfons J.M. Debets[1]
[1]Laboratory of Genetics, Plant Sciences, Wageningen University Arboretumlaan 4, 6703 BD Wageningen, The Netherlands
[2]CBS Fungal Biodiversity Centre, Uppsalalaan 8, 3584 CT Utrecht, The Netherlands
[3]University of Szeged, Faculty of Sciences, Dept. of Microbiology, Szeged, H-6701 Hungary

Abstract

This chapter contains both an overview of what is known about mycoviruses and of which *Aspergillus* species were found to be infected, as well as two detailed studies in the mycovirus dynamics in two different *Aspergillus* species. We start by giving a general review of viruses infecting fungi: these mycoviruses can contain either single-stranded (ss) or double-stranded RNA or double stranded DNA genomes. So far at least nine virus families have been identified that infect fungal species. For a long time these mycoviruses were generally considered to be cryptic, but nowadays more evidence points to the fact that on close examination most infections are deleterious to their hosts. Next, an overview is given of mycoviruses identified in *Aspergillus* species: most of the mycoviruses were identified in asexual species, but several sexual *Aspergillus* species also proved to be infected. Then two special cases of mycovirus infections and their dynamics in *Aspergillus* species are discussed: the natural infection with double-stranded RNA mycoviruses of approximately ten percent of asexual *A. niger* and other black Aspergilli and the artificial infection of both sexually and asexually reproducing *Emericella nidulans* (anamorph: *A. nidulans*). For these species horizontal and vertical transmission of mycoviruses was studied as well as the effect of heterokaryon incompatibility and sexual recombination on this transmission. The effect of virus infection on different fitness traits was quantified using isogenic virus free and infected lines. A population model based on transfer rates and fitness costs is presented to predict the stability of the mycovirus infections. We discuss the possibility of interspecies virus transfer between *Aspergillus* species and further relatives. Finally, we conclude with the possibilities the genomics era has for research of mycoviruses and their hosts.

Keywords: double-stranded RNA, mycoviruses, dynamics

1. An introduction to mycoviruses

1.1. A history of mycoviruses

Viruses are infectious agents that are invisible with the light microscope, are small enough to pass through a bacterial filter, lack a metabolism of their own and thus depend on living host cells for their replication, but they encode some of the genes necessary for their own replication. Often, but not necessarily, the RNA or DNA genome of the virus is protected by a protein coat. Since the first discovery of a virus in a fungal species, it has become clear that mycoviruses and virus-like replicons occur commonly in fungi (e.g. Buck, 1986, 1998; Ghabrial, 1998).

Some of the first discovered mycoviruses were found in *Aspergillus* species. In 1970 Banks *et al.* described the isolation of isometric virus particles from *Aspergillus foetidus*. These particles were approximately 40 nm in diameter and proved to contain double stranded RNA (Banks *et al.*, 1970). Further experiments showed that two distinct classes of particles were present, that could be separated electrophoretically indicating multiple infection with two distinct viruses. One class of particles contained four main dsRNA components and the other contained two that were separately encapsidated (Ratti and Buck, 1972). Most of the mycoviruses found since in fungal species have single or segmented double stranded RNA (dsRNA) genomes, but single stranded (ss) RNA and DNA genomes have also been described (Buck, 1986, 1998; Fauquet *et al.*, 2005). Often the mycovirus particles can be seen by electron microscopy as isometric particles (see for example Figure 1), but also other forms like rod-shapes or non-encapsidated mycoviruses are found. For a current classification of mycoviruses see paragraph 1.2.

Lhoas (1970) was the first to describe the transmission of virus-like particles (VLPs) in *Aspergillus niger* to previously virus free strains through heterokaryosis. Mycoviruses are exceptional in that except for the mycoreoviruses they do not seem to have an extracellular phase in their multiplication cycle and depend on transmission by intracellular routes (Buck, 1986, 1998). Protein coats may be essential for viruses in general to survive outside the host cell, and as mycoviruses seem to lack an extracellular phase completely, not all mycoviruses have retained capsid proteins or may use them for other functions. For example Wickner (1996) described a yeast virus capsid protein that provides both protection in the form of subcellular compartmentalisation for transcription and replication, but that also has a catalytic function in decapping host messenger RNA (mRNA) in favour of viral mRNAs.

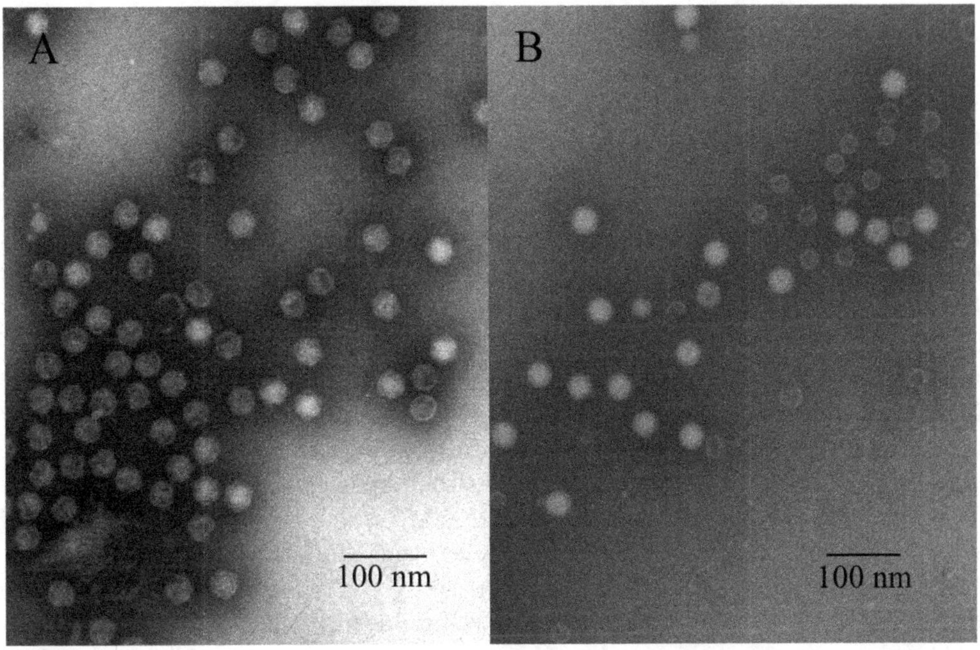

Figure 1. Electron microscopy photographs of mycoviruses in cytoplasmic fractions of two black Aspergillus *strains. A. isometric particles of about 34-38 nm diameter in a subsamples of* Aspergillus foetidus *strain N076 (CBS 618.78). B. Small isometric particles of 23-25 nm and larger particles of 31-35 nm in size from* Aspergillus tubingensis *strain Ind 1.6.19.*

In some fungi viruses have been noted because of their effects on their host. These effects vary and can be caused by virally coded products or simply by disturbances of the host cell metabolism. The killer toxins encoded by some viruses or satellite RNAs can have a selective advantage for their host in crowded environments (e.g. Wickner, 1996). Especially plant pathogenic fungi are known to become hypovirulent due to infections with mycoviruses, which can affect a large number of cell processes (e.g. Nuss, 1992, 1996). Other viruses are known because they cause serious crop losses as for instance in *Agaricus bisporus* or *Pleurotus* spp. (e.g. Go *et al.*, 1992; Rinker *et al.,* 1993; Van der Lende *et al.*, 1994; Romaine and Schlagnhaufer, 1995). But, because the effects of many mycoviruses on their host are not known or not easily visible, mycovirus infections are often considered 'cryptic'.

1.2. Taxonomy of viruses infecting fungi

According to modern virus taxonomy, at least members of nine virus families and of an unassigned genus infect fungi (Table 1 and Figure 2; Fauquet *et al.*, 2005). Most mycoviruses carry dsRNA genomes enclosed into nonenveloped isometric VLPs. These viruses carry one or several dsRNA segments, and replicate by the action of an RNA dependent RNA polymerase (RdRp). Five families of dsRNA mycoviruses are currently recognised: the Partitiviridae, Totiviridae, Chrysoviridae, Hypoviridae, and Reoviridae (Enebak *et al.*, 1994; Ghabrial, 1998; Hong *et al.*, 1999; Dawe and Nuss, 2001; Suzuki *et al.*, 2004; Wickner *et al.*, 2000b). The Partitiviridae family comprises viruses with bipartite genomes with sizes between 1.4 and 3.0 kbp. The two segments code for an RdRp and a capsid protein for the formation of isometric particles respectively (Ghabrial, 1998; Ghabrial *et al.*, 2000). Members of the family Totiviridae are dsRNA viruses with a linear uncapped genome of 4.6 to 7 kbp in size (Wickner *et al.*, 2000b). The type species for the Chrysoviridae is the *Penicillium chrysogenum* virus (PcV) which was first described by Banks *et al.*, (1969) and contains four double-stranded RNA fragments. Recently this virus has been cloned and sequenced, and the different fragments of 2.9 to 3.6 kbp in size were shown to code for a putative RNA-dependent RNA polymerase, a capsid protein and a virion-associated protein. All four dsRNAs were found to have extended regions of conserved terminal sequences at both ends (Jiang and Ghabrial, 2004). Other chrysovirus-like viruses were identified in the plant pathogen *Cochliobolus (Helminthosporium) victoriae* (Ghabrial *et al.*, 2002) and in several other *Penicillium* species (Jiang and Ghabrial, 2004). Members of the Hypoviridae family have pleiomorphic morphology, and distinguish themselves by attenuation of virulence and the alteration of developmental processes after infection of the fungal host. The Hypoviridae contain one segment of dsRNA between 10 and 13 kbp in length. The type virus is the *Cryphonectria* hypovirus 1 (Shapira *et al.*, 1991). The last family is Reoviridae. The *Mycoreovirus* genus has recently been accepted as a virus taxon to accommodate reoviruses of *Cryphonectria parasitica* (Enebak *et al.*, 1994; Hillman *et al.*, 2004; Suzuki *et al.*, 2004) and *Rosellinia necatrix* (Osaki *et al.*, 2002; Kanematsu *et al.*, 2004). Mycoreoviruses contain 11 dsRNA segments in approximately equimolar amounts and are the only mycoviruses which have an extracellular phase during their life cycle (Fauquet *et al.*, 2005).

Other mycoviruses carry single stranded RNA or dsDNA genomes. The Mushroom Bacilliform virus has a genome of positive-sense single-stranded RNA and as its name describes its virion is bacilliform in shape (Revill *et al.*, 1994). It belongs to the Barnaviridae family. A mycovirus with similar morphology and genome organisation has been found in *Botrytis cinerea* (Howitt *et al.*, 1995). The Narnaviridae family contains two positive stranded RNA mycovirus genera (Wickner *et al.*, 2000a). The

Table 1. Taxonomical outline of viruses infecting fungi (Fauquet et al., *2005).*

Nature of the genome	Family	Morphology	Genome structure	Genus	Type species	Closest relative (based on sequence data)
dsDNA	-	isometric	1 circular	*Rhizidiovirus*	*Rhizidiomyces* virus	Phycodnaviridae
dsRNA	Hypoviridae	pleiomorphic	1 segment	*Hypovirus*	*Cryphonectria* hypovirus 1-EP713	Potyviridae
	Chrysoviridae	isometric	4 segments	*Chrysovirus*	*Penicillium chrysogenum* virus	Totiviridae
	Partitiviridae	isometric	2 segments	*Partitivirus*	*Atkinsonella hypoxylon* virus	Alphacryptovirus
	Totiviridae	isometric	1 segment	*Totivirus*	*Saccharomyces cerevisiae* L-A virus	Giardia-, Leishmaniaviruses, *Ribes*, Totivirus
	Reoviridae	isometric	11 segments	*Mycoreovirus*	Mycoreovirus 1	Coltivirus
ss(+)RNA	Barnaviridae	bacilliform	1 ss(+)RNA	*Barnavirus*	Mushroom bacilliform virus	Luteo-, carmoviruses
	Narnaviridae	RNP complex	1 + segment	*Narnavirus*	*Saccharomyces cerevisiae* 20SRNA narnavirus	Leviviridae
				Mitovirus	*Cryphonectria parasitica* mitovirus-1 NB631	Leviviridae, RdRps of plant mtDNAs
ss(+)RNA-RT	Pseudoviridae	isometric/ spherical	1 + segment	*Pseudovirus*	*Saccharomyces cerevisiae* Ty1 virus	Hemivirus, Retroviridae, Metaviridae
				Hemivirus	*Drosophila melanogaster* copia virus[1]	Pseudovirus, Retroviridae, Metaviridae
	Metaviridae	spherical	1 + segment	*Metavirus*	*Saccharomyces cerevisiae* Ty3 virus	Caulimoviridae, Retroviridae, Pseudoviridae

[1]Includes *Saccharomyces cerevisiae* Ty5 virus.

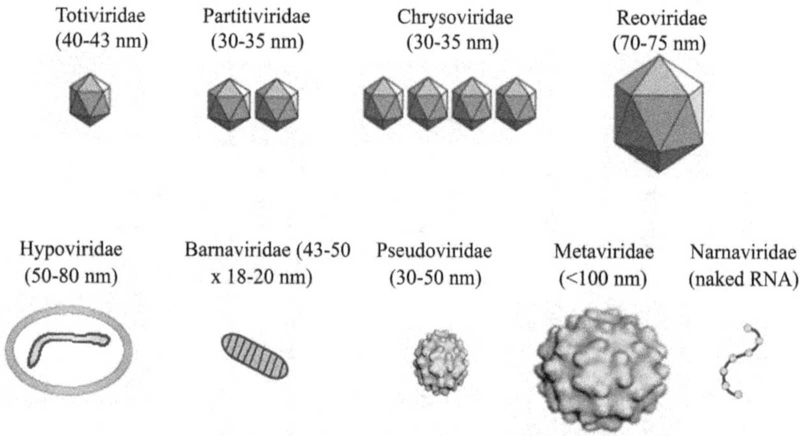

Figure 2. Diagram of the recognised virus families infecting fungi with their morphology and size range (modified after Nuss, 2005).

Narnavirus genus includes the *Saccharomyces cerevisiae* 20SRNA narnavirus as type species (Rodriguez-Cousino *et al.*, 1998), while the *Mytovirus* genus contains the *C. parasitica* mitovirus-1 and several *Ophiostoma* mitoviruses (e.g. Polashok and Hillman, 1994; Hong *et al.*, 1998, 1999). Two families of ssRNA mycoviruses resemble retrotransposons, and replicate by reverse transcription (Pseudoviridae, Metaviridae) (e.g. Boeke *et al.*, 1988 for the *Pseudovirus* type species, and Hansen and Sandmeyer, 1990 for the Metavirus type species). An isometric dsDNA mycovirus infecting *Rhizidiomyces* species has also been described, although this species belongs to the *Hyphochytriomycota phylum*, which is not treated as member of the fungal kingdom (Alexopoulos *et al.*, 1996).

Besides these viruses, a large number of mycoviruses are still uncharacterised. The genome sequence and organisation of some mycoviruses not fitting to any of the above families or genera have recently been characterised. Among these, an ssRNA mycovirus (DaRV) of *Diaporthe ambigua* could not be assigned to any of the above families; its genome sequence exhibited relatively high levels of homology to carmoviruses of the Tombusviridae family (Preisig *et al.*, 2000). Flexuous rod-shaped ssRNA viruses BvF and BvX of *B. cinerea* seem to be closely related to plant potex- and tymoviruses (Howitt *et al.*, 2001, 2006), a spherical mycovirus with ssRNA genome was found to be related to Tymoviruses (Yu *et al.*, 2003), while isometric ssRNA mycoviruses SmVa and SmVb found in *Sclerophthora macrospora* were related to the Nodaviridae family which occurs in animals, and to the plant pathogen *Sobemovirus* genus, respectively (Yokoi *et al.*, 1999, 2003). Some fungal viruses, including those infecting *Helicobasidium mompa* and *Phytophthora* species have been found to be related to the *Endornavirus* genus, a newly established dsRNA plant virus group (Horiuchi and Fukuhara, 2004; Osaki *et al.*, 2006; Hacker *et al.*, 2005).

Killer viruses of *S. cerevisiae* and of other yeasts also could not be assigned to any virus families; these satellite viruses need a helper virus for replication and encapsidation, since their genomes lack an RdRp gene (Buck, 1986). Other still uncharacterised mycoviruses include the rigid rod-shaped VLPs of rusts and powdery mildews, Herpesvirus-like viruses of lower fungi, enveloped bacilliform particles of the entomopathogenic zygomycete *Strongwelsea magna*, bacteriophage-like VLPs of yeasts and *Penicillium* species (Buck, 1986, 1998), a ssRNA virus of *Pleurotus eryngii* (Ro *et al.*, 2007), and mycoviruses responsible for MVX disease of *A. bisporus* (Grogan *et al.*, 2003; Rao *et al.*, 2007).

1.3. Multiple infections

One fungal isolate can be infected by a number of different mycoviruses simultaneously. For instance, *A. foetidus* CBS 618.78 is infected with two viruses, AfV-S belonging to the Totiviridae, and AfV-F, an unassigned mycovirus giving rise to a 6-band dsRNA profile on agarose gels (Buck and Ratti, 1975). Observation of 1–8 different dsRNA segments in individual isolates of black Aspergilli indicates that perhaps even more mycoviruses can be present in a single isolate at the same time (Varga *et al.*, 2001, Van Diepeningen *et al.*, 2006). Multiple infections have also been observed in other fungi including *Gremmeniella abietina* (Tuomivirta and Hantula, 2003b), *Penicillium stoloniferum* (Bozarth *et al.*, 1971), *Gaeumannomyces graminis* (Buck *et al.*, 1981), *Fusarium graminearum* (Chu *et al.*, 2004), *Chalara elegans* (Park *et al.*, 2005) and *A. bisporus* (Grogan *et al.*, 2003; Rao *et al.*, 2007). A single species can be the host of various mycoviruses belonging to different virus families, e.g. different *Cryphonectria parasitica* isolates may carry viruses belonging to either Hypoviridae, Narnaviridae, or a virus belonging to the Reoviridae family. *A. bisporus* may carry bacilliform ssRNA viruses (Barnaviridae) as well as isometric viruses of the Partitiviridae family, either 25 nm or 34 nm isometric viruses carrying 2 and 9 dsRNAs, respectively (Van der Lende *et al.*, 1994; Rao *et al.*, 2007; Buck, 1986).

Next to infection with multiple independent viruses, two types of extra dsRNA fragments, the so-called satellite and defective RNAs, can be detected in association with so-called helper viruses, on which they depend for their replication. Both satellite and defective RNAs can potentially interfere with the replication of their helper virus and they are likely to be widespread in populations of dsRNA mycoviruses. They contribute to the complexity of observed dsRNA profiles from individual fungi

(Buck, 1998). Satellite RNAs are comprised largely of sequences that are distinct from those of their helper viruses (Mayo *et al.*, 1995). Some protein toxins are encoded for by satellite RNAs as for example in *S. cerevisiae* (Buck, 1986; Wickner, 1996) or are found in association with hypovirulence (Deng and Boland, 2004) But, also novel dsRNA fragments originating from host' genomic DNA have been found (Lakshman and Tavantzis, 1994). The defective RNAs are derived from their helper viruses and occur generally after internal deletions as for example happened in *C. parasitica* hypovirus (Tartaglia *et al.*, 1986; Shapira *et al.*, 1991).

1.4. Transmission of mycoviruses

The only mycoviruses known to have an extracellular phase are the mycoreoviruses, all other mycoviruses are considered to be strictly bound to intracellular life within their host (Fauquet *et al.*, 2005). The location of these viruses within their host varies per virus species. Some are for example found in the mitochondria like the mitoviruses in *C. parasitica* and *Ophiostoma novo-ulmi* (Polashok and Hillman, 1994; Hong *et al.*, 1998, 1999), others in association with the trans-Golgi network (Jacob-Wilk *et al.*, 2006), and for many of them the exact intracellular location is not known. In the Aspergilli one of the dsRNA segments observed in *Petromyces alliaceus* was found in the mitochondria, but all other examined dsRNA segments of *Aspergillus* species have been found to be localised in the cytoplasm (Varga *et al.*, 1998, 2001). Thus, bound to intracellular life the mycoviruses depend on vertical transmission through asexual or sexual spores and for new infection on horizontal transmission after hyphal contact.

Vertical transmission through asexual spores is quite efficient in most cases. For example in the black Aspergilli, *Aspergillus flavus*, and *Emericella nidulans* (anamorph: *Aspergillus nidulans*) 100% of the tested conidiospores was infected (Van Diepeningen *et al.*, 1997, 2006; Coenen *et al.*, 1997; Wood *et al.*, 1974). However, in some species lower vertical transfer rates to asexual offspring were observed. For example, the 7 kb dsRNA fragment of *F. graminearum* was found to be transmitted to about 50% of conidia (Chu *et al.*, 2002), though in their next study isolates that showed a pattern of a mixed infection with two different mycoviruses were transmitted to all progeny, both in conidia and ascospores (Chu *et al.*, 2004). In *Heterobasidion annosum*, vertical transmission of dsRNA through conidia was found to be 3% and 55% in the two examined isolates (Ihrmark *et al.*, 2002). Transmission rates also varied in *A. flavus* isolates (Elias and Cotty, 1997) and were also far from absolute in *Penicillium* and *Magnaporthe* species (Buck, 1986).

Vertical transmission through sexual spores is usually much less effective than via asexual spores. For nearly thirty years of mycovirus research, mycoviruses have even only been found in asexual Aspergilli and not in any species able to undergo a sexual cycle. In other fungal genera mycoviruses were also more often found in asexuals than in sexual species. Mycoviruses have never been observed yet in natural isolates of some *Aspergillus* species which are able to undergo a sexual cycle, including *A. nidulans, Fennellia nivea* and *Eurotium* species (Coenen *et al.*, 1997; Varga *et al.*, 2001, 2005; unpublished observations). However, isolates of two *Neosartorya* species were found to carry dsRNA elements, and a *Petromyces alliaceus* isolate was also found to be infected (Varga *et al.*, 1998). Mycoviruses do occur in a number of other fungi which are able to reproduce both mitotically and meiotically (e.g. *Ophiostoma ulmi, C. parasitica, G. graminis*; for references see Buck, 1986), and also in some strains of *Neurospora crassa* (Myers *et al.*, 1988).

Exclusion of dsRNA segments from sexual spores was observed in *G. graminis* strains (McFadden *et al.*, 1983), in *O. ulmi* (Brasier, 1983), in *Epichloë festucae* (Romo *et al.*, 2007) and in *Helicobasidium mompa* and *R. necatrix* (Ikeda *et al.*, 2004). On the contrary, mycovirus transfer through ascospores was found to be effective in *Neosartorya hiratsukae*, while the stromata embedding the asci in *Petromyces alliaceus* were found not to transmit one of the two observed dsRNA segments (Varga *et al.*, 1998). This observation indicates that some mechanisms exist in Aspergilli to exclude cytoplasmically located dsRNA molecules from stromatal structures. Mycoviruses were transmitted very efficiently through basidiospores in *Agaricus brunnescens* (Buck, 1986), *Ustilago maydis* (Koltin and Day, 1976) and *Phaffia rhodozyma* (Pfeiffer *et al.*, 1996), *H. annosum* (Ihrmark *et al.*, 2004) and through ascospores in *S. cerevisiae* (Brewer and Fangman, 1980), and in *F. graminearum* (Chu *et al.*, 2004).

Horizontal transmission of mycoviruses usually takes place through hyphal anastomosis. This process is under the control of vegetative (*vic*) or heterokaryon (*het*) incompatibility genes which may have evolved to limit the spread of deleterious elements like mycoviruses (Caten, 1972). Lhoas (1970) was the first to describe the transmission through heterokaryosis of VLPs in *A. niger* to previously virus-free strains. In *C. parasitica*, transmission occurs readily between isolates carrying the same *vic(het)* genes, while the efficiency of transmission decreases with the increasing number of different *vic* genes in the partners (Liu and Milgroom, 1996). Similar findings were reported in *A. nidulans* (Coenen *et al.*, 1997). On the contrary, transmission of mycoviruses has been found to be effective between intersterility groups of *H. annosum* (Ihrmark *et al.*, 2002), and by anastomosis between individuals of *Beauveria bassiana* (Dalzoto *et al.*, 2006).

In the laboratory, mycoviruses can also be transferred from one isolate to another by protoplast fusion (Buck, 1986; Varga *et al.*, 1994a). This technique could be used to transfer mycoviruses between *A. niger*, *A. tubingensis*, *A. oryzae* and *A. nidulans,* and even from *Fusarium poae* to *A. niger* (Varga *et al.*, 1994a; Coenen *et al.*, 1997; Liang and Chen, 1987; Van Diepeningen *et al.*, 1998, 2000). Transfer via protoplast fusion circumvents some of the barriers formed by the *het*-genes and transfer by protoplast fusions seems to be more efficient than by the co-cultivation of incompatible strains or species (Coenen *et al.*, 1997; Van Diepeningen *et al.*, 1998, 2000).

Mycovirus transmission through introducing purified virus particles into the hyphae or protoplasts of fungi has rarely been reported; examples are from yeasts (El-Sherbeini and Bostian, 1987), *G. graminis* (Stanway and Buck, 1984), the hypovirus of *C. parasitica* (Sasaki *et al.*, 2002), and the mycoreoviruses of *C. parasitica* and *R. necatrix* (Hillman and Suzuki., 2004; Sasaki *et al.*, 2007). Sasaki *et al.* (2006) also succeeded in transfecting VLPs of a Partitivirus to *R. necatrix* protoplasts.

Some evidence has also been found that horizontal transmission may even occur between different species in nature. For instance the occurrence of conspecific mitoviruses in *Sclerotinia homeocarpa* and *O. novo-ulmi* has been treated as evidence for horizontal transmission of this dsRNA mycovirus (Deng *et al.*, 2003, Deng and Boland, 2004). Liu *et al.* (2003) found evidence for interspecies transfer in the genus *Cryphonectria*. Several weeks of co-cultivation of virus free *A. nidulans* and infected *A. niger* strains in the laboratory also led to relatively high number of successful transfers (Coenen *et al.*, 1997), where similar experiments with heterokaryon incompatible black Aspergilli yielded very little successful transfers (Van Diepeningen *et al.*, 1997). Also the similarity of mycoviruses in *A. niger*, other black *Aspergillus* species and *F. poae,* and the ease with which these mycoviruses could be transferred from one species to the other indicate that interspecies transfer may also occur in nature (Van Diepeningen *et al.*, 2000).

1.5. Effects of mycoviruses on their hosts

The presence of mycoviruses is usually not associated with any phenotypic traits in fungi and thus they are generally considered to cause latent or so-called cryptic infections. But, one can expect that even harbouring non-virulent, cryptic viruses must have a cost to the host simply by the resources the virus uses. Intraspecies variance between strains probably often obscures the true positive or negative effects of a virus infection. To quantify such effects of virus infections, one would need isogenic virus-free and infected strains. However, in both ascomycetes and basidiomycetes more and more fungal species are found with clear and strong phenotypic effects caused by mycovirus infection. One could classify these infections by their effects as mycoviruses with beneficial effects on their host (e.g. killer phenomena in yeast), mycoviruses that cause hypovirulence in their pathogenic hosts, mycoviruses that cause yield losses in economically important fungi, and mycoviruses having other debilitation effects on their host.

Mycoviruses could have evolved in concert with their hosts to limited detrimental effect or to mutual benefit. In structured habitats the benefits from interference competition can be strong. Viruses linked with killer phenotypes in yeasts and smuts confer a strong advantage to their hosts under these conditions by encoding toxins and protecting their host cells from the toxins produced (for references, see e.g. Buck 1986; Varga *et al.*, 2001; D. Wloch and J.A.G.M. de Visser, unpublished results).

The best studied example of mycoviruses causing hypovirulence is that of the chestnut blight fungus (e.g. Nuss, 1992, 1996; Anagnostakis *et al.*, 1998; Hoegger *et al.*, 2003; Hillman and Suzuki, 2004). Recently, hypovirulence has also been found to be caused by the presence of mycoviruses in a number of other fungi including *Sclerotinia sclerotiorum* (Xie *et al.*, 2006), *C. victoriae* (Ghabrial *et al.*, 2002), *B. cinerea* (Castro *et al.*, 2003), *F. graminearum* (Chu *et al.*, 2002), *Monosporascus cannonballicus* (Batten *et al.*, 2000), *O. novo-ulmi* (Hong *et al.*, 1999), *S. homeocarpa* (Deng *et al.*, 2003; Deng and Boland, 2004), *B. bassiana* (Dalzoto *et al.*, 2006) and *R. necatrix* (Kanematsu *et al.*, 2004). Interestingly, the presence of a dsRNA mycovirus was found to upregulate fungal virulence in *Nectria radicicola* (Ahn and Lee, 2001).

Especially in several edible mushrooms, virus effects were quantified as yield losses. In the oyster mushroom *Pleurotus florida* comparisons between virus infected and virus free strains showed reduced growth, an increase in growth abnormalities, increased infections in culture beds and a reduction in fruit body yield of approximately 30% (Go *et al.*, 1992). Rinker *et al.* (1993) compared isogenic infected and virus free lines of *Pleurotus pulmonarius* and found a reduced growth, no changes in either carpophore morphology or colour, and a total reduction in yield of 50% due to the presence of the mycovirus. Also in *P. eryngii* mycovirus infection causes severe epidemic symptoms (Ro *et al.*, 2007). In the white button mushroom *A. bisporus* several mycoviruses with segmented dsRNA genomes are known. The virus causing La France disease causes brown spots throughout the fruiting body (Hollings, 1962), virus X turns the outer layer of the white button mushroom brown instead of white (Rao *et al.*, 2007; Grogan *et al.*, 2003) and both viruses render the harvest unsellable.

Other phenotypic, mostly debilitating, effects have also been observed in mycovirus infected fungi. Differences were observed in growth between hypovirus infected and dsRNA free strains of *C. parasitica* (Elliston, 1985). Kazmierczak *et al.* (1996) attributed the cause of this reduced virulence, but also reduced sporulation and pigmentation to the reduced accumulation of mRNAs and proteins of four specific host genes under the influence of the virus infection. The presence of mycoviruses also affected spore production, growth rate, endochitinase production and infectivity of *Metarhizium anisopliae* isolates to the cattle tick *Boophilus microplus* (De La Paz Gimenez-Pecci *et al.*, 2002; Frazzon *et al.*, 2000). DsRNAs also may affect ceratoulmin production (Sutherland and Brasier, 1995) and were suggested to be responsible for the *de novo* generation of mitochondrial DNA plasmids in *O. novo-ulmi* (Charter *et al.*, 1993). In *B. cinerea* the infection with mycovirus lead to an important degree of cellular degeneration (Castro *et al.*, 1999) and the mycovirus infecting *F. graminearum* strain DK21 induced pronounced morphological changes including a reduction in mycelial growth, increased pigmentation, reduced virulence towards its host plant wheat and a 60-fold decrease in the production of trichothecene mycotoxins (Chu *et al.*, 2002). Recently, Marquez *et al.* (2007) found that the presence of a mycovirus in the endophyte *Curvularia protubera* is responsible for the heat tolerance of the host plant *Dichanthelium lanuginosum*.

Regarding Aspergilli, phenotypic characteristics have not been reliably attributed to the presence of VLPs in Aspergilli, with two exceptions. Elias and Cotty (1996), for example, found no virus effects on mycotoxin production of their section *Flavi* strains. Also Coenen *et al.* (1997) did not see an effect of mycovirus infection on its *A. nidulans* host. But, Varga *et al.* (1994a) observed that *A. niger* strains which were cured of dsRNAs lost not only their VLPs but also their 'arginine/proline leaky' phenotype and became prototrophic. In another case, Van Diepeningen *et al.* (2006) observed that in one *A. niger* strain the presence of mycovirus showed clear visible effects on its host with non-sporulating sectors. Besides, it was also demonstrated that otherwise cryptic mycovirus infections in *A. niger* and *A. tubingensis* did have detrimental effects on mycelial growth rate and spore production, and on competitive abilities with respect to other strains under different growth conditions (see below paragraph 3.3).

2. Mycoviruses identified in different *Aspergillus* species

Extensive screening of mycovirus infected isolates of several sections of the *Aspergillus* genus have been carried out in the last decades (Table 2). These studies clarified that most *Aspergillus* species are prone to mycovirus infection. The *Aspergillus* genus is divided into 18 sections (Varga *et al.*, 2006). In the following, the occurrence of mycoviruses in different *Aspergillus* sections will be detailed (Table 2).

Table 2. Aspergillus *species, their mode of reproduction, infection rates with mycoviruses, and references.*

Section	*Aspergillus* species	Sexual reproduction	Infection rate (# infected/# tested)	References
Candidi	*A. candidus*	-	0% (0/3)	Varga *et al.*, 1998
Circumdati	*A. ochraceus*	-	18.5% (5/27)	Kim and Bozarth, 1985; Varga *et al.*, 1998
Clavati	*A. clavatus*	-	20% (2/10)	Varga *et al.*, 1998, 2003
Flavi	*A. flavus*	-	8.9%(5/56)	Schmidt *et al.*, 1986; Elias and Cotty, 1996
	A. nomius	-	14.3% (1/7)	Elias and Cotty, 1996
	A. parasiticus	-	16.7% (1/6)	Elias and Cotty, 1996
	A. tamarii	-	20% (3/15)	Elias and Cotty, 1996
	A. oryzae		0% (0/4)	Elias and Cotty, 1996
	A. sojae		0% (0/4)	Elias and Cotty, 1996
	A. leporis		50% (1/2)	Varga *et al.*, 2001
	Petromyces alliaceus	+	20% (1/5)	Varga *et al.*, 1998
Fumigati	*A. fumigatus*	-	2.7% (2/74)	Varga *et al.*, 2001; Warn *et al.*, 2006; Anderson *et al.*, 1996
	Neosartorya hiratsukae	+	100% (3/3)	Varga *et al.*, 1998
	N. quadricincta	+	33% (1/3)	Varga *et al.*, 1998
	N. primulina	+	100% (1/1)	Varga *et al.*, 1998
	N. fischeri	+	0% (0/8)	Varga *et al.*, 1998
Nidulantes	*Emericella nidulans*	+	0% (0/112)	Coenen *et al.*, 1997
Nigri	*A. carbonarius*	-	5.3% (1/19)	Varga *et al.*, 1994a, 2001; Van Diepeningen *et al.*, 2006
	A. foetidus	-	33% (1/3)	Ratti and Buck, 1972; Buck and Ratti, 1975
	A. heteromorphus	-	100% (2/2)	Varga *et al.*, 1994a
	A. japonicus/A. aculeatus	-	8.6% (12/140)	Varga *et al.*, 1994a, 2001; Van Diepeningen *et al.*, 2006
	A. niger	-	12.8% (79/615)	Buck *et al.*, 1973; Varga *et al.*, 1994a, 2001; Van Diepeningen *et al.*, 2006

Table 2. Continued.

Section	*Aspergillus* species	Sexual reproduction	Infection rate (# infected/# tested)	References
	A. tubingensis	-	7.4% (29/393)	Varga *et al.*, 1994a, 2001; Van Diepeningen *et al.*, 2006
	A. ellipticus	-	0% (0/2)	Varga *et al.*, 2001
	A. brasiliensis	-	0% (0/7)	Varga *et al.*, 2001
Aspergillus	*Eurotium* spp.	+	0% (0/25)	Varga *et al.*, unpublished
Restricti	*A. restrictus*	-	0% (0/3)	Varga *et al.*, unpublished
Terrei	*A. terreus*	-	7.5% (3/40)	Varga *et al.*, 2005
	A. carneus	-	0% (0/3)	Varga *et al.*, 2005
Flavipedes	*Fennellia nivea*	+	0% (0/4)	Varga *et al.*, 2005

In *Aspergillus* section *Clavati*, two *Aspergillus clavatus* isolates were found to be infected with mycoviruses (isometric, 35-40 nm) with sole 6 kb double-stranded RNA genomes (Varga *et al.*, 2003). The presence of an unsegmented genome enclosed into isometric particles indicates that these mycoviruses belong to the genus *Totivirus* (Fauquet *et al.*, 2005). Although patulin was earlier found to inhibit mycovirus replication (Detroy and Still, 1976), one of the mycovirus carrying isolates also produced patulin, and both carried the iso- epoxydon dehydrogenase (IDH) gene fragment. These results indicate that patulin production and mycovirus replication take place in different compartments of the fungal cell (Varga *et al.*, 2003).

Regarding *Aspergillus* section *Fumigati*, Adler-Moore (1985) detected isometric particles of 20 nm in diameter in an *Aspergillus fumigatus* isolate, which was suggested to be a virus-like particle. Later, Anderson *et al.* (1996) identified dsRNA elements in three clinical *A. fumigatus* isolates, while Varga *et al.* (2001) also detected a dsRNA element of >10 kb in another *A. fumigatus* isolate. Interestingly, *A. fumigatus* isolate Af293, the whole genomic sequence of which has recently been determined, also carries dsRNA elements (Warn *et al.*, 2006). Regarding other species of section *Fumigati*, dsRNA size ranges of 1.3-3.4 kb were observed in one *Neosartorya quadricincta* and 3 *Neosartorya hiratsukae* isolates (Varga *et al.*, 1998).

In section *Terrei*, three *Aspergillus terreus* isolates carried dsRNA elements indicative of mycovirus infection. The sizes of the dsRNAs varied between 1.3–3.6 kbp. None of the isolates carrying dsRNA elements produced lovastatin (Varga *et al.*, 2005). In section *Flavipedes,* none of the examined *Fennellia flavipes* or *Fennellia nivea* isolates carried dsRNA elements (Varga *et al.*, 2005).

None of the three examined *Aspergillus candidus* strains examined harboured dsRNAs (Varga *et al.*, 1998). In *Aspergillus* section *Circumdati,* Kim and Bozarth (1985) detected a mycovirus complex (AoV) in *Aspergillus ochraceus* ATCC 28706. They observed three structural proteins and nine dsRNA segments in AoV. Serological analysis indicated that AoV has three unrelated components, and cross-reacted with the antiserum against PsV-S, a mycovirus of *P. stoloniferum*. Recently, Kim *et al.* (2003, 2005, 2006) determined the complete genomic sequences of the *P. stoloniferum* genomes of mycovirus PsV-S and the serologically and electrophoretically unrelated mycovirus PsV-F, and compared them to AoV. The authors found that one of the dsRNAs of AoV cross-hybridised to a PsV-S dsRNA component. Sequence analysis revealed the same genome organisation in AoV to that found in PsV-S. The RdRp gene sequence of AoV exhibited 81% homology to members of the *Partitivirus* genus, and grouped together with PsV-S in the phylogenetic tree. Since the capsid protein of AoV did not reacted against PsV-S antiserum, the authors suggested the cross-reaction between AoV and PsV-S was due to the homology between their RdRps.

In another study, four of the 27 *A. ochraceus* isolates examined were found to contain dsRNAs (Varga *et al.*, 1998). The dsRNA pattern observed earlier in *A. ochraceus* ATCC 28706 was not found in any of the *A. ochraceus* strains examined. DsRNA sizes ranged between 0.6-6.0 kb in *A. ochraceus* isolates. The dsRNA profiles of a number of strains, e.g. those of *A. ochraceus* strains and *A. foetidus* CBS 618.78 share a number of segments of similar mobilities. The differences observed in the relative intensities of some dsRNA bands within single isolates indicated that these strains are infected with two or more mycoviruses as observed earlier in the case of an *A. foetidus* strain (Ratti and Buck 1972). DsRNAs were not detected in any of the other 28 strains tested from *Aspergillus* section *Circumdati.* Strict correlation was not observed between ochratoxin production and dsRNA content of the strains, although none of the dsRNA infected strains produced ochratoxins (Varga *et al.*, 1996, 1998).

Regarding *Aspergillus* section *Flavi,* research focused on the possible negative correlation between aflatoxin production and presence of mycoviruses in *A. flavus* cultures. The absence of aflatoxin production in *A. flavus* strain NRRL A-12268 was originally found to be correlated with the occurrence of VLPs (Mackenzie and Adler, 1972). However, more extensive investigations failed to establish a relationship between absence of VLPs and aflatoxin production in other *A. flavus* strains (Wood *et al.*, 1974; Gussack *et al.*, 1977). Double stranded RNA particles in an *A. flavus* strain NRRL A-12268 were found to have a genome identical in size to a dsRNA found in *P. chrysogenum* (Schmidt *et al.*, 1986). When this strain was cured of dsRNA infection by exposing it to an RNA synthesis inhibitor (cycloheximide), it developed the ability to produce aflatoxins (Schmidt *et al.*, 1983). Additionally, artificial infection with virus isolated from *P. chrysogenum* arrested aflatoxin production. However, Elias and Cotty (1996) detected dsRNA elements in 5 of the examined 56 *A. flavus* isolates, and observed that aflatoxin-producing ability of the isolates was not affected by dsRNA infection. Strains that produced high levels of aflatoxins were as likely to be infected by dsRNA as strains that produced no aflatoxins. Moreover, curing strains infected with dsRNA did not result in altered aflatoxin-producing ability (Elias and Cotty, 1996). In another study, a lack of correlation was observed between production of aflatoxins and presence of VLPs in another *A. flavus* strain (Silva *et al.*, 2001).

Regarding other species of *Aspergillus* section *Flavi*, Elias and Cotty (1996) examined the presence of dsRNAs in several of them. Altogether 10 of the examined 92 isolates were found to be infected (Table 2). DsRNAs have been found in *A. flavus*, *Aspergillus parasiticus*, *Aspergillus nomius* and *Aspergillus tamarii* isolates. However, none of the isolates of the domesticated species *A. oryzae* and *Aspergillus sojae* were infected. The authors found that collection strains are more frequently infected than field isolates. The number of dsRNAs detected in a single isolate varied between 1-9, in the size range of <1-10 kb. One of the examined *Petromyces alliaceus* and one of the *Aspergillus leporis* isolates have also been found to carry dsRNA elements in other studies (Varga *et al.*, 1998, 2001). *A. leporis* harboured a 6 kb dsRNA element, while the *P. alliaceus* isolate FRR 4340 carried a 3.50 and a 0.60 kb dsRNA element. The stromata embedding the asci in *P. alliaceus* were found not to transmit the 3.50 kb dsRNA segment. The 0.6-kb dsRNA segment that was present in the single-stromatal cultures was found to be located in the mitochondrial fraction of this strain. Association of dsRNA molecules with the mitochondria was also detected in several other fungi, including *Puccinia* species (Kim and Klassen 1989), *S. cerevisiae* (Beilharz *et al.* 1982), *C. parasitica* (Polashock and Hillman 1994), *G. abietina* (Tuomivirta and Hantula, 2003a), *H. mompa* (Osaki *et al.*, 2005), *C. elegans* (Park *et al.*, 2006) and *O. ulmi* (Rogers *et al.* 1987; Hong *et al.*, 1999).

Among Aspergilli, black *Aspergillus* species have been the most extensively studied for the presence and variability of dsRNA elements. In the early seventies, Kenneth W. Buck and his co-workers identified a multicomponent virus system both in *A. foetidus* and *A. niger* isolates (Banks *et al.*, 1970; Ratti and Buck, 1972; Buck *et al.*, 1973; Buck and Ratti, 1975; Buck, 1986). More recently, altogether more than 1000 black *Aspergillus* isolates have been screened in our laboratories, and about 10%

(Van Diepeningen *et al.*, 2006) of them were found to carry dsRNA elements (Varga *et al.*, 2001; Van Diepeningen *et al.*, 2006). Interestingly, strains held in culture collections for a long time were found to be infected more frequently then isolates recently collected from nature (Varga *et al.*, 2001). Occurrence, variability and effects of mycoviruses of black Aspergilli will be described in detail in section 3.

The dsRNA genomes detected in Aspergilli have been found to be enclosed in isometric VLPs in the size-range of 25-40 nm (Varga *et al.*, 1994a; 2001). With the exception of one of the dsRNA segments observed in *P. alliaceus*, the other examined dsRNA segments of *Aspergillus* species have been found to be localized in the cytoplasm.

DsRNAs have not been observed in any of the examined *Emericella, Eurotium* or *A. restrictus* isolates (Coenen *et al.*, 1997; Varga *et al.*, unpublished results), although Hollings and Stone (1971) previously claimed that 25 nm isometric particles have been found in an *Aspergillus glaucus* isolate (teleomorph: *Eurotium herbariorum*). To our knowledge, the presence of mycoviruses has not been examined so far in isolates belonging to *Aspergillus* sections *Cervini, Cremei, Ornati, Sparsi, Usti, Versicolores* and *Ochraceorosei.*

From some of the *Aspergillus* species tested for the presence of mycoviruses only a low number of strains was checked, sometimes making it hard to draw definite conclusions. Overall eight different *Aspergillus* species which have the ability to reproduce sexually have been tested for the presence of mycoviruses. In four of these species mycoviruses were indeed detected. Of the twelve tested strains from these four species, six were found to be infected. But, on the whole of 161 tested sexual strains only 3.9% were infected. In the asexual species a much higher percentage of infected strains was found. On average the asexual infection rate was 9.7%. The largest group of asexuals tested were the black Aspergilli of section *Nigri*, where even 9.9% of the strains was found to be infected. *A. fumigatus* had a significantly lower infection rate than the other asexual Aspergilli.

3. Mycovirus dynamics in the black Aspergilli

3.1. Virus presence and variance

A. niger was first described by Van Tieghem in 1876 as a fungus capable of using tannic acid as carbon source. Growing on low concentrations of tannic acid is a feature many Aspergilli share, but the black Aspergilli are the only ones which are able to grow on high doses of tannic acid, up to 80% concentration of the polymer

(Van Diepeningen *et al.*, 2004). Rippel reported in 1939 the exclusive selection of *A. niger* on concentrations of 20% tannic acid. Nowadays *A. niger* is known to be part of a complex group of black spored Aspergilli within section *Nigri*, where all species share the ability of degrading tannic acids efficiently (Van Diepeningen *et al.*, 2004). Using the tannic acid degrading trait, many strains have been isolated from samples all over the world, but also large collections of black Aspergilli exist which were isolated by other means.

In the early 1970s the first mycoviruses were detected in some *A. foetidus* and *A. niger* isolates. The mycoviruses consisted of dsRNA genomes and formed isometric particles (Banks *et al.*, 1970; Ratti and Buck, 1972; Buck *et al.*, 1973; Buck and Ratti, 1975; Buck, 1986). Further experiments showed that two distinct classes of particles were present in *A. foetidus*, each with its own multiple dsRNA components. Thus, the possibility of multiple infection was demonstrated (Ratti and Buck, 1972). Mycovirus particles can be detected via electron microscopy (see Figure 1), but an easier screening for the potential presence of mycoviruses is the isolation of total nucleic acids from a strain and separation of DNA, dsRNA and ssRNA with gel electrophoresis.

Even isolation of minute amounts of nucleic acids from small amounts of mycelium (approx. 0.1 gram; for a protocol see Van Diepeningen *et al.*, 2004) suffices for the detection of mycoviruses. In untreated samples large molecular weight nuclear and mitochondrial DNA has a low mobility through the gel, small messenger RNAs and other single-stranded RNAs have the highest mobility, while dsRNA fragments have an intermediate mobility. The dsRNA nature of the detected fragments can easily be confirmed by treating the nucleic acid solution with DNAse, and RNAse under low and high salt concentrations (Varga *et al.* 1994a; Figure 3). Easy classification of isolates based on mitochondrial RFLP patterns can also be done with the miniprep isolations. RFLP haplotypes were described for *A. niger*, *A. tubingensis*, *A. brasiliensis*, *A. japonicus*, and *A. carbonarius* (Varga *et al.*, 1993, 1994b; Kevei *et al.*, 1996; Hamari *et al.*, 1997).

More than 1000 representatives of different black *Aspergillus* species were tested for the presence of dsRNA mycoviruses in different studies (Varga *et al.*, 1994a; Van Diepeningen *et al.*, 2006). An example of infected strains and one partially cured strain is given in Figure 4. From all the tested strains approximately 10% was found to be infected with mycoviruses (with 95% confidence the true infection rate lies between 8.8% and 12.4%). Strains examined originated from different countries. Most were isolated from soil samples, but also strains from other substrates were examined. From some countries only low numbers of strains were isolated, like

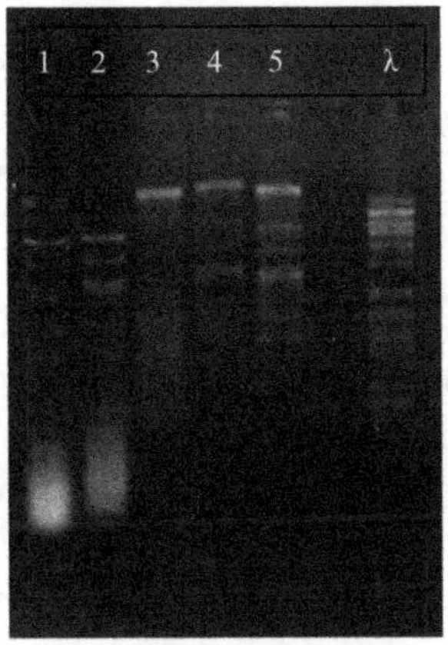

Figure 3. Confirmation of mycoviral dsRNA nature: 1. untreated total nucleic acid preparation, 2. DNase treated, 3. RNAse treated in milliQ water, 4: RNase treated 1 × SSC and 5. RNase treated 2 × SSC.

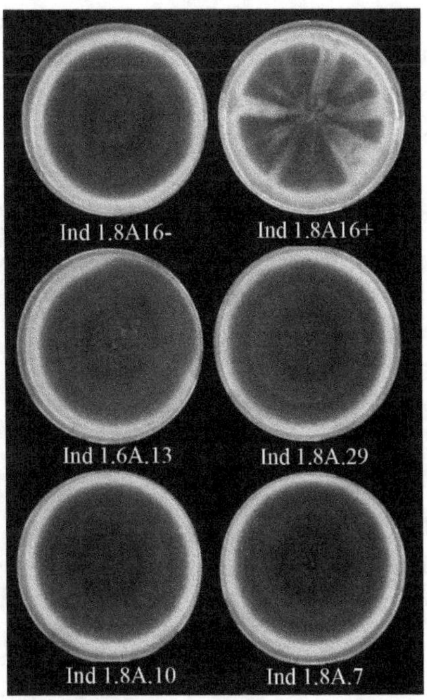

Figure 4. Photograph of different black Aspergilli with and without viruses. On top A. niger *strain Ind 1.8.16 is shown with and without the fitness decreasing effects caused by mycovirus infection, namely sectorial growth with sectors with reduced growth rate and spore production. The other four* A. niger *strains all harbour viruses with different dsRNA fragments that do not cause any clearly visible phenotypic effect.*

for example Canada, Egypt and Malaysia, and in these no viruses were detected. However, as these samples were small they do not differ significantly from the overall infection rate. Also the type of substrate the strains were isolated from does not affect the infection rate.

Moreover, the mycovirus infection rate does not seem to vary largely from the 10% in the different black *Aspergillus* species. Though, in species like *Aspergillus ellipticus* and *A. brasiliensis* no mycoviruses were detected and both *Aspergillus heteromorphus* strains were infected, the tested numbers of these species are too low to say that they are either always virus-free or always infected. Therefore it seems 10% infection with mycoviruses is true for black *Aspergillus* populations and species world-wide.

In the different infected black *Aspergillus* strains from one up to eight dsRNA fragments could be detected. These fragments ranged from approximately 0.5 to 6 kbp and were packed in isometric particles 25-40 nm in size, localised in the cytoplasm (Varga *et al.*, 1994a; 2001; Van Diepeningen *et al.*, 2006). The virus-infected black *Aspergillus* isolates exhibited at least 35 different dsRNA banding patterns (Figure 5). Hybridisation experiments revealed that electrophoretic banding patterns may even underestimate the diversity of the mycovirus genomes, satellite dsRNA fragments and defective fragments (Varga *et al.*, 2001). It is not determined whether the viruses

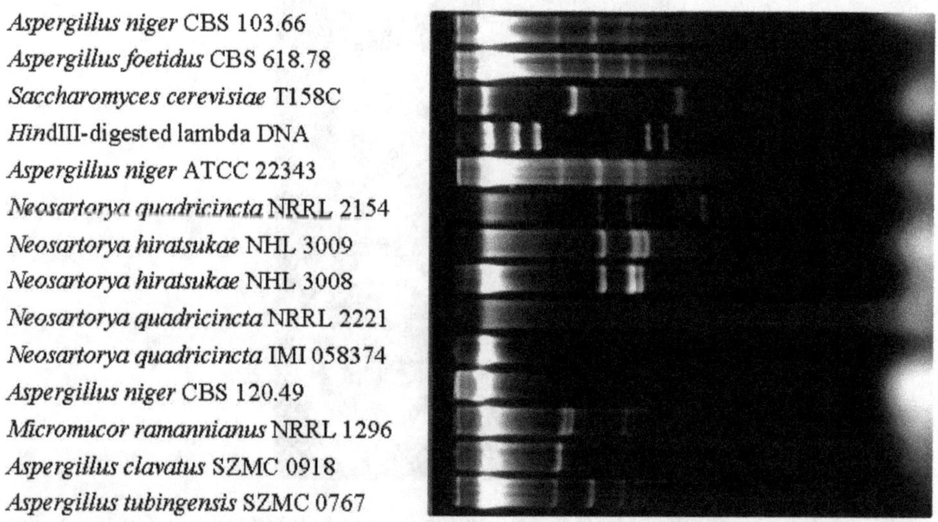

Figure 5. DsRNA profiles of mycoviruses found in various Aspergilli. Isolated total nucleic acid of different culture collection strains was examined by agarose gel electrophoresis. Different dsRNA viral patterns are visible in the infected strains, while some of the isolates do not carry mycoviruses.

belong to the families of the Totiviridae, Partitiviridae, Chrysoviridae or other known or yet unknown families.

3.2. Transfer of viruses in the black Aspergilli

As in most Aspergilli, the mycovirus particles with their dsRNA genomes are located in the cytoplasm. Except for the mycoreoviruses, the mycoviruses do not have a (known) extracellular phase. Thus, they depend on transmission to formed offspring (vertical transfer) and to other strains via cytoplasmic contact (horizontal transfer). The black Aspergilli occur world-wide with a slight preference for tropical and subtropical areas (Rippel, 1939; Raper and Fennel, 1965; Domsch *et al.*, 1980). In the tropics locally high densities of up to hundreds of black *Aspergillus* conidia per gram soil can be found which would give strains a chance of horizontal virus transfer when mycelia are formed. In moderate climates the spore densities are much lower (Van Diepeningen *et al.*, 2004).

The black Aspergilli are presumably asexual species. During all the years of research on the species no sexual reproduction has ever been observed. Also direct screening of the nuclear DNA showed no evidence for (para)sexual recombination (Van Diepeningen, 1999). In another presumed asexual, the opportunistic pathogen *A. fumigatus*, also never sexual reproduction was observed. But, in this species recently both mating types, *mat-1* and *mat-2* and meiosis related genes were identified (Pöggeler, 2005; Varga, 2003; Dyer and Paoletti, 2005; Paoletti *et al.*, 2005). Pál (2007) did a similar screening for mating type and meiosis related genes in the two recently sequenced *A. niger* genomes and checked more than 160 wild-type strains for the presence of *mat-1* and *mat-2* homologues. He did find some homologues of the *mat1* gene, but the putative *mat-2* sequence lacked the necessary key features. Quite a lot of meiosis related genes are conserved in the *A. niger* genome, but they may have pleiotropic functions beside meiosis. In wild-type strains Pál (2007) could detect the *mat-1* locus in about 85% of the strains, a *mat*-2-like sequence was detected very rarely. Some strains did not give a signal for either of the two mating type genes. Thus, as the black Aspergilli are truly asexual, vertical transfer of mycoviruses can only take place through conidia. This vertical transmission is very efficient in the black Aspergilli. Hundreds of single spore colonies of different virus infected strains were tested for their virus content and all contained the complete parental virus dsRNA patterns (Van Diepeningen *et al.*, 1997, 2006).

Horizontal transmission of mycoviruses needs to take place through hyphal anastomosis. This process of anastomosis is under the control of so-called vegetative (*vic*) or heterokaryon (*het*) incompatibility genes, which may have evolved to limit the spread of deleterious elements like some mycoviruses (Caten, 1972) or to prevent resource plundering (Debets and Griffiths, 1998). Lhoas (1970) was the first to describe the transmission through heterokaryosis of VLPs in *A. niger* to previously virus free strains. In asexual species like *A. niger* it is not easy to study the exact genes involved in heterokaryon incompatibility reactions. Pál (2007) did show that many homologous sequences to known *het* genes from *Podospora anserina* and *N. crassa* are present in the sequenced *A. niger* genomes, but we do not know whether these genes also have a function in incompatibility reactions in *A. niger*. However, using non-autonomous colour expression between different colour mutants, or complementation between different auxotrophic strains, dominant resistant strains and/or nitrate non-utilising strains, one can check for the ability of strains to form anastomoses and subsequently viable heterokaryons (Van Diepeningen *et al.*, 1997).

Hundreds of combinations of *A. niger, A. tubingensis,* and other black *Aspergillus* strains, mainly from one sampling area in Indonesia, were tested for their ability to form stable heterokaryons. Between different black *Aspergillus* species no heterokaryosis was observed. Within a species the average percentage for two random strains to be heterokaryon compatible was approximately 2% (Van Diepeningen *et al.*, 1997). Taking into account that the average infection rate of the black Aspergilli is only 9.9%, the chances on two compatible strains of the same species meeting of which at least one is infected with mycoviruses is only about 0.2%. Combinations of infected and virus free *A. niger* and some *A. tubingensis* strains have been tested in co-culture for the spontaneous transfer of mycoviruses between them. In the case of combinations of heterokaryon compatible strains the virus was rapidly transferred. In the case of heterokaryon incompatible strains that were co-cultured for six weeks, in less than one percent of the strains was virus transfer observed (Van Diepeningen *et al.*, 1997). Thus it seems that the chances on virus infection when two black *Aspergillus* strains meet are very low.

A way to avoid - at least part of - the heterokaryon incompatibility reactions in the lab is the use of protoplasts for fusion experiments. Thus, in the laboratory, mycoviruses have successfully been transferred from one *A. niger* isolate to another by protoplast fusion (Buck, 1986; Varga *et al.*, 1994a; Van Diepeningen *et al.*, 1998). Van Diepeningen *et al.* (1998) tested both single acceptor colonies after transfer experiments and pooled acceptor protoplasts together. Many of the single protoplasts proved to be virus free, though the infection ratio differed per combination of strains: in some cases all tested protoplasts were infected and in others none or a few. Even, when large pooled samples of protoplasts were tested, where occasional infecting viruses are expected to be able to spread throughout the resulting mycelium, transfer was not found in all combinations of strains. Also the transfer to another black

Aspergillus species, *A. tubingensis,* proved limited successful. In total approximately 55-60% of all inter *Aspergillus* niger transfers via protoplast fusion proved to yield new infections (Figure 6).

The protoplast fusion technique has also been used to successfully transfer mycoviruses from *A. niger* to other related species like *A. oryzae* and *A. nidulans* (Coenen *et al.*, 1997; Liang and Chen, 1987; Van Diepeningen *et al.*, 1998). Also transfer to *A. niger* from infected *A. nidulans* and even *F. poae* proved possible (Van Diepeningen *et al.*, 1998, 2000). Interestingly, all natural isolates of *F. poae* were found to be infected, some of the viral dsRNA fragments resembling those found in black Aspergilli (Fekete *et al.*, 1995). These interspecies transfers even seem easier then the intraspecies transfer in *A. niger.* For example the transfer to and from *A. nidulans* to black Aspergilli, tested in pooled samples of protoplasts, proved always (100%) successful, whereas the success within *A. niger* with similar experiments was only 55-60% (Figure 6). Even *A. niger* strains that could not be infected through other *A. niger* strains proved to be infectable via other species and not resistant

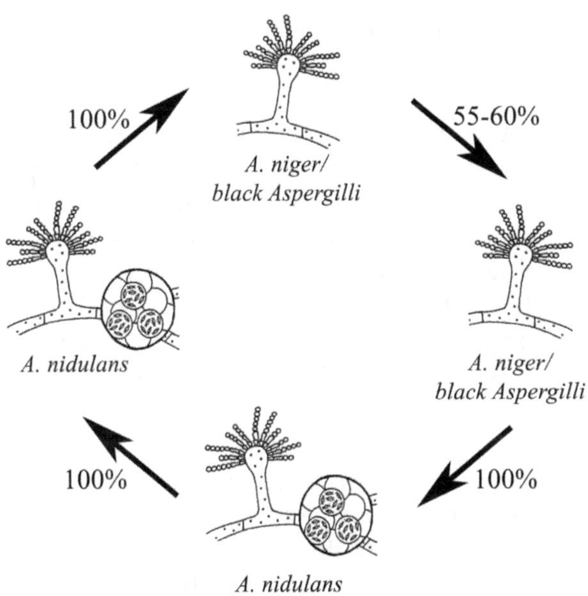

Figure 6. Diagram of the efficiency of inter and intraspecies virus transfer via protoplast fusion between sexual A. nidulans *and asexual* A. niger *and other black Aspergilli. Within the black Aspergilli transfer via protoplast fusions was successful in approximately 55-60% of the cases. However, transfer to and from* A. nidulans *and between* A. nidulans *strains was successful in 100% of the combinations (data from Van Diepeningen* et al., *1998).*

to infection. Thus, transfer via protoplast fusion circumvents not only some of the strong barriers formed by the *het*-genes within the black *Aspergillus* species, but even barriers between them and distantly related species.

Interspecies mycovirus transfer via co-culturing of species was only tested in four combinations between two *A. niger* and two *A. nidulans* strains. However, in contrast to the above mentioned transfer experiments within and between *A. niger* and *A. tubingensis* species, where transfer by co-culturing was successful in less than 1%, the transfer to *A. nidulans* was successful in half of the combinations (Coenen *et al.*, 1997).

3.3. Effects of virus infections in the black Aspergilli

As a rule, mycoviruses cause latent and persistent infections, while only a few of them are associated with variable phenotypic effects. Mycoviruses could have evolved in concert with their hosts to limited detrimental effects or to mutual benefit, like, for instance, the viruses linked with killer phenotypes in yeast and smuts that confer a strong advantage to their hosts in interference competition by encoding (costly) toxins. But also deleterious mycovirus infections are found (Ghabrial, 1998; Hillman and Suzuki, 2004).

In all 124 infected black *Aspergillus* strains examined (see Table 2), only one infected *A. niger* strain was found with a clear phenotypic effect (see Figure 4). Strain Ind 1.8.16 shows sectorial growth with locally reduced growth rates and reduced spore production, while all other strains do not show clear signs of mycovirus infections (Van Diepeningen *et al.*, 2006). However, when 0.01 mM of the protein synthesis inhibitor cycloheximide is added to the growth medium also in strains with no phenotypic effect on normal medium, sectors lighter or darker with conidiospores can sometimes be observed in mycovirus infected colonies (Van Diepeningen, unpublished results).

As strains can vary largely in growth characteristics like spore production and growth rate, one needs isogenic virus free and infected lines to quantify the effect of the mycovirus on its host. To cure infected *A. niger* and *A. tubingensis* strains of their mycovirus infections, different methods have been tested. Hundreds of single conidium-derived colonies have been tested for their virus contents, but all of them proved to be infected. Sequential hyphal tip isolations have been done on normal media and media with cycloheximide, but only in the case of the strain Ind 1.8.16 with clear phenotypic effect a partial curing of some of the dsRNA fragments occurred (Van Diepeningen *et al.*, 2006). Introducing viruses into previously virus free strains by co-culturing strains on agar plates proved nearly only possible when strains were heterokaryon compatible. Protoplast fusion could successfully be used to bypass most of the barriers formed by heterokaryon incompatibility and newly infected *A. niger* and *A. tubingensis* strains could thus be obtained. Further co-cultivation of newly infected strains with their selection markers (*fwn* colour and *nia* or *cnx* chlorate resistance and nitrate deficiency markers) and the original wild type black parental strains further spread the mycoviruses through heterokaryosis and thus virus free and infected isogenic lines were obtained (Van Diepeningen *et al.*, 2006).

Isogenic mycovirus infected and virus-free strains were compared for three different fitness traits: linear growth rate, conidium production and interference competition with a reference strain. *A. niger* strain Ind 1.8.16 showed the costly effect of mycovirus infection clearly (Figure 7). The effects were stronger on minimal medium than on rich medium, but in all three fitness traits, linear growth rate, spore production, and in interference competition the infected line did worse than the (partially) cured strain. Also all newly infected strains did worse than their virus-free counterparts.

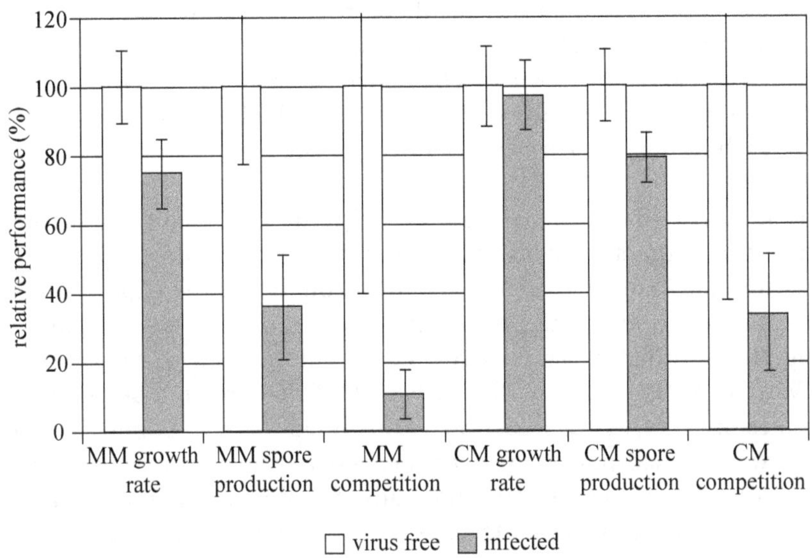

Figure 7. The fitness cost of virus infection depicted as the relative performance of virus free and infected lines of A. niger *strain Ind 1.8.16. Lines were tested on minimal (MM) or rich (CM) medium and compared for the fitness components linear growth rate, spore production and competition with heterokaryon incompatible A. niger strain Ind 1.8.9. Standard deviations are given. Except for the growth rate on CM, all fitness components differ significantly between the infected and virus free lines (p-values < 0.05).*

On average the infected lines grew 3.3% slower (p=0.001), produced 24% less spores (p=0.033) and in interference competition even lost 59% in comparison the virus-free lines (Van Diepeningen *et al.*, 2006). Thus one can conclude that there are serious costs to harbouring mycoviruses under the tested conditions.

3.4. Model for balanced infection in the black Aspergilli

Estimates of the fitness effects and the rates of the different modes of transfer of parasites, like the mycoviruses, would allow model predictions of implications for a host population. Fine (1975) specified that in an asexual host a parasite limited to vertical transmission could not persist in a population if it lowers the fitness. Lipsitch *et al.* (1995) analysed a model for a parasite transmitted both vertically and horizontally. Their model predicts that if prevalence is high, most transmission will be vertical, but that the horizontal transmission rates must be sufficiently high to reach and maintain stable equilibrium prevalence.

For a balanced and stable infection rate (prevalence) of the mycoviruses of approximately 10 percent as seen in the black Aspergilli, loss of viruses and new infections should occur at a similar rate if the mycoviruses are neutral to their host. Taking into account fitness effects of mycoviruses to their host this balance depends on the curing and loss of mycoviruses from mycelium and spores together with the fitness costs of harbouring viruses on one side and the infection rate and potential benefits of harbouring mycoviruses on the other side.

Thus, the negative effects of dsRNA mycoviruses on linear growth rate, spore production and competitiveness of *A. niger* strains as found suggest that novel infections should occur at a sufficient rate to counteract the expected decline of infected strains. However, the previously previous *in vitro* experiments showed little or no direct transfer of mycoviruses between non-isogenic black *Aspergillus* strains (Van Diepeningen *et al*, 1997, 1998, 2000). To obtain a better insight into the dynamics of dsRNA virus infection, we analysed a simple population model. In this model we can incorporate the fitness effects resulting from viral infections as estimated from our measurements, and study the effect of the rates of novel virus infections and of the production of virus-free spores by infected strains.

We assume a population will consist of two types of colonies, either infected with dsRNA mycovirus (relative frequency x) or uninfected (relative frequency 1-x). Furthermore, we suppose that all fungal colonies are subject to the following life cycle (see Figure 8): upon germination they may encounter close contact with an isogenic conspecific during vegetative growth; following this they sporulate and give

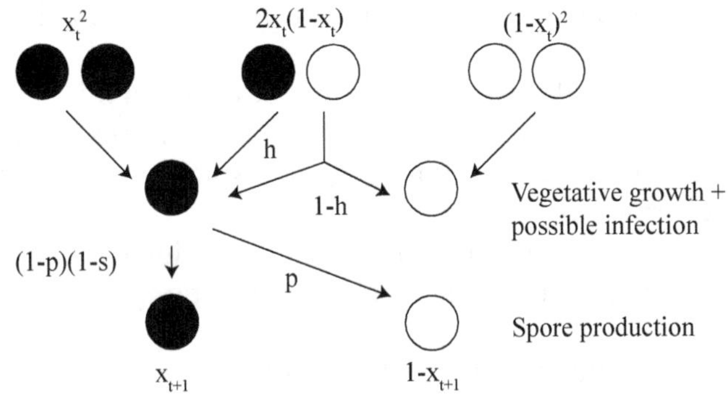

Figure 8. Schematic model of the black Aspergillus *population dynamics. Black = virus infected, white = virus-free, xt = fraction infected strains at generation t, h = rate of horizontal transmission to virus-free strains, s = selective disadvantage, p = spontaneous loss of virus.*

rise to the next generation. Pair wise contacts occur randomly, therefore two infected strains will meet with a probability proportional to x^2, two uninfected ones with probability $(1-x)^2$ and an infected strain will merge with an uninfected one with a probability proportional to $2x(1-x)$ (as under Hardy-Weinberg equilibrium). We assume that the latter category contact results in the infection of a virus-free colony with probability h (horizontal transfer), either as a consequence of anastomosis (in case of vegetative compatibility), or otherwise. Upon sporulation an infected colony produces $1-s$ times the number of spores from an uninfected colony (s the deleterious effect of virus infection). Finally, we assume that a fraction p of the spores produced by an infected colony will carry no dsRNA virus (spontaneous loss).

From these assumptions we can deduce the following equation for the change of the relative frequency of infected strains over one generation:

$$x_{t+1} = (1-p)(1-s)x_t[1+h(1-x_t)] \,/\, [(1-s)(1+h)(1-p)\, x_t(1-x_t)] \qquad (1)$$

Instead of solving this equation analytically we provide the following summarising conclusions based on linearisation at sufficiently small values of x_t and standard stability analysis.

1. If $p=0$ (there is no spontaneous curing and dsRNA mycovirus is included in all spores produced by an infected colony), then a stable coexistence of infected and uninfected strains is not possible. Eventually there will only be virus-free strains (if $h < s[1-s]^{-1}$) or only infected strains (if $h > s[1-s]^{-1}$).

2. If $p>0$ (there is at least some virus-free progeny from infected colonies), the two outcomes are possible:
   (a) If $(1-p)(1-s)(1+h) > 1$ (if the horizontal rate of infection h is sufficiently high to compensate for the virus loss caused by spontaneous curing and by an impaired fitness of infected colonies), then a stable coexistence of infected and virus-free strains is possible;
   (b) If $(1-p)(1-s)(1+h) < 1$ (if the horizontal rate of infection h is too low), then the viruses are expected to disappear from the population.

Based on the measured estimates of the fitness costs caused by virus infection, this model can be used to predict the population level consequences. In the laboratory experiments no transfer was found between heterokaryon incompatible strains (Van Diepeningen *et al.*, 1997). The great majority of strains isolated from nature are incompatible with one another and virus resistant strains have not been found. Based on the infection frequencies in different haplotypes of black Aspergilli and the frequencies of the haplotypes in a sampled population, the chance for a strain to meet a compatible strain with virus infection can be estimated to be less than 1 percent (Van Diepeningen *et al.*, 1997). Spontaneous loss of virus was not detected in progeny of the examined black Aspergilli (Van Diepeningen *et al*, 1997), nor was it easy to cure strains with sequential isolation of hyphal tips, indicating that spontaneous loss of virus, '*p*' in our model, will be very small. However, Elias and Cotty (1996) did find cured strains in their examined strains from the *Aspergillus* section *Flavi* (from 0 to 100% depending on strain and method used). Our black *Aspergillus* strains seem to have no active mechanisms to dispose of their virus(es) as was suggested for some stromatal structures in some (sexual) *Aspergillus* species (Varga *et al.*, 1998). If the horizontal transfer rate h should equal the selective disadvantage (ranging from on average –3% in growth rate to –59% in competition), far more transfer should take place than observed.

3.5. Dynamics of mycoviruses in the black Aspergilli

The infection rates of the different black *Aspergillus* species appear to be a stable 9.9% of all natural isolates (Varga *et al.*, 1994a, 2001; Van Diepeningen *et al.*, 2006). With the sizes of the different examined populations, also world wide the infection rate in the black Aspergilli does not seem to vary from the observed 9.9% (Van Diepeningen *et al.*, 2006). Spontaneous curing of strains seems very unlikely. Heterokaryon incompatibility between strains can be partly overcome by protoplast fusion, but in normal co-culturing of strains heterokaryon incompatibility efficiently blocks virus transfer in *A. niger* and related species. The chances of two heterokaryon compatible strains within all the black Aspergilli meeting, based on experiments with mostly isolates from one Indonesian location, are less than 1 percent. Thus with the observed infection rate, chances on transfer via heterokaryosis are less than 0.1%. Also the chances on transfer in heterokaryon incompatible combinations are less than 0.1%. However, the costs of virus infections range from 3% to 59% depending on the examined fitness traits. Thus, one would expect a rapid decline of mycoviruses in the population.

But in the face of a stable 9.9% of the population infected, perhaps higher rates of intraspecies transfer occur in nature than were observed under laboratory conditions. They may be facilitated by the natural conditions and/or perhaps with high densities of different black *Aspergillus* strains per location (though this would also intensify competition). Another possibility would be that interspecies virus transfer is more frequent than so far considered, either after direct contact or via other vectoring organisms. Notably, transfer via protoplasts between black *Aspergilli* and their sexual relative *A. nidulans* and vice versa proved more efficient than similar transfer between black *Aspergillus* strains (Coenen *et al.*, 1997; Van Diepeningen *et al.*, 1998). Also from non-related species like *F. poae* mycoviruses proved transferable to black *Aspergillus* strains via protoplast fusion (Van Diepeningen *et al.*, 2000). Regular intra- and interspecies transfer could explain the similarities in some infection patterns in different haplotypes and even in members of different species. Thus, it seems that the black *Aspergillus* mycoviruses are not cryptic with regard to (deleterious) fitness effects on their host, but may be cryptic in their modes of transfer.

4. Mycovirus dynamics in the *Aspergillus nidulans*

E. nidulans (anamorph: *A. nidulans)* is a homothallic filamentous ascomycete (Raper and Fennel, 1965). It can reproduce both sexually by the formation of ascospores in cleistothecia and asexually by the formation of conidia (Alexopoulos and Mims, 1979; Burnett, 1976). Heterokaryon incompatibility, and thus possibly the transfer of cytoplasmatic elements in *A. nidulans* is controlled by allelic nuclear *het* genes (Jinks and Grindle, 1963). So far at least eight *het* genes have been identified in *A. nidulans* of which some are multi-allelic (Croft, 1985; Anwar *et al.*, 1993; Dales *et al.*, 1993). Coenen *et al.* (1994) also identified a *partial het* gene, a gene that either causes a reduced form of the heterokaryon incompatibility response, modifies other *het* genes or is expressed less strongly depending on the environment. In nature, *A. nidulans* probably propagates in a primarily clonal fashion either by the formation of conidiospores or by selfed ascospores. However, Geiser *et al.* (1994) have shown that outcrossing and recombination events are frequent enough in nature to disrupt the stable maintenance of clonal genotypes. Hoffmann *et al.* (2001) observed that in *A. nidulans* diploid nuclei are preferentially formed between unlike haploid nuclei (the so-called 'relative heterothallism') which might contribute to the recombining population structure of this organism. The observed vegetative incompatibility groups probably occur due to recombination of (*partial*) *het* genes rather than by novel mutation (Geiser *et al.*, 1994). One advantage of sexual reproduction is that it slows down the accumulation of deleterious mutations (Bruggeman *et al.*, 2003).

So far mycoviruses have never been found in natural isolates or culture collection strains of *A. nidulans*. Coenen *et al.* (1997) screened a collection of 112 *A. nidulans* isolates for the presence of dsRNA mycovirus fragments, but found none of the strains to be infected. The screened collection consisted for the majority of strains isolated from different parts of the United Kingdom, but also included strains from (South) Africa, Barbados, Hungary, India, Trinidad, and the USA.

A. nidulans proved not to be resistant to infection with mycoviruses as strains could become readily infected with mycoviruses from different black *Aspergillus* strains via protoplast fusion (Coenen *et al.*, 1997; Van Diepeningen *et al.*, 1998). Surprisingly, all tested protoplast fusion experiments with different combinations of black Aspergilli and *A. nidulans* proved successful for transferring the mycoviruses between the species. Also the reciprocal protoplast fusion experiments with infected *A. nidulans* strains as donors and black *Aspergillus* strains as acceptor were 100% successful (Van Diepeningen *et al.*, 1998). Interspecies transfer by hyphal anastomosis between co-cultured strains was tested in combinations of two *A. niger* donor strains and two *A. nidulans* acceptor strains. Spontaneous interspecies transfer was observed with both combinations of one of the *A. niger* donor strains (Coenen *et al.*, 1997).

With the artificially infected *A. nidulans* strains the horizontal mycovirus transfer between different *A. nidulans* strains could be studied. Intraspecies virus transfer between *A. nidulans* strains proved always successful when strains where cocultured and large numbers of pooled spores from the acceptor strains were tested. The transfer was even successful when the strains were heterokaryon incompatible (Coenen *et al.*, 1997). Looking in closer detail, a quantitative analysis of somatic virus transmission by analysing single conidiospores showed that the more the two strains differ in their *het* genes, the chance on virus transfer gets smaller (Coenen *et al.*, 1997). This supports the theory that heterokaryon incompatibility genes limit the transfer of cytoplasmic elements (Caten, 1972), though they do not absolutely block transfer.

Vertical virus transfer did occur to 100% of the asexually produced conidia of *A. nidulans*, but usually not to the sexually produced ascospores (Coenen *et al.*, 1997). Thus, there seems to be a mechanism in the formation of the cleistothecia or the ascospores themselves that blocks the mycoviruses from entering in the sexual offspring. Due to the many *het* genes in *A. nidulans* isolates, crosses between two somatically incompatible strains will often yield offspring that belongs to different vegetative compatibility groups and that will be incompatible with both parental strains and other offspring. The thus sexually produced virus free offspring will not be infected as easily as the somatically compatible asexual offspring or ascospores formed after selfing. Thus, heterokaryon incompatibility in *A. nidulans* in combination with sexual reproduction and especially outcrossing limits the spread of possible deleterious elements like mycoviruses.

Four different *A. nidulans* strains were tested for the effect of viruses on their fitness. For each strain a combination of either a virus free line or the isogenic virus infected line with the same reference strain was made. Equal mixtures of the conidia of reference strains were made with those of either of the isogenic strains to be tested. After four days conidia were harvested and ratios between reference and test strains were determined. Ratios of the virus free and infected strains were compared, but with the relatively few data no fitness effects on the host strains were discovered (Coenen *et al.*, 1997). Since the experiments of Coenen *et al.* (1997) the infected *A. nidulans* strains have been tested further. These unpublished experiments show that harbouring a mycovirus is not without costs for the *A. nidulans* host strain, and show that the costs are similar to the costs of infection observed in black Aspergilli. This is clearly demonstrated when the mycovirus that gives a strong, visible phenotypic effect in *A. niger* is transferred to *A. nidulans* and gives a similar phenotype of sectors with reduced growth rates and spore production (Figure 9).

For a stable infection rate of mycoviruses in a population, the rate of new infections should be in balance with the fitness costs of infection and the loss of these viruses. As *A. nidulans* is not resistant to mycovirus infection and interspecies virus transfer does happen upon co-culturing in the lab, in principle mycovirus infections could occur in nature. Between two different mycelia heterokaryon incompatibility is not an absolute barrier for horizontal infection to occur, though these transfer rates are lower the more the strains differ in their *het* genes. Thus, mycovirus infection could occur in nature and could spread through the *A. nidulans* population upon hyphal contact. On the other hand, harbouring the mycoviruses does have a negative effect on the host' fitness components like growth rate and spore production. All conidia are infected, but usually both outcrossed and selfed ascospores are virus free. Besides, the ascospores produced by outcrossing will less probably become infected horizontally as they may belong to (many) different vegetative incompatibility groups. Obviously as no mycovirus infected *A. nidulans* strains were found so far, the inter- and intraspecies infection rates in nature are lower than the possible fitness costs and lower than the chances that viruses get lost during sexual reproduction.

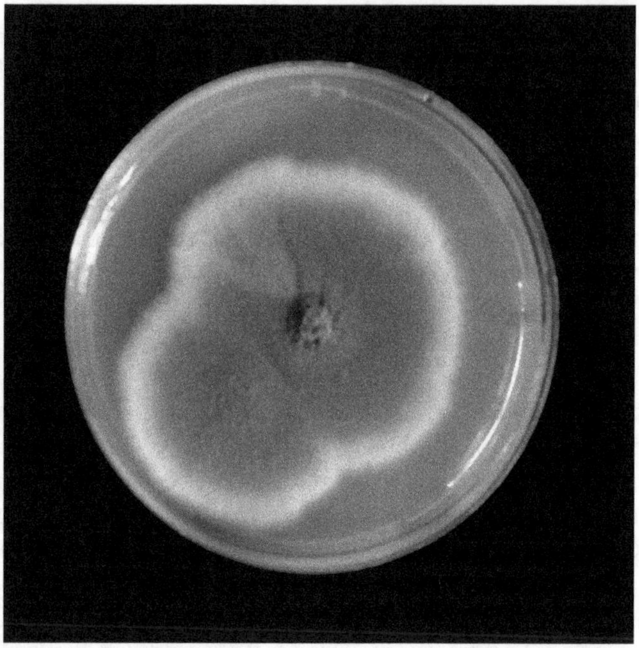

Figure 9. Photograph of A. nidulans *strain 702 infected with the mycovirus of* A. niger *Ind 1.8.16. After infection the* A. nidulans *strain forms sectors with reduced mycelial growth rate and reduced conidium production.*

5. Conclusions, perspectives and future research

Mycoviruses have been observed in many asexual and some sexual *Aspergillus* species, but some sections have been little examined or not yet investigated at all. So far, little is known about what species of mycoviruses exactly infect the different *Aspergillus* species or how they got infected. Research on mycovirus dynamics in asexual *A. niger* and sexual *A. nidulans* show that mycoviruses are generally very persistent through the asexual route, but are rapidly excluded when ascospores are produced. Virus infection in nature thus will depend on reproduction mode of the host and whether horizontal transfer occurs frequently. This horizontal transfer may well take place between species as well as within species, though heterokaryon incompatibility reactions may strongly influence virus transfer. Mycoviruses are not as cryptic as they were thought to be previously. In general, they have an adverse effect on their host' fitness. But, still a lot of research can be done on mycoviruses in the Aspergilli.

In the current genomics era it is much easier to sequence dsRNA fragments and to determine what essential genes like RNA polymerase and protein capsids they harbour. Do they have a similar organisation as the so far identified viruses? Comparison with known mycovirus families would tell us which mycovirus genera are infecting the different *Aspergillus* species in nature. Analysis of the multiple dsRNA fragments observed in some isolates will tell whether they are the result of infection with one multipartite virus or of multiple infections. Also satellite or defective fragments could be identified by comparison to the available host genome sequences, to the sequences of relatives or to known virus sequences. The appearance of novel dsRNA fragments could be explained by two mechanisms: recombination/ evolution inside the host or horizontal transmission (Ikeda *et al.*, 2005). Fragments have been identified that were the result of internal deletions from larger dsRNA fragments (Hiremath *et al.*, 1986; Tartaglia *et al.*, 1986; Shapira *et al.*, 1991), but novel dsRNA fragments originating from host genomic DNA have also been identified (Lakshman and Tavantzis, 1994).

Comparison of (some of the) 35 different dsRNA patterns observed in the black Aspergilli could tell us whether these viruses all have a common infection history or are the result of many different and perhaps regularly occurring infections. For example in *Fusarium poae* the dsRNA elements in vegetatively compatible strains were always related, while incompatible strains of the fungus contained either homologous or non-homologous dsRNAs of the same size (Compel *et al.*, 1998). Comparison of black *Aspergillus* viruses with those mycoviruses with similar organisation in other fungal species may give a better insight into the mechanism and frequency of intra- and interspecies transfers. In the monopartite *U. maydis* virus H1 analysis of the sequences of viruses from different geographic isolates gave insight into divergence and the geographic spread of the virus together with its host (Voth *et al.*, 2006). Characterisation of the dsRNA fragments infecting *A. niger* strain Ind 1.8.16 with the strong phenotypic effect of reduction in both mycelial growth and spore production may tell us why this particular mycovirus is much more virulent than the other *A. niger* mycoviruses, even when transferred to a new host like *A. nidulans*.

Also studies on the host' response to virus infection are possible, because in response to virus infection the host metabolisms may change. For example in infected *C. victoriae* the cellular protein Hv-p68 with alcohol oxidase and RNA-binding activity was found to interact with viral dsRNA and act as minor component of the virus capsids. Hv-p68 accumulates in infected strains and is proposed to play a role in the viral RNA packaging and replication (Soldevilla *et al.*, 2000, 2001). Overexpression experiments with the *Hv-p68* gene showed that it is not the accumulation of toxic aldehydes that induces the disease-phenotype associated with this virus in infected isolates (Zhao *et al.*, 2006). cDNA micro arrays have been used to monitor the transcriptional responses of the chestnut blight fungus *C. parasitica* to infection by virulence-attenuating hypoviruses (Allen *et al.*, 2003).

More advanced diagnostic techniques based on ELISA or quantitative PCR techniques could be used for the detection and quantification of small amounts of viruses inside the host. For example for the PeSV mycovirus in *P. eryngii* a diagnostic system has been developed based on a triple antibody sandwich-ELISA, that enables the detection of 0.5 μg of the virus particles in 1 g diseased mushroom tissue (Ro *et al.*, 2007). The exact location of a mycovirus within its host could be determined and hence what effect that location has on e.g. transmission or virulence.

Finally, in future introduced viruses might be used to cause directed changes in its host like the hypovirulence observed in pathogenic species like *Cryphonectria* and *Ophiostoma*. Besides, mycovirus infection could also be used for specific altering of expression of certain host genes and thus gene products.

References

Adler-Moore, J., 1985. Subcellular particles in pathogenic fungi. In: D.H. Howard (ed.) Fungi pathogenic to humans and animals, part B, Pathogenicity and detection: II. Marcel Dekker, New York, pp. 103-121.

Ahn, I.P. and Y.H. Lee, 2001. A viral double-stranded RNA upregulates the fungal virulence of *Nectria radicicola*. Molecular Plant-Microbe Interactions 14: 496-507.

Alexopoulos, C.J. and C.W. Mims, 1979. Introductory mycology. John Wiley and sons, New York, USA, 868 p.

Allen, T.D., A.L. Dawe and D.L. Nuss, 2003. Use of cDNA microarrays to monitor transcriptional responses of the chestnut blight fungus *Cryphonectria parasitica* to infection by virulence-attenuating hypoviruses. Eukaryotic Cell 2: 1253-1265.

Anagnostakis, S.L., B.S. Chen, L.M. Geletka and D.L. Nuss, 1998. Hypovirus transmission to ascospore progeny by field-released transgenic hypovirulent strains of Cr*yphonectria parasitica*. Phytopathology 88: 598-604.

Anderson, M.J., K. Gull and D.W. Denning, 1996. Molecular typing by random amplification of polymorphic DNA and M13 Southern hybridization of related paired isolates of *Aspergillus fumigatus.* Journal of Clinical Microbiology 34: 87-93.

Anwar, M.M., J.H. Croft and R.B.G. Dales, 1993. Analysis of heterokaryon incompatibility (h-c) groups R and GL provides evidence that at least eight *het* loci control somatic incompatibility in *Aspergillus nidulans*. Journal of General Microbiology 139: 1599-1603.

Banks, G.T., K.W. Buck, E.B. Chain, J.E. Darbyshire and F. Himmelweit, 1969. Virus-like particles in penicillin producing strains of *Penicillium chrysogenum*. Nature 221: 89-90.

Banks, G.T., K.W. Buck, E.B. Chain, J.E. Darbyshire, F. Himmelweit, G. Ratti, T.J. Sharpe and D.N. Planterose, 1970. Antiviral activity of double-stranded RNA derived from a virus isolated from *Aspergillus foetidus*. Nature 227: 505.

Batten, J.S., K.B.G. Scholthof, B.R. Lovic, M.E. Miller and R.D. Martyn, 2000. Potential for biocontrol of Monosporascus root rot/vine decline under greenhouse conditions using hypovirulent isolates of *Monosporascus cannonballus*. European Journal of Plant Pathology 106: 639-649.

Beilharz, M.W., G.S. Cobon and P. Nagley, 1982. A novel species of double-stranded RNA in mitochondria of *Saccharomyces cerevisiae*. Nucleic Acids Research 10: 1051-1070.

Boeke, J.D., D. Eichinger, D. Castrillon and G.R. Fink, 1988. The *Saccharomyces cerevisiae* genome contains functional and nonfunctional copies of transposon Ty1. Molecular and Cellular Biology 8: 1432-1442.

Bozarth, R.F., H.A. Wood and A. Mandelbrot, 1971. The *Penicillium stoloniferum* virus complex: two similar double-stranded RNA virus-like particles in a single cell. Virology 45: 516-523.

Brasier, C.M., 1983. A cytoplasmically inherited disease of *Ceratocystis ulmi*. Nature 305: 220-223.

Brewer, B.J. and W.L. Fangman, 1980. Preferential inclusion of extrachromosomal genetic elements in yeast meiotic spores. Proceedings of the National Academy of Sciences USA 77: 5380-5384.

Bruggeman, J., A.J.M. Debets, P.J. Wijngaarden, J.A.G.M. De Visser and R.F. Hoekstra, 2003. Sex slows down the accumulation of deleterious mutations in the homothallic fungus *Aspergillus nidulans*. Genetics 164: 479-485.

Buck, K.W., 1986. Fungal virology - an overview. In: K.W. Buck (ed.). Fungal virology. CRC Press, Boca Raton, Florida, USA, pp. 1-84.

Buck, K.W., 1998. Molecular variability of viruses of fungi. In: Bridge, P., Y. Couteaudier and J. Clarkson (eds). Molecular variability of fungal pathogens. CAB International, London, UK, pp. 53-72.

Buck, K.W. and G. Ratti, 1975. Biophysical and biochemical properties of two viruses isolated from *A. foetidus*. Journal of General Virology 27: 211-224.

Buck, K.W., M.R. Almond, J.J.P. McFadden, M.A. Romanos and J.C. Rawlinson, 1981. Properties of thirteen viruses and virus variants obtained from eight isolates of the wheat take-all fungus, *Gaeumannomyces graminis* var. *tritici*. Journal of General Virology 53: 235-245.

Buck, K.W., R.F. Girvan and G. Ratti, 1973. Two serologically distinct double-stranded ribonucleic acid viruses isolated from *A. niger*. Biochemical Society Transactions 1: 1138-1140.

Burnett, J.H., 1976. Fundamentals of mycology. Edward Arnold, London, UK, 673 p.

Castro, M., K. Kramer, L. Lorena, S. Ortiz, J. Benavente and J. Castillo, 1999. A new double-stranded RNA mycovirus from *Botrytis cinerea*. FEMS Microbiology Letters 175: 95-99.

Castro, M., K. Kramer, L. Valdivia, S. Ortiz and A. Castillo, 2003. A double-stranded RNA mycovirus confers hypovirulence-associated traits to *Botrytis cinerea*. FEMS Microbiology Letters 228: 87-91.

Caten, C.E., 1972. Vegetative incompatibility and cytoplasmic infection in fungi. Journal of General Microbiology 72: 221-229.

Charter, N.W., K.W. Buck and C.M. Brasier, 1993. De-novo generation of mitochondrial-dna plasmids following cytoplasmic transmission of a degenerative disease in *Ophiostoma novo-ulmi*. Current Genetics 24: 505-514.

Chu, Y.E., W.S Lim, S.J. Yea, J.D. Cho, Y.W. Lee and K.H. Kim, 2004. Complexity of dsRNA mycovirus isolated from *Fusarium graminearum.* Virus genes 28: 135-143.

Chu, Y.M., J.J. Jeon, S.J. Yea, Y.H. Kim, S.H. Yun, Y.W. Lee and K.H. Kim, 2002. Double-stranded RNA mycovirus from *Fusarium graminearum.* Applied and Environmental Microbiology 68: 2529-2534.

Coenen, A., F. Kevei and R.F. Hoekstra, 1997. Factors affecting the spread of double-stranded RNA viruses in *Aspergillus nidulans.* Genetical Research Cambridge 69: 1-10.

Coenen, A., J.H. Croft, M. Slakhorst, F. Debets and R.F. Hoekstra, 1994. Additive action of partial heterokaryon incompatibility (*partial-het*) genes in *Aspergilus nidulans.* Current Genetics 26: 233-237.

Compel, P., I. Papp, M. Bibo, C. Fekete and L. Hornok, 1998. Genetic interrelationships and genome organization of double stranded RNA elements in *Fusarium poae.* Virus Genes 18: 49-56.

Croft, J.H., 1985. Protoplast fusion and incompatibility in *Aspergillus.* In: J.F. Peberdy and L. Ferenczy (eds). Fungal protoplasts. Marcel Dekker Inc, New York, USA, pp 225-240.

Dales R.B.G., J. Moorhouse and J.H. Croft, 1993. Evidence for a multi-allelic heterokaryon incompatibility (*het*) locus detected by hybridization among three heterokaryon compatibility (h-c) groups of *Aspergillus nidulans.* Heredity 70: 537-543.

Dalzoto, P.R., C. Glienke-Blanco, V. Kava-Cordeiro, J.Z. Ribeiro, E.W. Kitajima and J.L. Azevedo, 2006. Horizontal transfer and hypovirulence associated with double-stranded RNA in *Beauveria bassiana.* Mycological Research 110: 1475-1481.

Dawe A.L. and D.L. Nuss, 2001. Hypoviruses and chestnut blight: exploiting viruses to understand and modulate fungal pathogenesis. Annual Review of Genetics 35: 1-29.

De La Paz Gimenez-Pecci, M., M.R. Bogo, L. Santi, C.K. Moraes, C.T. Correa, M. Henning-Vainstein and A. Schrank, 2002. Characterization of mycoviruses and analyses of chitinase secretion in the biocontrol fungus *Metarhizium anisopliae.* Current Microbiology 45: 334-339.

Debets, A.J.M. and A.J.F. Griffiths, 1998. Polymorphism of *het*-genes prevents resource plundering in *Neurospora crassa.* Mycological Research 102: 1343-1349.

Deng, F. and G.J. Boland, 2004. A satellite RNA of *Ophiostoma novo-ulmi* mitovirus 3a in hypovirulent isolates of *Sclerotinia homeocarpa.* Phytopathology 94: 917-923.

Deng, F., R. Xu and G.J. Boland, 2003. Hypovirulence-associated double-stranded RNA from *Sclerotinia homeocarpa* is conspecific with *Ophiostoma novo-ulmi* mitovirus 3a-Ld. Phytopathology 93: 1407-1414.

Detroy, R.W. and P.E. Still, 1976. Patulin inhibition of mycovirus stranded RNA replication in *Penicillium stoloniferum.* Journal of General Microbiology 92: 167-174.

Domsch, K.H., W. Gams and T.H. Anderson, 1980. Compendium of soil fungi vol 1 and 2. Academic Press, London, UK.

Dyer, P.S. and M. Paoletti, 2005. Reproduction in *Aspergillus fumigatus*: sexuality in a supposedly asexual species? Medical Mycology Supplement 1: 43.

Elias, K.S. and P.J. Cotty, 1996. Incidence and stability of infection by double stranded RNA genetic elements in *Aspergillus* section *Flavi* and effects on aflatoxigenicity. Canadian Journal of Botany 74: 716-725.

Elliston, J.E., 1985. Characteristics of dsRNA-free and dsRNA-containing strains of *Endothia parasitica* in relation to hypovirulence. Phytopathology 72: 151-158.

El-Sherbeini, M. and K.A. Bostian, 1987. Viruses in fungi: infection of yeast with the K1 and K2 killer viruses. Proceedings of the National Academy of Sciences USA 84: 4293-4297.

Enebak, S.A., B.I. Hillman and W.L. MacDonald, 1994. A hypovirulent isolate of *Cryphonectria parasitica* with multiple genetically unique dsRNA elements. Molecular Plant-Microbe Interactions 7: 590-595.

Fauquet, C.M., M.A. Mayo, J. Maniloff, U. Desselberger and L.A. Ball, 2005. Virus Taxonomy. VIIIth Report of the International Committee on Taxonomy of Viruses. Academic Press, New York. 1162p.

Fekete C., G. Giczey, I. Papp, L. Szabó and L. Hornok, 1995. High-frequency occurrence of virus-like particles with double-stranded RNA genome in *Fusarium poae*. FEMS Microbiology Letters 131: 295-299.

Fine, P.E.M., 1975. Vectors and vertical transmission: an epidemiologic perspective. Annuals of the New York Academy of Sciences 266: 173-194.

Frazzon, A.P., I. da Silva Vaz Junior, A. Masuda, A. Schrank and M.H. Vainstein, 2000. In vitro assessment of *Metarhizium anisopliae* isolates to control the cattle tick *Boophilus microplus*. Veterinary Parasitology 94: 117-125.

Geiser, D.M., M.L. Arnold and W.E. Timberlake, 1994. Sexual origins of British *Aspergillus nidulans* isolates. Proceedings of the National Academy of Sciences of the USA 91: 2349-2352.

Ghabrial, S.A., 1998. Origin, adaptation and evolutionary pathways of fungal viruses. Virus Genes 16: 119-131.

Ghabrial, S.A., A.I. Soldevila and W.M. Havens, 2002. Molecular genetics of the viruses infecting the plant pathogenic fungus *Helminthosporium victoriae*. In S. Tavantzis, (ed), Molecular biology of double-stranded RNA: Concepts and applications in agriculture, forestry and medicine. CRC Press, Boca Raton. pp. 213-236.

Ghabrial, S.A., R.F. Bozarth, K.W. Buck, G.P. Martelli and R.G. Milne, 2000. Partitiviridae. In: M.H.V. van Regenmortel, C.M. Fauquet, D.H.I. Bishop, E.B. Carstens, M.K. Estes, S.M. Lemon, J. Maniloff, M.A. Mayo, D.J. McGeoch, C.R. Pringle and R.B. Wickner (eds.). Virus taxonomy: seventh report of the international committee on taxonomy of viruses. Academic Press, New York, USA, pp. 503-513.

Go, S.J., D.Y. Cha and J.G.H. Wessels, 1992. Symptoms of virus infected oyster mushrooms, *Pleurotus florida*. Korean Journal of Mycology 20: 229-233.

Grogan, H.M., B.A.T. Adie, R.H. Gaze, M.P. Challen and P.R. Mills, 2003. Double-stranded RNA elements associated with the MVX disease of *Agaricus bisporus*. Mycological Research 107: 147-154.

Gussack, G., J.W. Bennett, S. Cavalier and L. Yatsu, 1977. Evidence for the parasexual cycle in a strain of *Aspergillus flavus* containing virus-like particles. Mycopathologia 61: 159-165.

Hacker, C.V., C.M. Brasier and K.W. Buck, 2005. A double-stranded RNA from a *Phytophthora* species is related to the plant endornaviruses and contains a putative UDP glycosyltransferase gene. Journal of General Virology 86: 1561-1570.

Hamari, Z., F. Kevei, É. Kovács, J. Varga, Z. Kozakiewicz and J.H. Croft, 1997. Molecular and phenotypic characterization of *Aspergillus japonicus* strains with special regard to their mitochondrial DNA polymorphisms. Antonie van Leeuwenhoek 72: 337-347.

Hansen, L.J. and S.B. Sandmeyer, 1990. Characterization of a transpositionally active Ty3 element and identification of the Ty3 integrase protein. Journal of Virology 64: 2599-2607.

Hillman, B.I. and N. Suzuki, 2004. Viruses of the chestnut blight fungus, *Cryphonectria parasitica.* Advances in Virus Research 63: 423-472.

Hillman, B.I., S. Supyani, H. Kondo and N. Suzuki, 2004. A Reovirus of the fungus *Cryphonectria parasitica* that is infectious as particles and related to the *Coltivirus* genus of animal pathogens. Journal of Virology 78: 892-898.

Hiremath, S., B. L'Hostis, S.A. Ghabrial and R.E. Rhoads, 1986. Terminal structure of hypovirulence-associated dsRNAs in the chestnut blight fungus *Endothia parasitica.* Nucleic Acids Research 14: 9877-9896.

Hoegger, P.J., U. Heiniger, O. Holdenrieder and D. Rigling, 2003. Differential transfer and dissemination of hypovirus and nuclear and mitochondrial genomes of a hypovirus-infected *Cryphonectria parasitica* strain after introduction into a natural population. Applied and Environmental Microbiology 69: 3767-3771.

Hoffmann, B., S.E. Eckert, S. Krappmann and G.H. Braus, 2001. Sexual diploids of *Aspergillus nidulans* do not form by random fusion of nuclei in the heterokaryon. Genetics 157: 141-147.

Hollings, M., 1962. Viruses associated with dieback disease of cultivated mushrooms. Nature 196: 962-965.

Hollings, M. and O.M. Stone, 1971. Viruses that infect fungi. Annual Reviews in Phytopathology 9: 93-118.

Hong, Y., T.E. Cole, C.M. Brasier and K.W. Buck, 1998. Evolutionary relationships among putative RNA-dependent RNA polymerases encoded by a mitochondrial virus-like RNA in the Dutch elm disease fungus, *Ophiostoma novo-ulmi*, by other viruses and virus-like RNAs and by the Arabidopsis mitochondrial genome. Virology 246: 158-169.

Hong, Y., S.L. Dover, T.E. Cole, C.M. Brasier and K.W. Buck, 1999. Multiple mitochondrial viruses in an isolate of the Dutch Elm disease fungus *Ophiostoma novo-ulmi.* Virology 258: 118-127.

Horiuchi, H. and T. Fukuhara, 2004. Putative replication intermediates in endornavirus, a novel genus of plant dsRNA viruses. Virus Genes 29: 365-375.

Howitt, R.L., R.E. Beever, M.N. Pearson and R.L. Forster, 2006. Genome characterization of a flexuous rod-shaped mycovirus, *Botrytis* virus X, reveals high amino acid identity to genes from plant 'potex-like' viruses. Archives of Virology 151: 563-579.

Howitt, R.L.J., R.E. Beever, M.N. Pearson and R.L.S. Forster, 1995. Presence of double-stranded RNA and virus-like particles in *Botrytis cinerea.* Mycological Research 99: 1472-1478.

Howitt, R.L.J., R.E. Beever, M.N. Pearson and R.L.S. Forster, 2001. Genome organisation of *Botrytis* virus F, a flexuous rod-shaped mycovirus resembling plant 'potex-like' viruses. Journal of General Virology 82: 67-78.

Ihrmark, K., E. Stenstrom E and J. Stenlid, 2004. Double-stranded RNA transmission through basidiospores of *Heterobasidion annosum*. Mycological Research 108: 149-153.

Ihrmark, K., H. Johannesson, E. Stenstrom and J. Stenlid, 2002. Transmission of double-stranded RNA in *Heterobasidion annosum*. Fungal Genetics and Biology 36: 147-154.

Ikeda, K.I., H. Nakamura, M. Arakawa and N. Matsumoto, 2004. Diversity and vertical transmission of double-stranded RNA elements in root rot pathogens of trees, *Helicobasidium mompa* and *Rosellinia necatrix*. Mycological Research 108: 626-634.

Ikeda, K.I., H. Nakamura, M. Arakawa, T. Koiwa and N. Matsumoto, 2005. Dynamics of double-stranded RNA segments in a *Helicobasidium mompa* clone from a tulip tree plantation. FEMS Microbiology Ecology 51: 293-301.

Jacob-Wilk, D., M. Turina and N.K. van Alfen, 2006. Mycovirus Cryphonectria Hypovirus 1 elements cofractionate with *trans*-Golgi network membranes of the fungal host *Cryphonectria parasitica*. Journal of Virology 80: 6588-6596.

Jiang, D. and S.A. Ghabrial, 2004. Molecular characterization of *Penicillium chrysogenum virus*: reconsideration of the taxonomy of the genus *Chrysovirus*. Journal of General Virology 85: 2111-2121.

Jinks, J.L. and M. Grindle, 1963. The genetical background of heterokaryon incompatibility in *Aspergillus nidulans*. Heredity 18: 407-411.

Kanematsu, S., M. Arakawa, Y. Oikawa, H. Osaki, H. Akamura, K. Ikeda, Y. Kuga-Uetake, H. Nitta, A. Sasaki, K. Suzuki, K. Yoshida and N. Matsumoto, 2004. A reovirus causes hypovirulence of *Rosellinia necatrix*. Phytopathology 94: 561-568.

Kazmierczak, P., P. Pfeiffer, L. Zhang and N.K. van Alfen, 1996. Transcriptional repression of specific host genes by the mycovirus *Cryphonectria* hypovirus 1. Journal of Virology 70: 1137-1142.

Kevei, F., Z. Hamari, J. Varga, Z. Kozakiewicz and J.H. Croft, 1996. Molecular polymorphism and phenotypic variation in *Aspergillus carbonarius*. Antonie van Leeuwenhoek 70: 59-66.

Kim, J.W. and R.F. Bozarth, 1985. Intergeneric occurrence of related fungal viruses: the *Aspergillus ochraceus* virus complex and its relationship to the *Penicillium stoloniferum* virus S. Journal of General Virology 66: 1191-2002.

Kim, J.W., E.Y. Choi and J.I. Lee, 2005. Genome organization and expression of the *Penicillium stoloniferum* virus F. Virus Genes 31: 175-183.

Kim, J.W., E.Y. Choi and Y.T. Kim, 2006. Intergeneric relationship between the *Aspergillus ochraceus* virus F and the *Penicillium stoloniferum* virus S. Virus Research 120: 212-215.

Kim, J.W., S.Y. Kim and K.M. Kim, 2003. Genome organization and expression of the *Penicillium stoloniferum* virus S. Virus genes 27: 249-256.

Kim, W.K. and G.R. Klassen, 1989. Double-stranded RNAs in mitochondrial extracts of stem rusts and leaf rusts of cereals. Current Genetics 15: 161-166.

Koltin, Y. and P.R. Day, 1976. Inheritance of killer phenotypes and double-stranded RNA in *Ustilago maydis*. Proceedings of the National Academy of Sciences USA 73: 594-598.

Lakshman, D.K. and S.M. Tavantzis, 1994. Sponaneous appearance of genetically distinct double-stranded RNA elements in *Rhizoctonia solani*. Phytopathology 84: 633-639.

Lhoas, P., 1970. Use of heterokaryosis to infect virus free strains of *Aspergillus niger*. *Aspergillus* Newsletter 11: 8.

Liang, P. and K. Chen, 1987. Virus transmission through interspecies protoplast fusion in *Aspergillus*. Transactions of the British Mycological Society 89: 73-81.

Lipsitch, M., M.A. Nowak, D. Ebert and R.M. May, 1995. The population dynamics of vertically and horizontally transmitted parasites. Proceedings of the Royal Society of London B 260: 321–327.

Liu, Y.C. and M.G. Milgroom, 1996. Correlation between hypovirus transmission and the number of vegetative incompatibility (vic) genes different among isolates from a natural population of *Cryphonectria parasitica*. Phytopathology 86: 79-86.

Liu, Y.-C., D. Linder-Basso, B.I. Hillman, S. Kaneko and M.G. Milgroom, 2003. Evidence for interspecies transmission of viruses in natural populations of filamentous fungi in the genus *Cryphonectria*. Molecular Ecology 12: 1619-1628.

Mackenzie, D.E. and J.P. Adler, 1972. Virus-like particles in toxigenic Aspergilli. Abstracts of the Annual Meeting of the American Society for Microbiology, ASM, Washington, p. 68.

Marquez, L.M., R.S. Redman, R.J. Rodriguez and M.J. Roossinck, 2007. A virus in a fungus in a plant: three-way symbiosis required for thermal tolerance. Science 315: 513-515.

Mayo, M.A., K.I. Berns, C. Fritsch, J.M. Kaper, A.O. Jackson, M.J.Leibowitz and J.M. Taylor, 1995. Subviral agents: satellites. In: F.A. Murphy, C.M. Fauquet, D.H.L. Bishop, S.A. Ghabrial, A.W. Jarvis, G.P. Martelli, M.A. Mayo and M.D. Summers (eds), Virus taxonomy, sixth report of the international committee on taxonomy of viruses. Springer-Verlag, Wien and New York, pp 487-492.

McFadden, J.J.P., K.W. Buck and C.J. Rawlinson, 1983. Infrequent transmission of double-stranded RNA virus particles but absence of DNA proviruses in single ascospore cultures of *Gaeumannomyces graminis*. Journal of General Virology 64: 927-937.

Myers, C.J., A.J.F. Griffiths, S.R. Kraus and R.R. Martin, 1988. Double-stranded RNA in natural isolates of *Neurospora*. Current Genetics 13: 495-501.

Nuss, D.L., 1992. Biological control of chestnut blight: an example of virus-mediated attenuation of fungal pathogenesis. Microbiology Reviews 56: 561-576.

Nuss, D.L., 1996. Using hypoviruses to probe and perturb signal transduction processes underlying fungal pathogenesis. Plant Cell 8: 1834-1853.

Nuss, D. L., 2005 Hypovirulence: mycoviruses at the fungal-plant interface. Nature Reviews Microbiology 3: 632-642.

Osaki, H., C.Z. Wei, M. Arakawa, T. Iwanami, K. Nomura, N. Matsumoto and Y. Ohtsu, 2002. Nucleotide sequences of double-stranded RNA segments from a hypovirulent strain of the white root rot fungus Rosellinia necatrix: Possibility of the first member of the Reoviridae from fungus. Virus Genes 25: 101-107.

Osaki, H., H. Nakamura, A. Sasaki, N. Matsumoto and K. Yoshida, 2006. An endornavirus from a hypovirulent strain of the violet root rot fungus, *Helicobasidium mompa*. Virus Research 118: 143-149.

Osaki, H., H. Nakamura, K. Nomura, N. Matsumoto and K. Yoshida, 2005. Nucleotide sequence of a mitochondrial RNA virus from the plant pathogenic fungus, *Helicobasidium mompa* Tanaka. Virus Research 107: 39-46.

Pál, K., 2007. Comparative analysis of genetic incompatibility in *Aspergillus niger* and *Podospora anserina*. PhD thesis, Wageningen University and Research Center, Wageningen, The Netherlands, 136p.

Paoletti, M., C. Rydholm, E.U. Schwier, M.J. Anderson, G. Szakacs, F. Lutzoni, J.P. Debeaupuis, J.P. Latgé, D.W. Denning and P.S. Dyer, 2005. Evidence for sexuality in the opportunistic pathogen *Aspergillus fumigatus*. Current Biology 15: 1242-1248.

Park, Y., D. James and Z.K. Punja, 2005. Co-infection by two distinct totivirus-like double-stranded RNA elements in *Chalara elegans* (*Thielaviopsis basicola*). Virus Research 109: 71-85.

Park, Y., X. Chen and Z.K. Punja, 2006. Molecular and biological characterization of a mitovirus in *Chalara elegans*. Phytopathology 96: 468-479.

Pfeiffer, I., J. Kucsera, J. Varga, A. Párducz and L. Ferenczy, 1996. Variability and inheritance of double stranded RNA viruses in *Phaffia rhodozyma*. Current Genetics 30: 294-297.

Pöggeler, S, 2002. Genomic evidence for mating abilities in the asexual pathogen *Aspergillus fumigatus*. Current Genetics 42: 153-160.

Polashock, J.J. and B.I. Hillman, 1994. A small mitochondrial double-stranded (ds) RNA element associated with a hypovirulent strain of the chestnut blight fungus and ancestrally related to yeast cytoplasmic T and W dsRNAs. Proceedings of the National Academy of Sciences USA 91: 8680-8684.

Preisig, O., N. Moleleki, W.A. Smit, B.D. Wingfield and M.J. Wingfield, 2000. A novel RNA mycovirus in a hypovirulent isolate of the plant pathogen *Diaporthe ambigua*. Journal of General Virology 81: 3107-3114.

Rao, J.R., D.W.A. Nelson and S. McClean, 2007. The enigma of double-stranded RNA (dsRNA) associated with mushroom virus X (MVX). Current Issues in Molecular Biology 9: 103-122.

Raper, K.B and D.I. Fennell, 1965. The genus *Aspergillus*. Williams & Wilkins, Baltimore, USA, 686 p.

Ratti, G. and K.W. Buck, 1972. Virus particles in *Aspergillus foetidus*: a multicomponent system. Journal of General Virology 14: 165-175.

Revill, P.A., A.D. Davidson and P.J. Wright, 1994. The nucleotide sequence and genome organization of mushroom bacilliform virus: a single stranded RNA virus of *Agaricus bisporus* (Lange) Imbach. Virology 202: 904-911.

Rinker, D.L., D.W. Stobbs and G. Aim, 1993. The effect of virus disease on oyster mushroom production. Micología Neotropical Aplicada 6: 73-79.

Rippel, A, 1939. Ueber die Verbreitung von *Aspergillus niger* ins besondere in Deutschland. Archiv für Mikrobiologie 11: 1-32.

Ro, H.S., E.J. Kang, J.S. Yu, T.S. Lee, C.W. Lee and H.S. Lee, 2007. Isolation and characterization of a novel mycovirus, PeSV, in *Pleurotus eryngii* and the development of a diagnostic system for it. Biotechnology Letters 29: 129-135.

Rodriguez-Cousino, N., A. Solorzano, T. Fujimura and R. Esteban, 1998. Yeast positive-stranded virus-like RNA replicons 20 S and 23 S RNA terminal nucleotide sequences and 3' end secondary structures resemble those of RNA coliphages. Journal of Biological Chemistry 273: 20363-20371.

Rogers, H.J., K. W. Buck and C.M. Brasier, 1987. A mitochondrial target for double-stranded RNA in diseased isolates of the fungus that causes Dutch elm disease. Nature 329: 558-560.

Romaine, C.P. and B. Schlagnhaufer, 1995. PCR analysis of the viral complex associated with La France disease of *Agaricus bisporus*. Applied and Environmental Microbiology 61: 2322-2325.

Romo, M., A. Leuchtmann, B. Garcia and I. Zabalgogeazcoa, 2007. A totivirus infecting the mutualistic fungal endophyte *Epichloë festucae*. Virus Research 124: 38-43.

Sasaki, A., M. Onoue, S. Kanematsu, K. Suzaki, M. Miyanishi, N. Suzuki, D.L. Nuss and K. Yoshida, 2002. Extending chestnut blight hypovirus host range within Diaporthales by biolistic delivery of viral cDNA. Molecular Plant-Microbe Interactions 15: 780-789.

Sasaki, A., S. Kanematsu, M. Onoue, Y. Oyama and K. Yoshida, 2006. Infection of *Rosellinia necatrix* with purified viral particles of a member of Partitiviridae (RnPV1-W8). Archives of Virology 151: 697-707.

Sasaki, A., S. Kanematsu, M. Onoue, Y. Oikawa, H. Nakamura and K. Yoshida, 2007. Artificial infection of *Rosellinia necatrix* with purified viral particles of a member of the genus *Mycoreovirus* reveals its uneven distribution in single colonies. Phytopathology 97: 278-286.

Schmidt, F.R., N.D. Davis, U.L. Diener and P.A. Lemke, 1983. Cycloheximide induction of aflatoxin synthesis in a nontoxigenic strain of *Aspergillus flavus*. BioTechnology 1: 794-795.

Schmidt, F.R., P.A. Lemke and K. Esser, 1986. Viral influences on aflatoxin formation by *Aspergillus flavus*. Applied Microbiology and Biotechnology 24: 248-252.

Shapira, R., G.H. Choi, B.I. Hillman and D.I. Nuss, 1991. The contribution of defective RNAs to the complexity of viral-encoded double-stranded RNA populations present in hypovirulent strains of the chestnut blight fungus *Cryphonectria parasitica*. EMBO Journal 10: 741-746.

Silva, V.N., E.L. Durigon, M.F.C. Pires, A. Lourenço, M.J. de Faria and B. Corrêa, 2001. Time-course of virus-like particles (VLPs) double-stranded RNA accumulation in toxigenic and non-toxigenic strains of *Aspergillus flavus*. Brazilian Journal of Microbiology 32: 56-60.

Soldevilla, A.I. and S.A. Ghabrial, 2001. A novel Alcohol Oxidase/RNA-binding protein with affinity for mycovirus double-stranded RNA from the filamentous fungus *Helminthosporium* (*Cochliobolus*) *victoriae*. Journal of Biological Chemistry 276: 4652-4661.

Soldevilla, A., W.M. Havens and S.A. Ghabrial, 2000. A cellular protein with an RNA-binding activity co-purifies with viral dsRNA from mycovirus-infected *Helminthosporium victoriae*. Virology 272: 183-190.

Stanway, C. and K.W. Buck, 1984. Infection of protoplasts of the wheat take-all fungus, *Gauemannomyces graminis* var. *tritici*, with double-stranded RNA viruses. Journal of General Virology 65: 2061-2066.

Sutherland, M.L. and C.M. Brasier, 1995. Effect of D-factors on in-vitro cerato-ulmin production by the dutch-elm disease pathogen *Ophiostoma novo-ulmi*. Mycological Research 99: 1211-1217.

Suzuki, N., S. Supyani, K. Maruyama and B.I. Hillman, 2004. Complete genome sequence of Mycoreovirus-1/Cp9B21, a member of a novel genus within the family Reoviridae, isolated from the chestnut blight fungus *Cryphonectria parasitica*. Journal of General Virology 85: 3437-3448.

Tartaglia, J., C.P. Paul, D.W. Fulbright and D.L. Nuss, 1986. Structural properties of double-stranded RNAs associated with biological control of chestnut blight fungus. Proceedings of the National Academy of Science of the USA 83: 9109-9113.

Tuomivirta, T.T. and J. Hantula, 2003a. *Gremmeniella abietina* mitochondrial RNA virus S1 is phylogenetically related to the members of the genus Mitovirus. Archives in Virology 148: 2429-2436.

Tuomivirta, T.T. and J. Hantula, 2003b. Two unrelated double-stranded RNA molecule patterns in *Gremmeniella abietina* type A code for putative viruses of the families Totiviridae and Partitiviridae. Archives of Virology 148: 2293-2304.

Van der Lende, T.R., M.C. Harmsen and J.G.H. Wessels, 1994. Double-stranded RNAs and proteins associated with the 34 nm virus particles of the cultivated mushroom *Agaricus bisporus*. Journal of General Virology 75: 2533-2536.

Van Diepeningen, A.D., 1999. Horizontal transfer of genetic elements in the black Aspergilli. PhD thesis, Wageningen University, Wageningen, The Netherlands, 144 p.

Van Diepeningen, A.D., A.J.M. Debets and R.F. Hoekstra, 1997. Heterokaryon incompatibility blocks virus transfer among natural isolates of black Aspergilli. Current Genetics 32: 209-217.

Van Diepeningen, A.D., A.J.M. Debets and R.F. Hoekstra, 1998. Intra- and interspecies virus transfer in Aspergilli via protoplast fusion. Fungal Genetics and Biology 25: 171-180.

Van Diepeningen, A.D., A.J.M. Debets and R.F. Hoekstra, 2006 Dynamics of dsRNA mycoviruses in black *Aspergillus* populations. Fungal Genetics and Biology 43: 446-452.

Van Diepeningen, A.D., A.J.M. Debets, J. Varga, M. van der Gaag and R.F. Hoekstra, 2004. Efficient degradation of tannic acid by black *Aspergillus* species. Mycological Research 108: 919-925.

Van Diepeningen, A.D., A.J.M. Debets, S.M. Slakhorst, C. Fekete, L. Hornok and R.F. Hoekstra, 2000. Interspecies virus transfer via protoplast fusions between *Fusarium poae* and black *Aspergillus* strains. Fungal Genetics Newsletter 47: 99-100.

Van Tieghem, Ph., 1867. Sur la fermentation gallique. Comptes rendus hebdomadaires des séances de l'Académie des Sciences ixv: 1091-1094.

Varga, J., 2003. Mating type homologs in *Aspergillus fumigatus*. Microbiology 149: 816-819.

Varga, J., B. Tóth, S. Kocsubé, B. Farkas and J. Téren, 2005. Evolutionary relationships among *Aspergillus terreus* isolates and their relatives. Antonie van Leeuwenhoek 88: 141-150.

Varga, J., B. Tóth, S. Szencz, J. Molnár, C.S. Fekete and L. Szabó, 2001. Double-stranded RNA elements and virus-like particles in Aspergilli. Acta Biologica Hungarica 52: 355-363.

Varga, J., E. Rinyu, F. Kevei, B. Tóth and Z. Kozakiewicz, 1998. Doublestranded RNA mycoviruses in species of *Aspergillus* sections Circumdati and Fumigati. Canadian Journal of Microbiology 44: 569-574.

Varga, J., F. Kevei, A. Vriesema, F. Debets, Z. Kozakiewicz and J.H. Croft, 1994b. Mitochondrial DNA restriction fragment length polymorphisms in field isolates of the *Aspergillus niger* aggregate. Canadian Journal of Microbiology 40: 612-621.

Varga, J., F. Kevei, C. Fekete, A. Coenen, Z. Kozakiewicz and J.H. Croft, 1993. Restriction fragment length polymorphisms in the mitochondrial DNA's of the *Aspergillus niger* aggregate. Mycological Research 97: 1207-1212.

Varga, J., F. Kevei, C. Vágvölgyi, A. Vriesema and J.H. Croft, 1994a. Double- stranded RNA mycoviruses in section *Nigri* of the *Aspergillus* genus. Canadian Journal of Microbiology 39: 325-329.

Varga, J., F. Kevei, E. Rinyu, J. Téren and Z. Kozakiewicz, 1996. Ochratoxin production by *Aspergillus* species. Applied and Environmental Microbiology 62: 4461-4464.

Varga, J., K. Rigó, J. Molnár, S. Szencz, J. Téren and Z. Kozakiewicz, 2003. Mycotoxin production and evolutionary relationships among species of *Aspergillus* section *Clavati*. Antonie van Leeuwenhoek 83: 191-200.

Varga, J., K. Rigó, S. Kocsubé, K. Pál, B. Tóth, R.A. Samson and Z. Kozakiewicz, 2006. Evolutionary relationships among economically important species of *Aspergillus* subgenera *Aspergillus* and *Fumigati*. In: A.K. Sharma and A. Sharma (eds). Plant genome: diversity and evolution Vol 2B. Cryptogams. Science Publishers, Inc., Enfield, pp. 285-332.

Voth, P.D., L. Mairura, B.E. Lockhart and G. May, 2006. Phylogeography of *Ustilago maydis virus H1* in the USA and Mexico. Journal of General Virology 87: 3433-3441.

Warn, P.A., A. Sharp, M.J. Anderson and D. W. Denning, 2006. Single-spored cultures of the sequenced strain of *Aspergillus fumigatus* (Af293) have significantly different mortality rates in a murine model of IPA. In: Proceedings of the second Advances Against Aspergillosis Conference, P031 (p. 134).

Wickner, R.B., 1996. Double-stranded RNA viruses of *Saccharomyces cerevisiae*. Microbial Reviews 60: 250-265.

Wickner, R.B., R. Esteban and B.I. Hillman, 2000a. *Narnaviridae*. In: Van Regenmortel, M.H.V., C.M. Fauquet, D.H.I. Bishop, E.B. Carstens, M.K. Estes, S.M. Lemon, J. Maniloff, M.A. Mayo, D.J. McGeoch, C.R. Pringle and R.B. Wickner (eds.). Virus taxonomy: seventh report of the international committee on taxonomy of viruses. Academic press, New York, USA, pp. 651-656.

Wickner, R.B., S.A. Ghabrial, J. Bruenn, K.W. Buck, J.K. Patterson, K.D. Stuart and C.V. Wang, 2000b. Totiviridae. Academic Press, San Diego, USA.

Wood, H.A., R.F. Bozarth, J.P. Adler and D.E. Mackenzie, 1974. Proteinaceous virus-like particles from an isolate of *Aspergillus flavus*. Journal of Virology 13: 532-534.

Xie, J., D. Wei, D. Jiang, Y. Fu, G.Li, S. Ghabrial and Y. Peng, 2006. Characterization of debilitation-associated mycovirus infecting the plant-pathogenic fungus *Sclerotinia sclerotiorum*. Journal of General Virology 87: 241-249.

Yokoi, T., S. Yamashita and T. Hibi, 2003. The nucleotide sequence and genome organization of *Sclerophthora macrospora* virus A. Virology 311: 394-399.

Yokoi, T., Y. Takemoto, M. Suzuki, S. Yamashita and T. Hibi, 1999. The nucleotide sequence and genome organization of *Sclerophthora macrospora* virus B. Virology 264: 344-349.

Yu, H.J., D. Lim and H.S. Lee, 2003. Characterization of a novel single-stranded RNA mycovirus in *Pleurotus ostreatus*. Virology 314: 9-15.

Zhao, T., W.M. Havens and S.A. Ghabrial, 2006. Disease phenotype of virus-infected *Helminthosporium victoriae is* independent of overexpression of the cellular alcohol oxidase/RNA-binding protein Hv-p68. Phytopathology 96: 326-332.

III.
Aspergilli in the biotechnology and agriculture

Aspergillus in grapes: ecology, biodiversity and genomics

Giancarlo Perrone[1], Antonia Gallo[1], Antonia Susca[1] and János Varga[2,3]
[1]Institute of Sciences of Food Production, CNR, Via Amendola, 122/O 70126 Bari, Italy
[2]Department of Microbiology, Faculty of Sciences, University of Szeged, H-6701 Szeged, P.O. Box 533, Hungary
[3]CBS Fungal Biodiversity Centre, Uppsalalaan 8, 3584 CT Utrecht, The Netherlands

Abstract

Black Aspergilli, which comprises species belonging to *Aspergillus* section *Nigri*, are distributed worldwide and have a significant impact on human society due to both their beneficial and harmful effects. Although black Aspergilli are mainly isolated from the soil, they are among the most common fungi causing food spoilage. Recently, several studies focused on this group of organisms due to their role as causative agents of black rot of grapes and subsequently cause ochratoxin A contamination. Herein an overview on the occurrence, biodiversity, ecology, toxigenic potential and genetic aspects relevant to ochratoxin production and molecular detection of black Aspergilli in grapes was presented. Nine different black *Aspergillus* species have been identified on grapes with different secondary metabolites profiles. These species are often difficult to identify with classical methods. The polyphasic approach used in our studies led to characterisation of 3 new species occurring on grapes: *A. brasiliensis*, *A. ibericus* and *A. uvarum*. However, the main source of ochratoxin A contamination in grapes is *Aspergillus carbonarius*, followed by *A. niger* and *A. tubingensis*. The genetic aspects of ochratoxin production evidenced that a polyketide synthase is likely to be linked to ochratoxin production, in a similar way as in the cases of other mycotoxins such as fumonisins and aflatoxins. Furthermore, it was found that genes in different ochratoxigenic fungi like *Penicillium* and *Aspergillus* are dissimilar. Although four genes of ochratoxin biosynthetic pathway have recently been characterised from *Penicillium nordicum*, no gene proved to be involved in ochratoxin biosynthesis has yet been identified in *A. carbonarius*. Management of ochratoxin risk in the wine food chain is a challenging task. Various biotic, abiotic, agronomical and environmental factors play an important role in relation to ochratoxin accumulation in grapes and wine. Possible prevention and reduction strategies in the wine food chain have been proposed, and main critical control points and preventive measures to be taken in account in high OTA risk areas is presented.

Keywords: ochratoxin A, black rot, black Aspergilli, molecular approach (detection), biosynthetic genes, risk management (CCPs)

1. Introduction

Aspergillus species occupy a wide spectrum of habitats in animal and plant environment, so they can result economically important both as harmful and useful microorganisms. They can contaminate foods and feeds at different stages including pre- and postharvest stages, processing and handling. On the contrary, they are also frequently used in the fermentation industry for the production of organic acids, enzymes, vitamins and antibiotics (Kozakiewicz, 1989). The most important aspect of food and feed spoilage is the formation of mycotoxins, which may have harmful effects on human and animal health. Several *Aspergillus* mycotoxins have been identified as contaminants in foods and feeds, the economically most important of which are the aflatoxins, ochratoxins and patulin. Recent studies indicate that these compounds can be produced by a number of *Aspergillus* species (Varga *et al.,* 2003b, 2004). However, only a few of these mycotoxin producers can be regarded as potential health hazards either because they produce only traces of the given mycotoxin, or are encountered rarely if at all in food products (e.g. aflatoxin producing *A. ochraceoroseus, Emericella venezuelensis* and *E. astellata* isolates) (Pitt, 2001; Abarca *et al.,* 2001). However, new data indicate that some species recently reported to be mycotoxin producers can be regarded as main sources of mycotoxin contamination in various food products (Pitt, 2001). Although they are not considered to be major cause of plant disease, *Aspergillus* species are responsible of various damages on plants such as chlorosis of almonds, albinism of citrus, black rot of onions, crown rot of peanuts, vine canker and black rot of grapes (Varga *et al.,* 2004).

In the last ten years the importance of the black rot of grapes is highly increased due to the association between this disease and the ochratoxin A (OTA) contamination of grapes and wine. OTA is a major mycotoxin produced by several species of *Aspergillus* and *Penicillium* naturally occurring in a variety of food commodities prior to harvest or more commonly during storage. Numerous animal studies have shown that OTA is a potent nephrotoxin with the degree of renal injury depending on both toxin dose and exposure time; decreasing nephrotoxic sensitivity was observed from pig to rat, to mice. OTA is immunotoxic, neurotoxic *in vitro* and *in vivo* in rats, teratogenic in mice, rats and rabbits (JECFA, 2001). OTA was detected in wine for the first time in 1996 (Zimmerli and Dick, 1996). Thereafter several surveys have been conducted, mainly in Europe, on the occurrence of the toxin in wine and related products, showing it as a problem mainly for Southern Europe. Studies conducted in the last lustre have focused on identifying the source of OTA in grapes and wine, data shows that OTA contamination in Southern Europe and in subtropical parts of Brazil, Argentina and Australia is due to grape berry rot caused by species belonging to *Aspergillus* Sect. *Nigri*, the so called black Aspergilli (Da Rocha Rosa *et al.*, 2002; Battilani *et al.*, 2003b; Serra *et al.*, 2003; Tjamos *et al.*, 2004; Bau *et al.*, 2005; Chulze *et al.*, 2006; Leong *et al.*, 2006a). In the following headings of this chapter different aspects of *Aspergillus* species occurring and infecting grapes berries, their ecology, toxigenicity and genomics are reported together with an analysis of the main critical control point (CCP) for the prevention of OTA contamination in wine.

2. *Aspergillus* black rot of grapes

Aspergillus bunch rots, typically caused by members of *Aspergillus* Section *Nigri*, were formerly considered among one of the many fungal bunch rots affecting grapes prior to harvest. Recently, the significance of this rot increased a lot when some species belonging to *Nigri* Section were recognised as potential source of OTA in grapes and wine. The disease appears on the berries as a black rot due to intense sporulation of the fungus after it has invaded and deprived the berry which looks completely empty and dry (Figure 1).

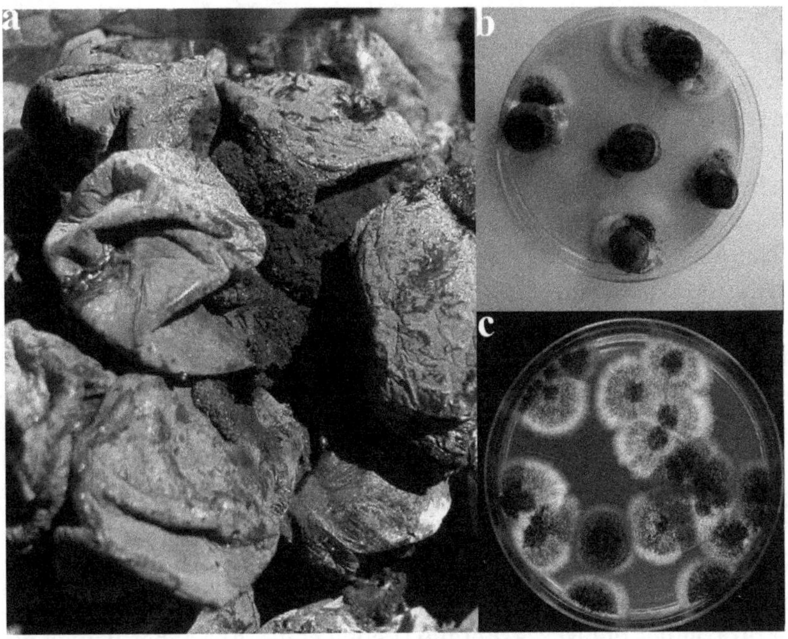

Figure 1. Black Aspergilli on grapes: (a) black rot of berries caused by black Aspergilli; (b) direct plating of berries on DRBC agar; (c) different black Aspergillus *colonies from berries' homogenate diluted and plated on DRBC.*

The berry skin is the primary barrier to fungal infection, the mycobiota on the skin comprises yeast and bacteria living at low metabolic rate on the nutrient leak from the berries, and dormant fungal and bacteria spores. When the skin is damaged, nutrients become more available and the microbial population increases dramatically. Damages to berries skin can be caused by many factors including fungal diseases (downy mildew, powdery mildew), pests (grape berry moth, bunch mites) and environmental factors (wind, hail, rain or sunburn injury). These factors, if don't cause a direct damage to skin, can cause localised hardening of the skin and increase the susceptibility of the berries to splitting. Berry splitting, which occurs when the influx of moisture in a berry is much higher than the efflux, is highly correlated with the bunch rot of grapes caused by *Aspergillus*, *Botrytis*, *Penicillium* and *Rhizopus*; or sour rot caused by a number of undesirable yeasts and bacteria. The incidence of these moulds on berry rot depends on climatic conditions and geographical exposure of the vineyards, generally *Aspergillus* spp. are favoured by warm to hot climates and *Penicillium* by cool and humid conditions (Hewitt, 1988). The black *Aspergillus* rot is usually associated with a wound and is at first tan to brown, but the area is soon covered with a dusty mass of brown or black spores. These rot areas are initially soft but later become firm and leathery. Under warm conditions (20-32 °C) in a drop of water, the fungus may infect mature fruit directly through the skin (Hewitt, 1988). The colonies of these fungi are present on the berry skin from setting and increase in amount from early veraison to harvest, with a peak at ripening. This situation is even more complex regarding the species composition of black Aspergilli occurring on grapes in relation to species identity, OTA production, and ecological conditions. *A. carbonarius* isolates produce the highest amounts of OTA, and although their incidence is 2-3 times lower than that of *A. niger* species aggregate on berries, it is considered to be the main causative agent of OTA contamination of grapes (Battilani *et al.*, 2003a; Abarca *et al.*, 2003; Leong *et al.*, 2006a). Its presence on grape berries increases from early veraison to harvest with a peak at ripening. *Aspergillus japonicus/aculeatus* are the least represented group, sporadically occurring in some countries like Portugal, Spain and Greece (Serra *et al.*, 2006b; Bellì *et al.*, 2006a; Tjamos *et al.*, 2006), more prevalent in France, Italy and Israel, being present in higher densities in the early stage of berry maturation (setting and pea size) and decreasing during the ripening (Battilani *et al.*, 2006b; Cozzi *et al.*, 2007). The occurrence of the different black *Aspergillus* species on grapes in various countries worldwide is summarised in Table 1.

Various surveys have evidenced that soil and vine trash on soil are the primary sources of propagules of black Aspergilli (Leong *et al.*, 2006a; Battilani *et al.*, 2006b; Cozzi *et al.*, 2007).

Table 1. Fungi responsible for ochratoxin contamination of grapes and wine in different countries.

Country	Species responsible	Reference
Argentina	*A. carbonarius, A. foetidus, A. aculeatus/ japonicus, A. niger* aggregate	Magnoli *et al.*, 2003; Magnoli *et al.*, 2004; Chulze *et al.*, 2006; Romero *et al.*, 2005; Ponsone *et al.*, 2007
Australia	*A. carbonarius, A. niger* aggregate, *A. aculeatus*	Heenan *et al.*, 1998; Leong *et al.*, 2004, 2006a
Brazil	*A. carbonarius, A. niger* aggregate	Da Rocha *et al.*, 2002
France	*A. carbonarius, A. niger* aggregate	Sage *et al.*, 2002; Pechavy *et al.*, 2003; Lataste *et al.*, 2004; Sage *et al.*, 2004; Bejaoui *et al.*, 2006
Greece	*A. carbonarius, A. niger* aggregate	Tjamos *et al.*, 2004, 2006
Hungary	*A. niger* aggregate	Varga *et al.*, 2005, 2007b
Israel	*A. carbonarius, A. niger* aggregate	Mitchell *et al.*, 2004; Guzev *et al.*, 2006
Italy	*A. carbonarius, A. niger* aggregate, *A. japonicus, A. tubingensis*	Battilani *et al.*, 2003b, 2006b, Perrone *et al.*, 2006a,b
Lebanon	*A. carbonarius*	El Khoury *et al.*, 2006
Portugal	*A. carbonarius, A. niger* aggregate, *A. japonicus, A. ochraceus*	Abrunhosa *et al.*, 2001, 2003; Serra *et al.*, 2003, 2006b
Spain	*A. carbonarius, A. japonicus, A. niger* aggregate, *A. tubingensis, A. ochraceus*	Abarca *et al.*, 2003; Bau *et al.*, 2005; Bellí *et al.*, 2004a, 2005a, 2006a; Cabañes *et al.*, 2002; Medina *et al.*, 2005; Gómez *et al.*, 2006; Valero *et al.*, 2005
Tunisia	*A. carbonarius, A. niger* aggregate	Lasram *et al.*, 2007, Fredj *et al.*, 2007

It has been postulated that air movement deposits spores from the soil onto berry surfaces, because the incidence of black Aspergilli in air samples increases approaching the soil. The role of air movement in the dispersal of spores from the soil is further highlighted by the increased incidence of *A. carbonarius* on berries after a severe dust storm, and on berries growing in crop systems closer to the soil (Leong *et al.*, 2006a; Cozzi *et al.*, 2007), although according to surveys carried out by Battilani *et al.* (2003a) the height of bunches above ground level cannot be considered relevant for fungal presence. Insects, birds or other fungal infections are also important factors affecting the development of the disease and OTA accumulation in berries. Rain prior to harvest is also a common cause of berry damage, and the severity of *Aspergillus* infection was highest in years with such meteorological conditions (Leong *et al.,* 2006a; Cozzi *et al.,* 2007). During rain fall, high osmotic pressure within the berries, combined with low evaporative water losses cause the berries to swell and if critical turgor pressure is reached, the berry skin splits. In this respect, development of *Aspergillus* rots requires both damage and the presence of spores on the berry surface and black Aspergilli were usually isolated as part of the natural mycoflora on the surface of grape berries. The increase in colony counts between veraison and pre-harvest, especially seen for *A. carbonarius,* reflects the change in composition of berry that initially (green, pea-sized) seems hostile for the survival of the *Aspergillus* spores, and successively, as berries start to mature, they become susceptible to *Aspergillus* rots due to increase in softness and sugar content and detriment of juice acidity of the berries (Leong *et al.,* 2006b). The grape variety and the cropping system may influence the susceptibility and the accumulation of OTA in grapes (Battilani *et al.,* 2004; Leong *et al.,* 2006a; Cozzi *et al.,* 2007). Differences in bunch structures and in berry splitting susceptibility may explain the varietal difference in susceptibility to *Aspergillus* bunch rots, despite intrinsic varietal differences in resistance to mould growth and OTA accumulation have been reported in an *in vitro* study (Battilani *et al.,* 2004). Compact bunches may trap more airborne spores of *Aspergillus* than loose bunches, and berries in a compact bunch are more likely to split due to reduced evaporative loss and to the squeezing between each other. Finally, the development of the *Aspergillus* bunch rot is completely dependent from the geographical areas and exposure (closeness to the sea); and the seasonal weather conditions (humidity, high temperature and rainfall during ripening). In this regard, a 3-year survey (2004-2006) performed on 8 vineyards located in the Salento peninsula of Southern Italy, showed a wide variability of OTA levels between different cultivation years. In particular, the 2005 crop was the most conducive to black Aspergilli contamination due to the higher relative humidity and rainfall levels, associated with hot temperatures at ripening and harvest time (Figure 2, Cozzi *et al.,* 2007). Finally, various data collected in Europe, have evidenced significantly the relation between incidence of berries infected by black Aspergilli at harvesting to latitude and longitude with a positive West-East and North-South gradient (Battilani *et al.,* 2006a).

Other mycotoxigenic species have also been identified on grapes. Although the two well-known producers of OTA, *Penicillium verrucosum* and *P. nordicum* have never been found in grapes, other *Penicillium* species were detected such as *P. expansum* producing citrinin and patulin in pure culture. Only one strain out of 51 of *P. expansum* could produce citrinin in grape juice, whereas 33 strains out of 51 could produce patulin (Abrunhosa *et al.,* 2001). Patulin contamination of wine have been detected by Scott *et al.* (1977). Some Penicillia are able to produce (-)-geosmin, a volatile compound giving fungal or earthy odor to wine. Two species able to produce this compound, *P. expansum* and *P. carneum* have been identified on

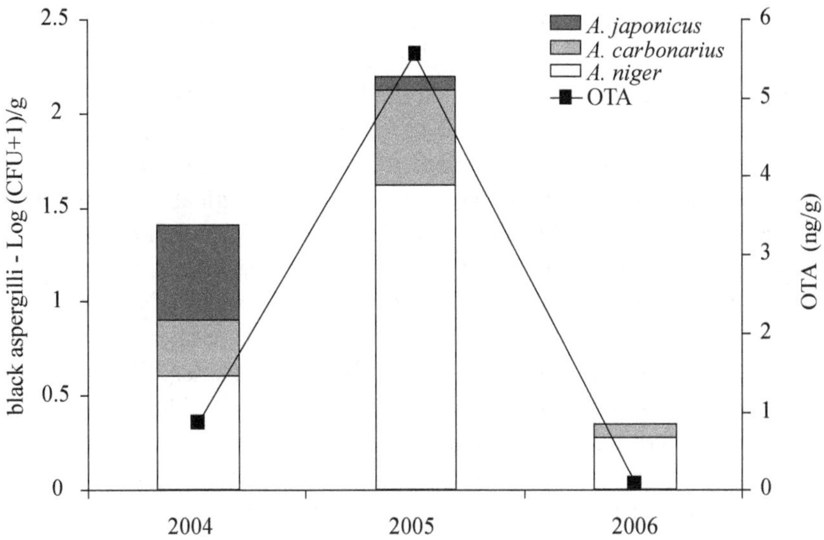

Figure 2. Occurrence of black Aspergilli and OTA in 8 vineyards in Apulia during three grape-harvest seasons (2004-2006).

grapes recently (La Guerche *et al.*, 2005). Recently, El Khoury *et al.* (2006) observed aflatoxin producing *A. flavus* isolates on Lebanese grapes, while Fredj *et al.* (2007) and Varga *et al.* (2007b) isolated *A. flavus* from Tunisian and Hungarian grapes, respectively, indicating that grape juices and wine could also be contaminated by aflatoxins. Indeed, several reports have been published on the occurrence of aflatoxins grape wine (Lemieszek-Chodorowska and Syncerski, 1971; Schuller *et al.*, 1967; Sripathomswat and Thasnakorn, 1981).

3. Ecology, toxigenicity and biodiversity of black Aspergilli

Ochratoxins are produced mostly by *Penicillium* species in colder temperate climates, and by a number of *Aspergillus* species in warmer and tropical parts of the world. *Aspergillus* isolates usually produce both OTA and ochratoxin B (dechlorinated analogue of OTA), while Penicillia produce only OTA. In *Aspergillus* section *Circumdati, A. cretensis, A. flocculosus, A. pseudoelegans, A. roseoglobulosus, A. westerdijkiae, A. sulphurous,* and *Neopetromyces muricatus* produce consistently high amounts of OTA (Frisvad *et al.*, 2004). Two other species, *A. ochraceus* and *A. sclerotiorum* produce large or small amounts of OTA, but less consistently, while four further species produce OTA inconsistently and in trace amounts according to the literature: *A. melleus, A. ostianus, A. petrakii,* and *A. persii*. The most important species regarding potential OTA production in coffee, rice, beverages and other foodstuffs are *A. ochraceus, A. westerdijkiae* and *A. steynii*. In *Aspergillus* section *Flavi, A. albertensis* was found to produce the highest amounts of OTA in synthetic media (Varga *et al.,* 1996; unpublished results). The latter species together with *A. alliaceus* has recently been transferred from section *Circumdati* to *Aspergillus* section *Flavi* based on molecular, biochemical, physiological and morphological studies (Varga *et al.,* 2000b). Fungi responsible for OTA accumulation in cereals, i.e., *Aspergillus ochraceus* and *Penicillium verrucosum,* were first thought to be involved also in OTA formation in grapes. However, a number of studies performed during the last decade evidenced a quite different situation showing that all fungi responsible for OTA accumulation in grapes belong to *Aspergillus* section *Nigri,* the so called black Aspergilli (Battilani *et al.,* 2003a). Most epidemiological surveys performed in Mediterranean, Australian and South American countries have shown that among black Aspergilli the biseriate species *A. carbonarius* and *A. niger* species aggregate, and the uniseriate species *A. aculeatus* and *A. japonicus,* are the prevalent ones occurring on grapes (Battilani *et al.,* 2003b; Da Rocha Rosa *et al.,* 2002; Leong *et al.,* 2006a).

The taxonomy of this Section is still not completely resolved, especially within the *A. niger* species aggregate (a group of morphologically indistinguishable species), leading often to misidentification of the species distribution in food. A comprehensive molecular characterisation of the black Aspergilli occurring in grape in Europe was performed using representative strains isolated from 107 vineyards of the Mediterranean basin (Bau *et al.,* 2006; Perrone *et al.,* 2006a,b). These studies led to the identification of four main populations, namely *A. carbonarius, A. tubingensis, A. niger,* and a group of *Aspergillus* 'uniseriate', that could be separated by using molecular methods including AFLP, RFLP and sequences analysis. The *Aspergillus* 'uniseriate' group was clearly separated from *A. japonicus* and *A. aculeatus* by molecular techniques but was morphologically indistinguishable (Perrone *et al.,* 2006a,b). Ecological and morphological differences between these species are summarised below.

A. carbonarius can easily be distinguished from other biseriate species due to its big and spiny conidia; a high percentage of strains of this species (98–100%) have been shown to produce OTA. Spore germination of *A. carbonarius* is very rapid, and occurs within 24 hours with water activity (a_w) 0.90–0.99 and temperatures between 25–35 °C. Optimal growth is at 32-35 °C and a_w 0.95–0.98 (min 10 °C and max 42 °C). Optimal conditions for OTA production by *A. carbonarius* are at 20 °C and a_w 0.95/0.98 (Bellí *et al.* 2005b). No OTA production is detected at a_w 0.90, some reports observed higher OTA production at 25 °C (Mitchell *et al.,* 2003). However atypical strains of *A. carbonarius*, unable to produce OTA, difficult to distinguish by morphological methods, and isolated from grapes in Spain and Portugal, were recently described as *A. ibericus*, a new species different from *A. carbonarius* for its slightly smaller conidia and faster growth at 37 °C (Serra *et al.*, 2006a).

Grape-derived isolates belonging to the *A. niger* species aggregate comprised four different species not distinguishable by morphological characteristics. The most frequent species isolated from grapes were *A. niger* and *A. tubingensis*, while *A. foetidus* and *A. brasiliensis* were detected at lower frequencies (Perrone *et al.* 2007b). Optimal growth conditions were 35–37 °C and a_w 0.93-0.98 (min 6–8 °C and max 47 °C) for these isolates. This is one of the most common group of species in a wide range of fresh and dry fruits, cereals, etc., and its enzymes and organic acids are used in food processing as 'GRAS' (generally regarded as safe). OTA production by *A. niger* species aggregate normally occurs at 20-25 °C, and a_w 0.95/0.98 (Esteban *et al.*, 2004). A low percentage of OTA producing strains (5–10%) were detected among the *A. niger* species aggregate isolates.

Among the uniseriate group, *A. aculeatus* and *A. japonicus* were often isolated from grapes, but were not proven to produce OTA although they grow at similar conditions as *A. carbonarius*. Recently the *Aspergillus* 'uniseriate' population from grapes in Europe was found to be different from *A. japonicus* and *A. aculeatus* based on molecular evidence (Perrone *et al.*, 2006b). This population is being described as a new species *A. uvarum* (Perrone *et al.* 2007c). This species grows more slowly at 37°C than *A. japonicus* or *A. aculeatus*, reaching only 16–22 mm diameter in 7 days.

Ecophysiological studies have been conducted to explain the dominance of OTA-producing fungi during ripening and sun-drying process analysing the presence of fungal mycoflora at different ripening stages and sun-dried grapes, and carrying out *in vitro* studies on synthetic grape medium at different a_w and temperatures. The results demonstrate that when relative moisture of grapes is high and temperature is between 20 and 30 °C that may favour most of the fungi, *Alternaria* was the most common fungus, followed by *Penicillium*, *Aspergillus* and others. When temperature was higher (35–40 °C), black Aspergilli, mainly *A. niger* aggregate became the predominant fungi as they are better adapted to high temperatures and low a_w (Valero *et al.*, 2005). This advantage of black Aspergilli for growing at hot and humid environmental conditions has also been reported by Leong *et al.* (2004). *A. niger* aggregate isolates were found to be predominant in respect to *A. carbonarius* because are capable of faster growth at higher temperatures and it is the only group with good growth over the 40 °C. As a very small percentage of *A. niger* aggregate isolates are able to produce OTA (2-5%) compared to *A. carbonarius* (90–100%), it is foreseeable that OTA accumulation takes places at temperatures ranging between 20–30°C and at a_w> 0.95. Further studies on the effect of photoperiod and day-night temperatures in simulating field conditions on growth and OTA production by *A. carbonarius* showed that the growth of the fungus is stimulated by field conditions but OTA production was not statistically affected by alternate photoperiod and temperature (Bellí *et al.*, 2006b). However, maximum amount of OTA was detected after 5 days of incubation (Bellí *et al.*, 2004b and 2006b), and incubation at night temperature was found to be beneficial for OTA production, while at higher temperatures OTA formation was restricted as metabolism was used for growing (Bellí *et al.*, 2006b).

OTA production of black Aspergilli occurring on grapes was widely studied in the last years with sometimes ambiguous reports on the toxigenicity and the percentage of toxigenic strains among the species. The OTA producing strains of *A. carbonarius* ranged between 70 and 100% when grown *in vitro* and tested using HPLC, while the range of producing strains was around 2–20% for *A. niger* and *A. tubingensis* (Battilani *et al.*, 2006b; Perrone *et al.*, 2006a). Some reports claimed the production of OTA also by *A. japonicus* but it has not yet been confirmed (Dalcero *et al.*, 2002; Battilani *et al.*, 2003b). Recently, Ponsone *et al.* (2007) studying the occurrence and toxigenicity of *Aspergillus* species in Argentinean vineyards found that *A. niger* aggregate was the most frequent species on grapes with 27% of the isolates producing OTA. The authors also confirmed the production of OTA by *A. japonicus* and *A. aculeatus* strains, but this work lacks molecular identification of the strains.

The difficulties in species recognition within the *Aspergillus niger* species aggregate and the fact that most of the studies carried out on black Aspergilli occurring on grapes lack molecular characterisation of the strains perplexed the extent of their natural occurrence and species distribution on grapes and food. In our survey of about 600 strains of black Aspergilli, representative of 3-year sampling, 5% of *A. niger* species aggregate strains (360) resulted OTA producers, while all *A. carbonarius* strains (200) and none of the *A. japonicus* strains (50) were positive to OTA production (Cozzi *et al.*, 2007). In order to establish a fully correct relationship between species and OTA production, the reported producing isolates and the chemical evidence need to be further confirmed as recommended by Frisvad *et al.* (2006).

The biodiversity of black Aspergilli studied by Perrone *et al.* (2006a) has evidenced that the genetic variability of the four main populations identified by the AFLP analysis ranged from 15 to 35% in *A. carbonarius*, *A. tubingensis* and the *Aspergillus* 'uniseriate' group and 45–55% in the *A. niger* group. The higher genetic diversity encountered in *A. niger* reflect the complexity of this taxon/group and the difficulties of identification at species level. The same grouping was obtained by phylogenetic analysis of partial calmodulin sequence data and part of the β-tubulin gene. These data indicate the need for molecular characterisation of these populations for a better and comprehensive identification of the complex of species involved in the *Aspergillus* black rot disease of grapes. In this respect, the molecular diversity of black Aspergilli using partial calmodulin gene sequence data was widely exploited in the last three years and led to the development of primer pairs and SSCP tools for the rapid and robust identification of the main species within the section (Perrone *et al.*, 2004, Susca *et al.*, 2007a, 2007b).

4. Structure and biosynthesis of ochratoxins

OTA contains a 7-carboxy-5-chloro-8-hydroxy-3,4-dihydro-3R-methylisocoumarin ring linked through the 7-carboxy group to L-β-phenylalanine by an amide bond (Figure 3). The isocoumarin ring of OTA is a pentaketide skeleton, to which a chlorine atom is introduced, and a C1 unit is added from methionine which is subsequently oxidised to carboxyl. An intact L-phenylalanine is linked to the isocoumarin ring through this carboxyl group (Moss 1996, Turner and Aldridge 1983). Several related compounds are also synthesised in the producing organisms. These are the dechlorinated analogue (ochratoxin B), the isocoumarin nucleus of OTA (ochratoxin α), its dechlorinated analogue (ochratoxin β), methyl and ethyl esters including ochratoxin C, which is an ethyl ester derivative of OTA, and several amino acid analogues (Moss, 1996, 1998; Xiao *et al.*, 1995).

Regarding OTA biosynthesis, early studies clarified that phenylalanine was incorporated into OTA, whereas ochratoxin α was constructed from five acetate units with one carbon at C-8 came from methionine (Ferreira and Pitout, 1969).

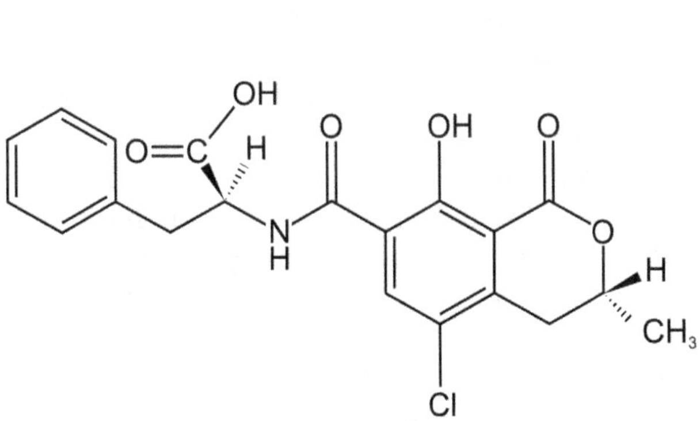

Figure 3. Chemical structure of ochratoxin A.

Wei *et al.* (1971) demonstrated incorporation of ^{36}Cl into OTA, possibly due to the action of a chloroperoxidase enzyme (Steyn and Holzapfel, 1967). Several schemes have been proposed for the OTA biosynthesis pathway, but the biosynthetic steps involved in OTA production are not as yet well established (Huff and Hamilton, 1979; Harris and Mantle, 2001). The isocoumarin group is a pentaketide skeleton formed from acetate and malonate via a polyketide pathway. A chlorine atom is incorporated most probably through the action of a chloroperoxidase to form the isocoumarin portion of OTA. A C1 unit is also added and oxidised to a carboxyl group at C-8. Phenylalanine derived from the shikimic acid pathway is linked through the additional carboxyl group (Moss, 1996, 1998). Precursor feeding experiments carried out by Harris and Mantle (2001) did not support an intermediary role for mellein in OTA biosynthesis, but could not rule out the role of 7-methyl-mellein. These results indicated that chlorination of both ochratoxin β and OTB can give rise to OTA. However, there was no evidence for the intermediate role of the ester ochratoxin C proposed by Huff and Hamilton (1979). These results suggest that one biosynthetic pathway involves ochratoxins $\beta \rightarrow \alpha \rightarrow$ OTA, with also a branch for ochratoxins $\beta \rightarrow$ OTB. According to the molecular structure of the mycotoxin, a number of enzymatic activities, besides the primary pathway for phenylalanine biosynthesis, are deduced to be required for the biosynthesis: a polyketide synthase for the synthesis of the polyketide dihydroisocoumarin, a cyclase, a chloroperoxidase, an esterase and a peptide synthetase for ligation of the phenylalanine to the dihydroisocoumarin. A scheme of the OTA biosynthetic pathway adapted from Huff and Hamilton (1979), O'Callaghan and Dobson (2005), Edwards *et al.* (2002) and Harris and Mantle (2001) is shown in Figure 4.

Since most of the identified genes of secondary metabolite pathways are clustered, like the aflatoxin, sterigmatocystin and trichothecene biosynthetic genes, it was supposed that the genes of the OTA biosynthetic pathway are clustered too. The synthesis of the isocoumarin group represents a crucial step. It is probably derived from a pentaketide skeleton formed from acetate and malonate via a polyketide synthesis pathway, therefore a polyketide synthase is likely to be linked to OTA production, in a similar way as fungal polyketide synthases are involved in the biosynthesis of other mycotoxins such as fumonisins and aflatoxins. Fungal polyketide synthases (PKSs) are multidomain enzymes using an iterative strategy to generate a polyketide from a single set of active sites. Bingle *et al.* (1999) identified two subclasses of fungal PKSs designated as WA type and MSAS type, based on sequence analysis of the genes: WA type are considered involved in pigment and aflatoxin biosynthesis, while MSAS type take part in the biosynthesis of 6-methyl-salicylic acid. Recently the existence of hybrid NRPS/PKS enzymes has also been demonstrated, where NRPS is a non-ribosomal peptide synthase. The domains of PKSs are higly conserved and

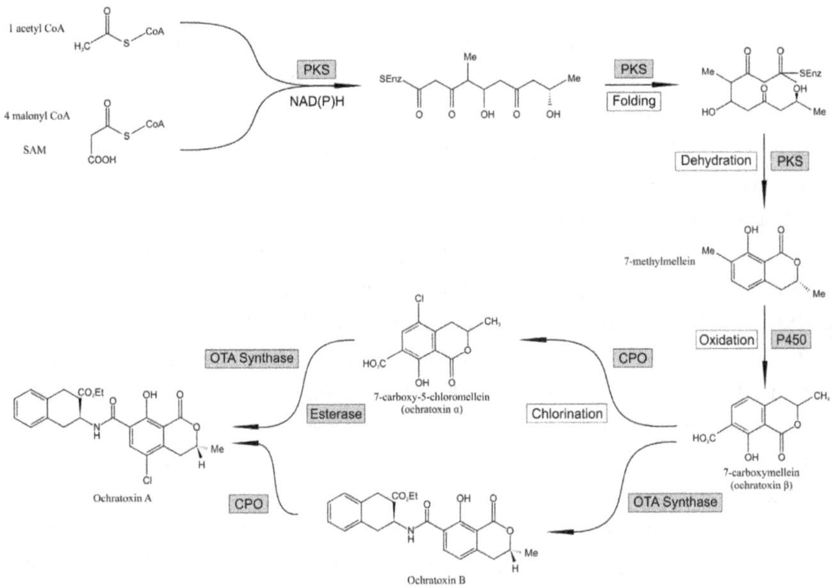

Figure 4. A proposed pathway of ochratoxin biosynthesis (adapted from Huff and Hamilton, 1979; O'Callaghan and Dobson, 2005, and Harris and Mantle, 2001). SAM: S-adenosyl methionine, PKS: polyketide synthase, P450: cytochrome P450-containing enzyme, CPO: chloroperoxidase.

their nucleotide sequences are homologous between different fungal species. Owing to this, primer pairs for polymerase chain reaction (PCR) have been developed for the ketosynthase, ketoreductase and C-methylation domains (Nicholson *et al.*, 2001). Mainly those developed for the KS domain are valuable to identify new gene sequences coding for PKS enzyme responsible for the biosynthesis of important mycotoxins. The analysis of PKS gene sequences in ochratoxin producing organisms represents an effective approach for the identification of OTA biosynthetic genes, even though a single fungal species may contain multiple polyketide synthase (*pks*) genes showing diversity of the nucleotide and corresponding amino acid sequences (Varga *et al.*, 2003a).

5. Biosynthetic genes in OTA producing fungi

Advanced findings on the genetic aspects of OTA biosynthetic pathway have been recently obtained from the fungus *Penicillium nordicum*. *Penicillium* species are responsible for the production of OTA in cereals and in some fermented products, especially in temperate regions. OTA producing *Penicillium* strains have recently been reclassified as either *P. verrucosum*, the exclusive ochratoxigenic species in cereals, or *P. nordicum*, which occurs mainly as a contaminant in protein-rich foods like fermented meats and cheeses (Lund and Frisvad, 2003). Initially a 10 kb genomic DNA region of *P. nordicum* carrying three genes was described (Karolewiez and Geisen, 2005), two of them coding for enzymes assumed to play a role in the OTA biosynthetic pathway, in particular the polyketide synthase (*otapks*PN) and the non-ribosomal peptide synthetase (*otanps*PN). The sequenced region of *otapks*PN gene coded for a ketoacyl synthetase (KS) and an acyltransferase (AT) domains and experiments of gene inactivaction showed an involvement in the pathway. From comparative analysis, the *otapksPN* gene exhibited high homology mainly to fungal polyketide synthases of MSAS type. The second gene characterised (*otanps*NP) displayed high homology to various non-ribosomal peptide synthetase and the product of this gene was supposed to be responsible for the formation of the peptide bond between the dihydroisocoumarin group and the amino acid phenylalanine. A second 4.3 kb gene cluster of putative biosynthetic genes was identified in the same fungus by Geisen *et al.* (2006). This cluster carried one incomplete open reading frame (ORF) with homology to an organic anion transporter from rat kidneys (*otatra*PN) and probably involved in ochratoxin export and another gene (*otachl*PN) showing some homology to a chloroperoxidase of *Gluconobacter oxidans*. There is no absolute evidence for the *otachl*PN gene encoding a chloroperoxidase linked to OTA biosynthesis but it appeared coordinately expressed with other genes known to be required. Both clusters might be organised as one in the chromosome of *P. nordicum*, such other mycotoxin biosynthetic genes and genes coding for possible regulators and transporters are in close proximity within the fungal genome. The genes in *P. nordicum* were expressed in a coordinated temporal manner suggesting a model for OTA production. After the production by the *otapks*PN gene, the polyketide dihydroisocoumarin would be ligated to phenylalanine by the activity of the *otanps*PN gene, subsequently the joined molecule would be chlorinated by *otachl*PN and then excreted out of the cell by the *otatra*PN gene product. The cluster resulted present only in the two OTA producing species *P. nordicum* and *P. verrucosum* which had significant identity between their biosynthetic genes, except for the polyketide synthase, while the producing *Aspergilli* species gave negative results when tested with primers based on the sequences of the *Penicillium* genes. Interestingly the ochratoxin *pks* gene of *A. ochraceus* which was identified by O'Callaghan *et al.* (2003) exhibits much lower homology to the *otapks*PN gene than other fungal polyketide synthases. The *pks* gene in *Aspergillus ochraceus*, the predominant OTA producer in tropical regions, was characterised using a suppression subtractive hybridisation PCR (SSH-PCR)-based technique to establish genes differentially expressed under OTA permissive conditions. The sequence of the *pks* gene included a highly conserved AT region typical of PKS protein. It was demonstrated to have a functional role in the biosynthesis since mutant strains of *A. ochraceus*, in which the *pks* gene was knocked out, lost their ability to produce the mycotoxin. Further studies revealed that the transcription level of the *pks* gene and production of OTA in *A. ochraceus* were regulated not only by nutritional availability but also by the pH of the growth medium (O'Callaghan *et al.*, 2006). The phylogenetic analysis, based on the alignment of the ketosynthase domains, showed that the OTA-PKS was most similar to MSAS type PKS proteins. In addition two putative p450 monooxygenase genes resulted upregulated during production of the toxin and their cooexpression together with the *pks* gene, under different physiological condition, was reported, indicating a possible role for these genes in the biosynthesis. Both genes showed a high degreee of similarity to other monooxygenase genes encoding enzymes involved in the production of polyketide secondary metabolites but their clustering has yet to be proven.

The attempt to identify a PKS related to OTA biosynthesis in *A. carbonarius*, considered to be responsible for contamination in grapes, wines and coffee, on the basis of the *A. ochraceus pks* gene sequence was unsuccessful. This result confirms the fact that the ochratoxin polyketide synthase genes in different ochratoxigenic fungi are dissimilar and are much less homologous to each other than expected, subsequently the OTA biosynthetic pathways may substantially differ between *Penicillium* and *Aspergillus* species.

In order to elucidate the genetic control of OTA biosynthesis in *A. carbonarius*, the attention was focused on PKS genes also in this *Aspergillus* species. PCR approaches with different sets of degenerate and specific primers for the KS domain were used to investigate the presence and multiplicity of *pks* genes. Atoui *et al.* (2006) obtained five different KS domain sequences in *A. carbonarius*. These sequences were submitted to similarity search and phylogenetic analyses, and were found to be highly diverse, probably because they represent PKSs responsible for different functions. Two of the KS sequences characterised in *A. carbonarius* were identified using the primers KS1/KS2 (Liou and Khosla, 2003) designed to target fungal and bacterial KS domains and were found to be closely related to the PKSs involved in the biosynthesis of reduced polyketides such as compactin, lovastatin and T-toxin. One of them displayed about 60% identity to a KS domain sequence which was found to join the AT domain sequences of the PKS involved in OTA biosynthesis in *A. ochraceus*. The different primer pair LC3/LC5 (Geisen, 2004), targeting MSAS type PKSs, allowed the amplification of two more KS sequences in *A. carbonarius*, with one of them showing a 44% identity to *otapks*PN of *P. nordicum*. Another sequence was obtained using specific primers deduced from the conserved motifs of the KS domain of the OTA *pks* gene in *A. ochraceus*. This sequence was used as target to detect specifically *A. carbonarius* by PCR methods and to design a specific primer pair to be employed in real time PCR for the direct quantification of this fungus in grape samples (Atoui *et al.*, 2007). However none of the *pks* sequences of *A. carbonarius* was proved to be undoubtedly connected to the biosynthesis of toxin. In our study on the molecular aspects of OTA production in *A. carbonarius*, a strain producing high levels of OTA was examined using degenerate and specific primers that allowed us to characterise an extra KS domain of a not yet identified *pks* gene. A comparative analysis displayed the highest similarity to a hypothetical protein of the known producer *A. niger*. Reverse transcription analysis (RT-PCR) indicated the correlation of its expression to the production in permissive conditions (Gallo *et al.*, 2006), but further studies, both physiological and molecular, are in progress to confirm the relationship. Still a lot remains to clarify about the genetic regulation of OTA biosynthesis in *A. carbonarius*.

The technological breakthroughs in genomic research, such as Expressed Sequence Tag (EST), chromosome sequencing and large-scale genome sequencing, will help the understanding of the genetics of OTA producing fungi. Valuable information about the set of genes responsible or related to toxin production could come from the genome sequencing of *A. carbonarius*, that is in progress. Such comparative analyses have permitted to establish the presence of a putative cluster for OTA synthesis in the sequenced genome of *Aspergillus niger* (Pel *et al.*, 2007).

6. Molecular tools for identification of black Aspergilli

Rapid and precise identification of spoilage fungi to species level is critical for improving the management of toxicological risk in foods and feeding stuffs. Moreover, sensitive detection tools are essential for tracking spoilage organisms in epidemiological investigations, and to clarify their toxigenic potential in foods. To address these requirements, numerous assays have been developed, based on molecular microbiological techniques, which have the potential to achieve rapid, sensitive, and specific identification of fungi from foods and feeds, to replace cumbersome and time consuming microbiological and chemical methods for detection and identification of fungi. The development of the polymerase chain reaction (PCR) in the mid 1980s precipitated a series of rapid technological advancements in DNA-based diagnostics as well as DNA sequencing technology. In particular, DNA sequencing has permitted elucidations of phylogeny in many difficult taxonomic groups, such as black Aspergilli, and led to the accumulation of numerous data which make it possible to identify molecular markers specific for either species or to their toxigenic properties. To date, many molecular genetic techniques have been devised to discriminate taxa or individuals. These methods include the classical hybridisation approach of Restriction Fragment Length Polymorphism (RFLP) as well as PCR based approaches such as PCR-RFLP and Single-Strand Conformation Polymorphism (SSCP) using specific primers, or anonymous primers such as Random Amplified Polymorphic DNA (RAPD), Amplified Fragment Length Polymorphism (AFLP) (Table 2). They could be essentially distinguished as two kinds of approaches: known gene and anonymous fragment based approaches.

6.1. Known gene based approaches

The identification of specific, polymorphic loci often requires a substantial initial effort to identify and characterise suitable loci. A category of site-specific polymorphisms exploits the sequences of introns (short stretches of non-coding sequences which interrupt many genes) of single copy metabolic and structural genes; another category exploits sequences of ribosomal RNA coding DNA regions and their spacers (ITS and IGS). Primers designed using highly conserved sequences flanking the introns have been described for a number of genes for filamentous fungi and yeasts (Glass and Donaldson, 1995; White *et al.*, 1990). The variability found within those genes has been exploited in designing a variety of PCR primers that amplify parts of

Table 2. Species-specific PCR assays for black Aspergillus *species.*

Species	Targeted sequence	Assay	Reference
A. carbonarius	RAPD fragment	PCR	Fungaro *et al.*, 2004
A. carbonarius	AFLP fragment	PCR	Schmidt *et al.*, 2004a
A. carbonarius	AFLP fragment	PCR	Schmidt *et al.*, 2004b
A. carbonarius/A. ochraceus	ITS	PCR	Patino *et al.*, 2005
A. niger, A. japonicus, A. heteromorphus, A. ellipticus	ITS	PCR	Gonzales-Salgado *et al.*, 2005
A. tubingensis/ A.niger	calmodulin gene	PCR	Susca *et al.*, 2007a
A. carbonarius/A. japonicus	calmodulin gene	PCR	Perrone *et al.*, 2004
A. ochraceus, A. niger, A. carbonarius	ITS	PCR	Haugland *et al.*, 2004
A. carbonarius/A. ochraceus/A niger	RAPD fragment	Multiplex-PCR	Sartori *et al.*, 2006
A. tubingensis/ A.niger	ITS	PCR-RFLP	Accensi *et al.*, 1999
black Aspergilli	ITS	PCR-RFLP	Martínez-Culebras and Ramón, 2007
A. carbonarius	RAPD fragment	RAPD	Voetz, 2002
A. carbonarius	calmodulin gene	Real-time PCR	Mulè *et al.*, 2006
A. carbonarius	PKS gene	Real-time PCR	Atoui *et al.*, 2007
black Aspergilli	calmodulin gene	SSCP	Susca *et al.*, 2007b

gene only in the fungus of interest or its close relatives. Specific primer pairs for *A. carbonarius* and *A.japonicus* were designed based on calmodulin gene by Perrone *et al.* (2004); for *A. ochraceus, A. niger, A. carbonarius* based on ITS by Haugland *et al.* (2004); for *A. carbonarius* and *A. ochraceus* based on ITS by Patino *et al.* (2005); for *A. japonicus, A. niger, A. tubingensis, A. ellipticus* and *A. heteromorphus* based on ITS by Gonzales-Salgado *et al.* (2005), and for *A. niger* and *A. tubingensis* based on calmodulin sequences by Susca *et al.* (2007a).

However, the variability of intron sequences might not necessarily involve length polymorphisms and may result solely from nucleotide substitutions between fragments of identical size, necessitating a more sophisticated method of detection. SSCP analysis is an adequate method to circumvent this problem, thus it allows to study low-level sequence variability in known PCR products. The electrophoretic migration of single strands is a function of secondary structure formed due to spontaneous self-annealing upon entry into the non-denaturing gel matrix. Single base variability may alter this secondary structure, hence changing the relative electrophoretic mobility of the molecule. Using this technique Susca *et al.* (2007b) developed a primer pair which allows amplification of all the black *Aspergillus* spp. occurring on grapes using the same PCR condition, and the discrimination of them in post-run amplification analysis, using capillary electrophoresis.

Alternatively, Bufflier *et al.* (2007) have developed an oligosorbent microarray (OLISA®) combining a conventional PCR assay, using primers targeting the calmodulin gene and the classical hybridisation approach, to identify and detect simultaneously *A. carbonarius* and other black Aspergilli in grapes.

The simplest method for screening sequence variability in PCR products is PCR-RFLP, by digesting them with restriction endonucleases. Variation is detected as change in fragment length or changes in base sequence resulting in the loss or gain of a restriction enzyme recognition site. Relatively little sequence information can be inferred from PCR-RFLP analysis since the probability of encountering a change in a specific site of four nucleotides is quite low, although the inclusion of additional restriction enzymes improves the analysis. It was proposed by Accensi *et al.* (1999) and recently re-proposed by Gonzales-Salgado *et al.* (2005) to discriminate *A. niger* and *A. tubingensis* and by Martínez-Culebras and Ramón (2007) to identify black Aspergilli.

6.2. Anonymous fragment based approaches

Using known gene sequences could cause a bias in the phylogenetic analysis since some gene sequences are under strong selection during evolution. Nucleotides in third codon positions and introns probably would not be the subjects to such selection, their frequency could be affected by their proximity to selected regions through genetic hitchhiking. In this case it is necessary to analyse the whole genome to find variations, a task that may involve the use of arbitrary regions.

RFLP analysis has the advantage of examining DNA variability directly, digesting the whole genomic or mitochondrial DNA (mtDNA) with restriction endonucleases. The resulting DNA fragments generate patterns which may represent polymorphic forms of a structural gene or noncoding regions. *Sma*I digested repetitive DNA profiles hybridised with the ribosomal repeat unit of *A. nidulans* were found to have distinctive value among black Aspergilli (Varga *et al.*, 1994, 2000a). Among the strains of the *A. niger* species complex, four profiles were observed, among which rDNA types I and III were shown by *A. niger* and *A. brasiliensis* strains, respectively, while rDNA types II and II' were characteristic of the *A. tubingensis* strains (Varga *et al.*, 1994). Regarding mtDNA polymorphisms, within the *A. niger* aggregate, most isolates were classifiable as *A. niger* or *A. tubingensis* according to their *Hae*III-*Bgl*II digested mtDNA patterns (Varga *et al.*, 1993, 1994). The *A. niger* and *A. tubingensis* species could be grouped into 5 and 6 mtDNA types, respectively. Six of the 13 Brazilian isolates examined exhibited mtDNA and rDNA types different from those of all the other strains. Later these strains have been assigned to the *A. brasiliensis* species (Varga *et al.*, 2007a). Among the 16 collection strains and field isolates of *Aspergillus carbonarius* examined, the *Hae*III-digested mtDNA profiles revealed only slight variations (Kevei *et al.*, 1996).

Although RFLP methodologies have been powerful analytical tools, they require large amounts of genomic DNA, are difficult to automate and require substantial time to complete. RAPD analysis uses a set of randomly designed primers allowing rapidly marker identification by screening a panel of candidate arbitrary oligonucleotide primers without *a priori* knowledge of the target sequence. Fungaro *et al.* (2004) and Voetz (2002) designed PCR primers based on sequences of RAPD fragments for the detection of *A. carbonarius* and OTA producing *Aspergillus* and *Penicillum* species. Also a multiplex PCR method which allows to detect amplicons of 809, 372 and 260 pb using primers based on a RAPD fragment has been developed to discriminate *A. ochraceus, A. carbonarius* and *A. niger* simultaneously (Sartori *et al.*, 2006). However, many investigators have found random primer fingerprinting methods to be hampered by problems related to reproducibility, due to the presence of non-specific bands resulting from PCR artefacts such as heteroduplexes. Therefore PCR-based fingerprinting techniques relying on random primers are not robust and generally unsuitable for use as population markers.

AFLP is a fingerprinting method which combines elements of both RFLP and random primed PCR. AFLP compares polymorphic patterns in fragments generated from the selected amplification of a subset of restriction fragments resulting from an enzymatic digestion. AFLP appears to be more robust than RAPD and related methods because the use of longer primers and known target sequence permit a higher stringency of hybridisation during the amplification procedure, producing fewer artefacts due to spurious priming events. To date, several studies have demonstrated the applicability of this methodology in resolving relatedness among black Aspergilli. Perrone *et al.* (2006a) demonstrated that *A. tubingensis,* which is morphologically indistinguishable from *A. niger,* can be distinguished from it based on their AFLP profiles. Moreover, AFLP profiles allowed to develop species-specific primers to identify *A. ochraceus* and *A. carbonarius* (Schmidt *et al.,* 2004a,b).

Molecular techniques can also be used for the direct detection of fungi from environmental samples, allowing the establishment of early detection and prevention strategies which can lead to the control of mycotoxin contamination of food. The simplicity of PCR technology together with its potential to detect small numbers of target micro-organisms without the need of culturing makes it an important method for monitoring toxigenic fungi. Despite this potential, technical limitations have continued to limit the large-scale use of PCR with environmental samples primarily because extraction techniques have been labor-intensive and often unreliable. The direct extraction of DNA from grapes contaminated by fungi without prior purification or culturing would clearly provide an attractive goal. An efficient method to remove PCR inhibitors from grape samples which contain many PCR-inhibiting substances such as tannins, polysaccharides and pigments was described by Mulè *et al.* (2006) for real-time PCR application and could be improved by nucleic acid purification based on magnetic beads.

Real-time PCR is currently a promising PCR method for quantifying DNA. The incorporation of probes increases the specificity relative to conventional PCR and allows the monitoring of several DNA targets in the same reaction tube (multiplex real-time PCR).

In the last years the availability of diagnostic assays for the detection of OTA producing fungi in raw cultures and to prevent entering mycotoxins to the food chain has become an urgent issue. Particular attention was given to *A. carbonarius* which is considered to be the main causative agent of OTA contamination in grapes and grape-derived products. The quantification of *A. carbonarius* in grapes at the early stage of maturation would permit to take necessary measures in order to reduce the final OTA contamination level. The first real-time PCR based detection/quantification assay for *A. carbonarius* using Taqman® chemistry has been developed by Mulè *et al.* (2006), targeting species-specific variability in the calmodulin gene. Subsequently the increasing knowledge about the genetic background of OTA production allowed to Atoui *et al.* (2007) to develop an assay, using SYBR Green® chemistry, also specific for *A. carbonarius*, targeting a putative polyketide synthase gene. Both methods, although target different genes, are equally informative about the occurrence of *A. carbonarius* in food products, not about their OTA production. Mycotoxin biosynthesis is highly dependent on several environmental conditions, therefore RNA-based methods, aimed at examininig expression levels of genes specifically involved in OTA biosynthesis, are preferable to those based on species-specific markers which detect the occurrence of certain mycotoxin-producing fungal species. The genes which are responsible for mycotoxin production were shown to be regulated and induced only under certain conditions and not expressed constitutively, so that the molecular data on relative RNAs levels give direct information about the environmental and nutritional parameters which leads to the induction of the genes and the biosynthesis of toxins. With this aim a real-time PCR system specific for the polyketide synthase gene of *P. nordicum* has been used to study the influence of different conditions on the gene expression and on ochratoxin production (Geisen, 2004). The sequences of OTA related genes were also employed for the development of a microarray (Schmidt-Heydt and Geisen, 2007) to study and evaluate the influence of environmental parameters like substrate, pH, temperature and water activity on the activation and regulation of the genes and thereby on mycotoxin biosynthesis. A more detailed knowledge of the genetic regulation of OTA production can help to develop further valuable tools for the control of intoxication of foods, by avoiding conditions favourable for OTA biosynthesis.

7. Managing the ochratoxin A risk in the grape-food chain

OTA is produced in vineyards and is normally absent up to early veraison. Bunches without visible symptoms can also contain OTA although berries with visible black moulds normally show higher contamination levels. The inoculum of black Aspergilli is always present in vineyards and the detection of *A. carbonarius* amount in the field is crucial as it is considered to be the major responsible of OTA presence in berrie unless some Argentinean and Brazilian report refers to *A. niger* as major responsible (Chulze *et al.*, 2006). The most important critical control points are related to identification of geographical areas with high risk of OTA contamination and monitoring of meteorological conditions in those areas from veraison to harvest including temperature, humidity and rainfall. In this respect, using a geostatistical approach described by Battilani *et al.* (2006a), data on incidence of *A. carbonarius* were run with ArcView and a predictive map was drawn. The incidence of *A. carbonarius* was significantly correlated with geographic coordinates showing a positive gradient going towards the South of Europe. Based on the combination of the degree-day and rainfall parameters in late August-early September in several countries of the Mediterranean basin, discriminant analysis gave promising perspectives for predicting OTA presence in vineyards by the development of thermo-wetness maps (Battilani *et al.*, 2006a). Looking at the biotic damages that may occurs to berries, while the fungal disease due to downy mildew and powdery mildew are usually well controlled in the field, pest like grape berry moth (*Lobesia botrana*) are mainly controlled during the first generations larvae, and not in the last period of the grape growing season when larvae can either contribute to spore dispersal or act as spore vectors, by trapping conidia in the cuticle ornamentation. In this respect, good correlation between pest damage and OTA content has been found in grape berries, due to the contribution of *L. botrana* to berry wounds and fungal spore dissemination (Cozzi *et al.*, 2006). Field trials, performed in 2004 and 2005 using both biological and conventional insecticidal treatments, confirmed that a successful control of the third generation of *L. botrana* reduces the inoculum of black Aspergilli and the formation of OTA in grapes (Perrone *et al.*, 2007a; Kappes *et al.*, 2005). However, various chemicals have been shown to be active in reducing both fungal growth and OTA levels in grape bunches to different extent: mepanipyrim, pyrimethanil, fluazinam, iprodione and the mixture cyprodinil and fludioxonil. The latter mixture was confirmed as effective in field trials carried out in several Mediterranean countries including France, Spain, Greece and Italy (Bellì *et al.*, 2007; Kappes *et al.*, 2005; Tjamos *et al.*, 2004). The most effective treatment was observed at 21 days before harvesting and a previous treatment at veraison was suggested in high risk conditions. This mixture of active ingredients was applied against black Aspergilli with the same combination and schedule, both in dosage and timing, as effectively used against grey mould, caused by *Botrytis cinerea*. Moreover, the insecticide treatment against *L. botrana* in combination with the fungicide contributes significantly to reduction of OTA level in the field particularly in crop years at high contamination risk (Kappes *et al.*, 2005). Promising results were also obtained by using biological control agents isolated from yeasts naturally occurring on authoctonous grapes in Greece and in Italy. In particular, good results were obtained with two strains of *Cryptococcus laurentii* and *Aureobasidium pullulans*, respectively, in Greece (Dimakopoulou *et al.*, 2005), and with a strain of *Hanseniaspora uvarum* in Italy using weekly or two-weekly treatments.

OTA occurring in grapes is transferred to wine during wine-making. As processing can play an important role in diminishing the potential risks of mycotoxin-contaminated food commodities, it is important to know OTA distribution in wine and relevant by-products to determine if the toxin level can be managed through post-harvest or corrective procedures. In this respect, the fate of OTA during vinification of grape has been studied by different authors with contrasting results as reviewed recently by Visconti *et al.* (2007 in press). In particular, some authors observed an increase of OTA concentration in must during maceration of crushed grapes and a consistent reduction of OTA during pomaces and lees separations and after spontaneous malo-lactic fermentation (Fernandes *et al.,* 2003; Grazioli *et al.,* 2006). In contrast, Rousseau (2004) reported that OTA content in must increases after crushing grapes, and reaches maximum levels during malo-lactic fermentation. Finally Leong *et al.* (2006d) reported that 24% of OTA originally present in crushed red grape passed into free run wine (must) and a 72% OTA reduction was recorded in wine after the first racking. The fate of OTA and its distribution in wine and winery by-products during vinification of naturally contaminated Negroamaro and Primitivo grapes has been recently reinvestigated at laboratory (microvinification) and industrial level by Solfrizzo *et al.* (2007). Results of microvinification experiments showed that only 4% of the OTA present in grapes remains in the wine whereas 95% of OTA originally present is retained on pressed grape pomaces (98% in the skin and 2% in the seeds) and 1% is retained on the lees. OTA concentration remained nearly constant in must collected after maceration, pressing, juice clarification, alcoholic fermentation and lees separations (after first and second racking). The same OTA concentrations were found in wine samples analysed after one year (Solfrizzo *et al.,* 2007).

The corrective actions actually available for reduction of OTA concentrations in contaminated wines are mainly based on several fining agents (Castellari *et al.,* 2001; Leong *et al.,* 2006c). The best results in terms of OTA removal were obtained with carbon or commercial preparations containing carbon but the removing of OTA resulted proportional to the reduction of polyphenol content of treated wines (Visconti *et al.,* 2007 in press). Treatment with oak wood fragments showed some efficacy in reducing OTA concentration by about 26-42% depending on the quantity of wood chips and powder used (Savino *et al.,* 2007). Some reports have reported the reduction of OTA in grape juice and wine by using oenological yeast strains (Bejaoui *et al.,* 2004; Garcia Moruno *et al.,* 2005; Cecchini *et al.,* 2006) but the removal of OTA during fermentation is based on adsorption mechanism other than degradation.

However, the efficacy of yeasts for OTA reduction at industrial level as well as their impact on wine quality parameters (phenol compounds) has not been shown. Visconti *et al.* (2007 in press) confirmed OTA reduction by yeasts or inactivated yeast walls while a consistent reduction of colour index (expressed in terms of Folin Ciocalteu index) occurred. Finally, the main source of OTA in the wine production chain is the infection by black Aspergilli occurring in the field. *A. carbonarius* is the main cause of OTA accumulation in grape berries occurring from early veraison to ripening. A summary of the main CCP and preventive measures to be taken in account in high OTA risk areas is presented in Table 3.

Table 3. Preventive and corrective measures to reduce OTA in wines[1].

Main critical control point	Preventive and corrective measures
Field and pre-harvest	
Mediterranean basin, closeness to the sea High temperature and relative humidity from veraison to harvest Rainfall during ripening period (berry splitting) Berry damage (high risk with grape berry moth infestation) Grape variety and training system susceptibility	Monitoring climatic conditions and black Aspergilli presence from veraison to harvest Monitoring with trap system the *Lobesia botrana* pressure in the vineyards Avoid excess of vigour and vegetation Combined fungicide/insecticide treatments (1 or 2) when favourable climatic conditions occur
Harvest – wine making	
Mechanical harvest without selection of bunches High incidence of rot bunches Long grape storage after harversting (>8 h)	Anticipate harvest time in high OTA risk areas when favouring conditions occur Segregate rot bunches at harvesting Minimise storage time before processing Control OTA contamination in must Use carbon preparations to reduce OTA contamination during fermentation

[1]Modified from Visconti *et al.* (2007).

8. Conclusions

In this review we wished to give an overview on the occurrence, biodiversity, toxigenic potential and detection of black Aspergilli in grapes. Black Aspergilli are the causative agents of black rot of grapes, and contaminate grapes with OTA. OTA contents of wine can reach high levels in some parts of Europe including South Italy, Greece, some parts of Spain and France. OTA production is influenced by: climatic conditions/geographic areas; grape varieties/crop systems; berry damages caused by insects, fungal infection or excessive irrigation/rainfall. Fungicidal and insecticidal treatments can reduce OTA contamination. Susceptibility to infection can vary from different years and regions. Several molecular methods have been developed for fast detection of black Aspergilli on grapes. The availability of these rapid methods for identification and quantification of OTA producing fungi in early stages of veraison in combination with the environmental and biotic factors can help to apply the right chemical treatment against black Aspergilli in the field. Methods targeting expression levels of genes specifically involved in OTA biosynthesis should be preferable to those based on species-specific markers. Further studies are necessary to identify these genes and to develop real-time PCR or microarray based approaches directly correlated to the level of OTA contamination of grapes in the field. Attempts to reduce fungal colonisation and OTA content of grapes including agronomic practices and biological and chemical treatments met with varying degrees of success, and the data obtained are sometimes controversial. Good Agriculture Practices (balanced soil tillage, irrigation, nitrogen fertilisation, pruning) and Good Manufacturing Practices (reduced harvest to vinification time, segregation of rot bunches) help considerably to reduce OTA contamination risk. Several attempts have been made to identify possible corrective actions for OTA decontamination of wine and grape juice. Activated carbon proved to reduce effectively OTA levels in wine but affected negatively wine quality. Further studies are necessary to find suitable practices for lowering fungal contamination and OTA levels in grapes and grape-derived products. The biodiversity of black Aspergilli on grapes has been widely studied in Europe by a polyphasic approach and several new species has been characterised. However, worldwide studies on biodiversity and population structure of black Aspergilli on grapes should be carried out. These studies are needed either to know the toxigenicity and species composition or to comprehend the biogeography and migration of black Aspergilli causing the bunch rot of grapes at a global level.

References

Abarca, M.L., F. Accensi, M.R. Bragulat and F.J. Cabanes, 2001. Current importance of ochratoxin A-producing *Aspergillus* spp. Journal of Food Protection 64: 903-906.

Abarca, M.L., F. Accensi, M.R. Bragulat, G. Castella and F.J. Cabañes, 2003. *Aspergillus carbonarius* as the main source of ochratoxin A contamination in dried vine fruits from the Spanish market. Journal of Food Protection 66: 504-506.

Abrunhosa, L., R.R.M. Paterson, Z. Kozakiewicz, N. Lima and A. Venâncio, 2001. Mycotoxin production from fungi isolated from grapes. Letters in Applied Microbiology 32: 240-242.

Abrunhosa, L., R. Serra, N. Lima and A. Venâncio, 2003. Ochratoxin A risk assessment in Portuguese wines: a one-year case study. Le Bulletin de l'OIV 76: 618-634.

Accensi, F., J. Cano, L. Figuera, M.L. Abarca and F.J. Cabanes, 1999. New PCR method to differentiate species in *Aspergillus niger* aggregate. FEMS Microbiology Letters 180: 191-196.

Atoui A., H.P. Dao, F. Mathieu and A. Lebrihi, 2006. Amplification and diversity analysis of ketosynthase domains of putative polyketide synthase genes in *Aspergillus ochraceus* and *Aspergillus carbonarius* producers of ochratoxin A. Molecular Nutrition & Food Research 50: 488-493.

Atoui, A., F. Mathieu and A. Lebrihi, 2007. Targeting a polyketide synthase gene for *Aspergillus carbonarius* quantification and ochratoxin assessment in grapes using real-time PCR. International Journal of Food Microbiology 115:313-318.

Battilani, P., P. Giorni and A. Pietri, 2003a. Epidemiology of toxin producing fungi and ochratoxin A occurrence in grape. European Journal of Plant Pathology 109: 715-722.

Battilani, P., A. Pietri, T. Bertuzzi, L. Languasco, P. Giorni and Z. Kozakiewicz, 2003b. Occurrence of ochratoxin A-producing fungi in grapes grown in Italy. Journal of Food Protection 66: 633-636.

Battilani, P., A. Logrieco, P. Giorni, G. Cozzi, T. Bertuzzi and A. Pietri, 2004. Ochratoxin A production by *Aspergillus carbonarius* on some grape varieties grown in Italy. Journal of the Science of Food and Agriculture 84: 1736-1740.

Battilani, P., C. Barbano, S. Marin, V. Sanchis, Z. Kozakiewicz and N. Magan, 2006a. Mapping of *Aspergillus* Section *Nigri* in Southern Europe and Israel based on geostatistical analysis. International Journal of Food Microbiology 111 S1: S72-S82.

Battilani, P., P. Giorni, T. Bertuzzi, S. Formenti and A. Pietri, 2006b. Black Aspergilli and ochratoxin A in Italy. International Journal of Food Microbiology 111S1: S53-S60.

Bau, M., M.R. Bragulat, M.L. Abarca, S. Minguez and F.J. Cabañes, 2005. Ochratoxigenic species from Spanish wine grapes. International Journal of Food Microbiology 98: 125-130.

Bau, M., G. Castellá, M.R. Bragulat and F.J. Cabañes, 2006. RFLP characterization of *Aspergillus niger* aggregate species from grapes from Europe and Israel. International Journal of Food Microbiology 111S1: S18-S21.

Bejaoui, H., F. Mathieu, P. Taillandier and A. Lebrihi, 2004. Ochratoxin A removal in synthetic and natural grape juices by selected oenological *Saccharomyces* strains. Journal of Applied Microbiology 97: 1083-1044.

Bejaoui, H., F. Mathieu, P. Taillandier and A. Lebrihi, 2006. Black Aspergilli and ochratoxin A in French vineyards. International Journal of Food Microbiology 111S1: S46-S52.

Bellí, N., E. Pardo, S. Marín, G. Farré, A.J. Ramos and V. Sanchis, 2004a. Occurrence of ochratoxin A and toxigenic potential of fungal isolates from Spanish grapes. Journal of the Science of Food and Agriculture 84: 541-546.

Bellí, N., A.J. Ramos, V. Sanchis and S. Marin, 2004b. Incubation time and water activity effects on ochratoxin A production by *Aspergillus* section *Nigri* strains isolated from grapes. Letters in Applied Microbiology 38: 72-77.

Bellí, N., D. Mitchell, S. Marín, I Alegre, A.J. Ramos, N. Magan and V. Sanchis, 2005a. Ochratoxin A-producing fungi in Spanish wine grapes and their relationship with meteorological conditions. European Journal of Plant Pathology 113: 233-239.

Bellí, N., A.J. Ramos, I. Coronas, V. Sanchis and S. Marín, 2005b. *Aspergillus carbonarius* growth and ochratoxin A production on a synthetic grape medium in relation to environmental factors. Journal of Applied Microbiology 98: 839-844.

Bellí, N., M. Bau, S. Marín, M.L. Abarca, A.J. Ramos and M.R. Bragulat, 2006a. Mycobiota and ochratoxin A producing fungi from Spanish wine grapes. International Journal of Food Microbiology 111 S1: S40-S45.

Bellí, N., A.J. Ramos, V. Sanchis and S. Marín, 2006b. Effect of photoperiod and day-night temperatures simulating field conditions on growth and ochratoxi A production of *Aspergillus carbonarius* strains isolated from grapes. Food Microbiology 23: 622-627.

Bellí, N., S. Marín, E. Argiles, A.J. Ramos and V. Sanchis, 2007. Effect of chemical treatments on ochratoxigenic fungi and common mycobiota of grapes (*Vitis vinifera*). Journal of Food Protection 70: 157-163.

Bingle, L.E.H., T.J. Simpson and C.M. Lazarus, 1999. Ketosynthase domain probes identify two subclasses of fungal polyketide synthase genes. Fungal Genetics and Biology 26: 209-223.

Bufflier, E., A. Susca, M. Baud, G. Mulè, K. Brengel and A. Logrieco, 2007. Detection of *Aspergillus carbonarius* and other black aspergilli from grapes by DNA OLISAtrade mark microarray. Food Additives and Contaminants 24. 1138-1147.

Cabañes, F.J., F. Accensi, M.R. Bragulat, M.L. Abarca, G. Castellá, S. Minguez and A. Pons, 2002. What is source of ochratoxin A in wine? International Journal of Food Microbiology 79: 213-215.

Castellari, M., A. Versari, A. Fabiani, G.P. Parpinello and S. Galassi, 2001. Removal of ochratoxin A in red wines by means of adsorption treatments with commercial fining agents. Journal of Agricultural and Food Chemistry 49: 3917-3921.

Cecchini, F., M. Morassut, E. Garcia Moruno and R. Di Stefano, 2006. Influence of yeast strain on ochratoxin A content during fermentation of white and red must. Food Microbiology 23: 411-417.

Chulze, S.N., C.E. Magnoli and A.M. Dalcero, 2006. Occurrence of ochratoxin a in wine and ochratoxigenic mycoflora in grape and dried vine fruits in South America. International Journal of Food Microbiology 111 S1: S5-S9.

Cozzi, G., M. Pascale, G. Perrone, A. Visconti and A. Logrieco, 2006. Effect of *Lobesia botrana* damages on black Aspergilli rot and ochratoxin A content in grapes. International Journal of Food Microbiology 111 S1: S88-S92.

Cozzi, G., G. Perrone, F. Epifani, M. Pascale and A. Visconti, 2007. Epidemiology of ochratoxin A producing fungi in Apulian vineyards. XII International IUPAC Symposium on Mycotoxins and Phycotoxins, 2007 May, 21-25; Istanbul. Poster 1422.

Da Rocha Rosa, C.A., V. Palacios, M. Combina, M.E. Fraga, A. De Oliveira Rekson, C.E. Magnoli and A.M. Dalcero, 2002. Potential ochratoxin A producers from wine grapes in Argentina and Brazil. Food Additives and Contaminants 19: 408-414.

Dalcero, A., C. Magnoli, C. Hallak, S.M. Chiacchiera, G. Palacio and C.A.R. Rosa, 2002. Detection of ochratoxin A in animal feeds and capacity to produce this mycotoxin by *Aspergillus* section *Nigri* in Argentina. Food Additives and Contaminants 19: 1065-1072.

Dimakopoulou, M., S.E. Tjamos, E.C. Tjamos and P.P. Antoniou, 2005. Chemical and biological control of sour rot caused by black Aspergilli in the grapevine variety agiorgitico of Korinth region. International Workshop on Ochratoxin A in Grapes and Wine: Prevention and Control held in Marsala (TP) Italy, 2005 October 20-21; p. 77 (Poster).

Edwards, S.G., J. O'Callaghan and A.D.W. Dobson, 2002. PCR-based detection and quantification of mycotoxigenic fungi, Mycological Research 106: 1005-1025.

El Khoury, A., T. Rizk, R. Lteif, H. Azouri, M.L. Delia and A. Lebrihi, 2006. Occurrence of ochratoxin A- and aflatoxin B1-producing fungi in Lebanese grapes and ochratoxin a content in musts and finished wines during 2004. Journal of Agricultural and Food Chemistry 54: 8977-8982.

Esteban, A., M.L. Abarca, M.R. Bragulat and F.J. Cabañes, 2004. Effects of temperature and incubation time on production of ochratoxin A by black Aspergilli. Research in Microbiology 155: 861-866.

Fernandes, A., A. Venâncio, F. Moura, J. Garrido and A. Cerdeira, 2003. Fate of ochratoxin A during a vinification trial. Aspects of Applied Biology 68: 73-80.

Fredj, S.M.B., S. Chebil, A. Lebrihi, S. Lasram, A. Ghorbel and A. Mliki, 2007. Occurrence of pathogenic fungal species in Tunisian vineyards. International Journal of Food Microbiology 113: 245-250.

Frisvad, J.C., J.M. Frank, J.A.M.P. Houbraken, A.F.A. Kuijpers and R.A. Samson, 2004. New ochratoxin A producing species of *Aspergillus* section *Circumdati.* Studies in Mycology 50: 23-43.

Frisvad, J.C., F.K. Nielsen and R.A. Samson, 2006. Recommendations concerning the chronic problem of misidentification of mycotoxinogenic fungi associated with foods and feeds. Advances in Experimental Medicine and Biology 571: 33-46.

Fungaro, M.H., P.C. Vissotto, D. Sartori, L.A. Vilas-Boa, M.C. Furlaneto and M.H. Taniwaki, 2004. A molecular method for detection of *Aspergillus carbonarius* in coffee beans. Current Microbiology 49: 123-127.

Gallo, A., G. Perrone, F. Epifani, G. Panzarini, A.D.W. Dobson and G. Mulè, 2006. Analysis of *pks* genes in OTA producer *Aspergillus carbonarius. **In:*** Book of Abstract of MYCO-GLOBE International Conference on 'Advances on genomics, biodiversity and rapid systems for detection of toxigenic fungi and mycotoxins' Monopoli (Bari) Italy 26-29 September 2006. p. 81.

Garcia Moruno, E., C. Sanlorenzo, B. Boccacino and R. Di Stefano, 2005. Treatment with yeast to reduce the concentration of ochratoxin A in red wine. American Journal of Enology and Viticulture 56: 73-76.

Geisen, R., 2004. Molecular monitoring of environmental conditions influencing the induction of ochratoxin A biosynthesis genes in *Penicillium nordicum.* Molecular Nutrition & Food Research 48: 532-540.

Geisen, R., M. Schmidt-Heydt and A. Karolewiez, 2006. A gene cluster of the ochratoxin A biosynthetic genes in *Penicillium.* Mycotoxin Research 22: 134-141.

Glass N.L. and G.C. Donaldson G.C., 1995. Development of primers sets designed for use with the PCR to amplify conserved genes from filamentous ascomycetes. Applied and Environmental Microbiology 61: 1323-30.

Gómez, C., M.R. Bragulat, M.L. Abarca, S. Minguez and F.J. Cabañes, 2006. Ochratoxin A-producing fungi from grapes intended for liqueur wine production. Food Microbiology 23: 541-545.

Gonzalez-Salgado A., B. Patino, C. Vazquez and M.T. Gonzalez-Jaen, 2005. Discrimination of *Aspergillus niger* and other *Aspergillus* species belonging to section *Nigri* by PCR assays. FEMS Microbiology Letters 245: 353-361.

Grazioli, B., M.D. Fumi and A. Silva, 2006. The role of processing on ochratoxin A content in Italian must and wine: A study on naturally contaminated grapes. International Journal of Food Microbiology 111 S1: S93-S96.

Guzev, L., A. Danshin, S. Ziv and A. Lichter, 2006. Occurrence of ochratoxin a producing fungi in wine and table grapes in Israel. International Journal of Food Microbiology 111 S1: S67-S71.

Harris, J.P. and P.G. Mantle, 2001. Biosynthesis of ochratoxins by *Aspergillus ochraceus*. Phytochemistry 58:709-716.

Haugland, R.A., M. Varma, L.J. Wymer and S.J. Vesper, 2004. Quantitative PCR analysis of selected *Aspergillus, Penicillium* and *Paecilomyces* species. Systematic and Applied Microbiology 27: 198-210.

Heenan, C.N., K.J. Shaw and J.I. Pitt, 1998. Ochratoxin A production by *Aspergillus carbonarius* and *A. niger* and detection using coconut cream agar. Journal of Food Mycology 1: 67-72.

Hewitt, W.B., 1988. Berry rots and raisin moulds. In: R.G. Pearson and A.C. Goohen (eds) Compendium of Grape Diseases. The American Phytopathological Society, Minnesota, USA, pp. 26-28.

Huff, W.E. and P.B. Hamilton, 1979. Mycotoxins-their biosynthesis in fungi: Ochratoxins-metabolites of combined pathways. Journal of Food Protection 42: 815-820.

JECFA 2001. Joint FAO/WHO Expert Committee on Food Additives, 56th Meeting, Geneva, 6-15 February 2001.

Kappes, ME, L. Serrati, J.B. Drouillard, J.M. Cantus and M. Kazantzidou, 2005. A crop protection approach to *Aspergillus* and OTA management in Southern European vineyards. International Workshop on Ochratoxin A in Grapes and Wine: Prevention and Control, Marsala (TP) Italy, 2005 October 20-21, p. 24.

Karolewiez, A. and R. Geisen, 2005. Cloning a part of the ochratoxin A biosynthetic gene cluster of *Penicillium nordicum* and characterization of the ochratoxin polyketide synthase gene. Systematic and Applied Microbiology 28: 588-595.

Kevei, F., Z. Hamari, J. Varga, Z. Kozakiewicz and J.H. Croft, 1996. Molecular polymorphism and phenotypic variation in *Aspergillus carbonarius*. Antonie van Leeuwenhoek 70: 59-66.

Kozakiewicz, Z., 1989. *Aspergillus* species on stored products. Mycological Papers 161: 1-188.

La Guerche, S., S. Chamont, D. Blancard, D. Dubourdieu and P. Darriet, 2005. Origin of (–)-geosmin on grapes: on the complementary action of two fungi, *Botrytis cinerea* and *Penicillium expansum*. Antonie van Leeuwenhoek 88: 131-139.

Lasram, S., N. Bellí, S. Chebil, Z. Nahla, M. Ahmed, V. Sanchis and A. Ghorbel, 2007. Occurrence of ochratoxigenic fungi and ochratoxin A in grapes from A Tunisian vineyard. International Journal of Food Microbiology 114: 376-379.

Lataste, C., L. Guérin, D. Solanet, J.L. Berger, P. Cottereau and B. Molot, 2004. Évolution de la contamination en ochratoxine A: du vignoble Français au vin. Progrès Agricole et Viticole 121: 57-64.

Lemieszek-Chodorowska, K. And A. Syncerski, 1971. Aflatoxin B1 in grape wines. Roczniki Państwowego Zakładu Higieny 22: 527-529. [Article in Polish].

Leong, S.L., A.D. Hocking and J.I. Pitt, 2004. Occurrence of fruit rot fungi (*Aspergillus* section *Nigri*) on some drying varieties of irrigated grapes. Australian Journal of Grape and Wine Research 10: 83-88.

Leong, S.L., A.D. Hocking, J.I. Pitt, B.A. Kazi, R.W. Emmett and E.S. Scott, 2006a. Australian research on ochratoxigenic fungi and ochratoxin A. International Journal of Food Microbiology 111 S1: S10-S17.

Leong, S.L., A.D. Hocking, and E.S. Scott, 2006b. Survival and growth of *Aspergillus carbonarius* on wine grapes before harvest. International Journal of Food Microbiology 111 S1: S83-S87.

Leong, S.L., A.D. Hocking and E.S. Scott, 2006c. The effect of juice clarification, static or rotary fermentation and fining on ochratoxin A in wine. Australian Journal of Grape and Wine Research 12: 245-251.

Leong, S.L., A.D. Hocking, P. Varelis, G. Giannikopoulos and E.S. Scott, 2006d. Fate of ochratoxin A during vinification of Semillon and Shiraz grapes. Journal of Agricultural and Food Chemistry 54: 6460-6464.

Liou, G.F. and C. Khosla, 2003. Building.block selectivity of polyketide synthases. Current Opinion in Chemical Biology 7: 279-284.

Lund, F. and J.C. Frisvad, 2003. *Penicillium verrucosum* in wheat and barley indicates presence of ochratoxin A. Journal of Applied Microbiology 95: 1117-1123.

Magnoli, C.E., M. Violante, M. Combina, G. Palacio and A.M. Dalcero, 2003. Mycoflora and ochratoxin-producing strains of *Aspergillus* section *Nigri* in wine grapes in Argentina. Letters in Applied Microbiology 37: 179-184.

Magnoli, C.E., A. Astoreca, L. Ponsone, M. Combina, G. Palacio, C.A.R. Rosa and A.M. Dalcero, 2004. Survey of mycoflora and ochratoxin A in dried vine fruits from Argentina markets. Letters in Applied Microbiology 39: 326-331.

Martinez-Culebras, P.V. and D. Ramon, 2007. An ITS-RFLP method to identify black *Aspergillus* isolates responsible for OTA contamination in grapes and wine. International Journal of Food Microbiology 113: 147-153.

Medina, A., M. Rufino, L. López-Ocaña, F.M. Valle-Algarra and M. Jiménez, 2005. Study of Spanish grape mycobiota and ochratoxin A production by isolates of *Aspergillus tubingensis* and other members of *Aspergillus* section *Nigri*. Applied and Environmental Microbiology 71: 4696-4702.

Mitchell, D., D. Aldred and N. Magan, 2003. Impact of ecological factors on the growth and ochratoxin A production by *Aspergillus carbonarius* from different regions of Europe. Aspect of Applied Biology 68: 109-116.

Mitchell, D., R. Parra, D. Aldred and N. Magan, 2004. Water and temperature relations of growth and ochratoxin A production by *Aspergillus carbonarius* strains from grapes in Europe and Israel. Journal of Applied Microbiology 97: 439-445.

Moss, M.O., 1996. Mode of formation of ochratoxin A. Food Additives and Contaminants 13 (Suppl.): 5-9.

Moss, M.O., 1998. Recent studies of mycotoxins. Journal of Applied Microbiology 84 (Suppl.): 62S-76S.

Mulè, G., A. Susca, A. Logrieco, G. Stea and A. Visconti, 2006. Development of a quantitative real-time PCR assay for the detection of *Aspergillus carbonarius* in grapes. International Journal of Food Microbiology 111: 28-34.

Nicholson, T.P, B.A.M. Rudd, M. Dawson, C.M. Lazarus, T.J. Simpson and R.J. Cox, 2001. Design and utility of oligonucleotide gene probes for fungal polyketide synthases. Chemistry & Biology 8:157-178.

O'Callaghan, J. and A.D. Dobson, 2005. Molecular characterization of ochratoxin A biosynthesis and producing fungi, *Advances in Applied Microbiology* 58: 227-243.

O'Callaghan, J., M.X. Caddick and A.D.W. Dobson, 2003. A polyketide synthase gene required for ochratoxin A biosynthesis in *Aspergillus ochraceus*. Microbiology 149: 3485-3491.

O'Callaghan, J., P.C. Stapleton and A.D.W. Dobson, 2006. Ochratoxin A biosynthetic genes in *Aspergillus ochraceus* are differentially regulated by pH and nutritional stimuli. Fungal Genetics and Biology 43: 213-221.

Patino, B., A. Gonzalez-Salgado, M.T. Gonzalez-Jaen and C. Vazquez, 2005. PCR detection assays for the ochratoxin-producing *Aspergillus carbonarius* and *Aspergillus ochraceus* species. International Journal of Food Microbiology 104: 207-214.

Pechavy, M., H. Bejaoui, F. Mathieu, B.L. Nguyen, M. Roset and A. Lebrihi, 2003. Evolution de la population des champignons toxinogènes et suivi de la production d'OTA pendant la maturation des grappes de raisin. VIIème Symposium International d'oenologie, 19–21th June 2003. Bordeaux.

Pel, H.J., J.H. de Winde, D.B. Archer, P.S. Dyer, G. Hofmann, P.J. Schaap, G. Turner, R.P. de Vries, R. Albang, K. Albermann, M.R. Andersen, J.D. Bendtsen, J.A. Benen, M. van den Berg, S. Breestraat, M.X. Caddick, R. Contreras, M. Cornell, P.M. Coutinho, E.G. Danchin, A.J. Debets, P. Dekker, P.W. van Dijck, A. van Dijk, L. Dijkhuizen, A.J. Driessen, C. d'Enfert, S. Geysens, C. Goosen, G.S. Groot, P.W. de Groot, T. Guillemette, B. Henrissat, M. Herweijer, J.P. van den Hombergh, C.A. van den Hondel, R.T. van der Heijden, R.M. van der Kaaij, F.M. Klis, H.J. Kools, C.P. Kubicek, P.A. van Kuyk, J. Lauber, X. Lu, M.J. van der Maarel, R. Meulenberg, H. Menke, M.A. Mortimer, J. Nielsen, S.G. Oliver, M. Olsthoorn, K. Pal, N.N. van Peij, A.F. Ram, U. Rinas, J.A. Roubos, C.M. Sagt, M. Schmoll, J. Sun, D. Ussery, J. Varga, W. Vervecken, P.J. van de Vondervoort, H. Wedler, H.A. Wosten, A.P. Zeng, AJ. van Ooyen, J. Visser and H. Stam, 2007. Genome sequencing and analysis of the versatile cell factory *Aspergillus niger* CBS 513.88. Nature Biotechnology 25: 221-231.

Perrone, G., A. Susca, G. Stea and G. Mule, 2004. PCR assay for identification of *Aspergillus carbonarius* and *Aspergillus japonicus, European Journal of Plant Pathology* 110: 641-649.

Perrone, G., G. Mulè, A. Susca, P. Battilani, A. Pietri and A. Logrieco, 2006a. Ochratoxin A production and AFLP analysis of *Aspergillus carbonarius, Aspergillus tubingensis,* and *Aspergillus niger* strains isolated from grapes in Italy. Applied and Environmental Microbiology 72: 680-685.

Perrone, G., A. Susca, F. Epifani and G. Mulè, 2006b. AFLP characterization of Southern Europe population of *Aspergillus* Section *Nigri* from grapes. International Journal of Food Microbiology 111 S1: S22-S27.

Perrone, G., G. Cozzi, M. Pascale, A. Logrieco and A. Visconti, 2007a. *Lobesia botrana* - an ochratoxin A risk factor in grapevine management. XII International IUPAC Symposium on Mycotoxins and Phycotoxins; 2007 May, 21-25 Istanbul. Poster 1423.

Perrone, G., A. Susca, F. Epifani, G. Mulè and A. Logrieco, 2007b. Biodiversity of *Aspergillus* Section *Nigri* from grapes in Europe. Book of Abstracts of the Aspergillus International Workshop: '*Aspergillus* systematics in the genomic era'; 2007 April 12-14; CBS, Utrecht, The Netherlands, 39-41.

Perrone, G., J. Varga, A. Susca, J.C. Frisvad, G. Stea, S. Kocsubé, B. Tóth, Z. Kozakiewicz and R.A. Samson, 2007c. *Aspergillus uvarum* sp. nov., an uniseriate black Aspergillus species isolated from grapes in Europe. International Journal of Systematic and Evolutionary Microbiology (in press).

Pitt, J.I., 2001. Toxigenic fungi: which are important? Medical Mycology 38 (Suppl): 17-22.

Ponsone, M.L., M. Combina, A. Dalcero and S. Chulze, 2007. Ochratoxin A and ochratoxigenic *Aspergillus* species in Argentinian wine trapes cultivated under organic and non-organic systems. International Journal of Food Microbiology 114: 131-135.

Romero, S.M., R.M. Comerio, G. Larumbe, A. Ritieni, G. Vaamonde and V. Fernández Pinto, 2005. Toxigenic fungi isolated from dried vine fruits in Argentina. International Journal of Food Microbiology 104: 43-49.

Rousseau J. 2004. Ochratoxin A in wines: current knowledge. Vinidea.net Wine Internet Technical Journal (5) http://www.infowine.com/default.asp?scheda=1038

Sage, L., S. Krivobok, É. Delbos, F. Seigle-Murandi and E.E. Creppy, 2002. Fungal flora and ochratoxin A production in grapes and musts from France. Journal of Agricultural and Food Chemistry 50: 1306-1311.

Sage, L., D. Garon and F. Seigle-Murandi, 2004. Fungal microflora and ochratoxin A risk in French vineyards. Journal of Agricultural and Food Chemistry 52: 5764-5768.

Sartori, D., M.C. Furlaneto, M.K: Martins, M.R. Ferriera de Paula, A.A. Pizzirani-Kleiner, M.H: Taniwaki and M.H. Pellegrinelli Fungaro, 2006. PCR method for detection of potential ochratoxin-producing *Aspergillus* species in coffee beans. Research in Microbiology 157: 350-354.

Savino, M., P. Limosani and E. Garcia-Moruno, 2007. Reduction of ochratoxin A contamination in red wines by oak wood fragments. American Journal of Enology and Viticulture 58: 97-101.

Schmidt, H., M. Bannier, R. F. Vogel and L. Niessen, 2004a. Detection and quantification of *Aspergillus ochraceus* in green coffee by PCR. Letters in Applied Microbiology 38: 464.

Schmidt, H., M.H. Taniwaki, R.F. Vogel and l. Niessen, 2004b. Utilization of AFLP markers for PCR-based identification of *Aspergillus carbonarius* and indication of its presence in green coffee samples. Journal of Applied Microbiology 97: 899-909.

Schmidt-Heydt, M. and R. Geisen, 2007. A microarray for monitoring the production of mycotoxins in food. International Journal of Food Microbiology 117: 131-140.

Schuller, P.L., T. Ockhuizen, J. Werringloer and P. Marquardt, 1967. Aflatoxin B1 and histamine in wine. Arzneimittelforschung 17: 888-890 [Article in German].

Scott, P.M., T. Fuleki and J. Harvig, 1977. Patulin content of juice and wine produced from moldy grapes. Journal of Agricultural and Food Chemistry 25: 434-437.

Serra, R., L. Abrunhosa, Z. Kozakiewicz and A. Venâncio, 2003. Black *Aspergillus* species as ochratoxin A producers in Portuguese wine grapes. International Journal of Food Microbiology 88: 63-68.

Serra, R., J. Cabanes, G. Perrone, Z. Kozakiewicz, G. Castellá, A. Venâncio and G. Mulè, 2006a. *Aspergillus ibericus:* a new species of the Section *Nigri* isolated from grapes. Mycologia 98 (2): 295-306.

Serra, R., C. Mendonça and A. Venâncio, 2006b. Ochratoxin A occurrence and formation in Portuguese wine grapes at various stages of maturation. International Journal of Food Microbiology 111 S1: S35-S39.

Solfrizzo, M., G. Panzarini and A. Visconti, 2007. Fate of ochratoxin A during vinification of naturally contaminated primitivo and negroamaro grapes. XIIth International IUPAC Symposium on Mycotoxins and Phycotoxins. May 21-25, 2007, Istanbul, Turkey, p. 1425.

Sripathomswat, N. and P. Thasnakorn, 1981. Survey of aflatoxin-producing fungi in certain fermented foods and beverages in Thailand. Mycopathologia 73: 83-88.

Steyn, P.S. and C.W. Holzapfel, 1967. The isolation of the methyl and ethyl esters of ochratoxins A and B, metabolites of *Aspergillus ochraceus* Wilh. Journal of the South African Chemical Institute 20: 186-189.

Susca, A., G. Stea, G. Mulè and G. Perrone, 2007a. Polymerase chain reaction (PCR) identification of *Aspergillus niger* and *Aspergillus carbonarius* based on the calmodulin gene. Food Additives and Contaminants 24: 1154-1160.

Susca, A., G. Stea and G. Perrone, 2007b. Rapid polymerase chain reaction (PCR)-single-stranded conformational polymorphism (SSCP) screening method for the identification of *Aspergillus* section *Nigri* species by the detection of calmodulin nucleotide variations. Food Additives and Contaminants 24: 1148-1153.

Tjamos, S.E., P.P. Antoniou, A. Kazantzidou, D.F. Antonopoulos, I. Papageorgiou and E.C. Tjamos, 2004. *Aspergillus niger* and *Aspergillus carbonarius* in Corinth raisin and wine-producing vineyards in Greece: population composition, ochratoxin A production and chemical control. Journal of Phytopatology 152: 250-255.

Tjamos, S.E., P.P. Antoniou and E.C. Tjamos, 2006. Aspergillus spp., distribution, population composition and ochratoxin A production in wine producing vineyards in Greece. International Journal of Food Microbiology 111 S1: S61-S66.

Turner, W.B. and D. C. Aldridge, 1983. Fungal metabolites II. Academic Press Ltd., London, UK

Valero, A., S. Marín, A.J. Ramos and V. Sanchis, 2005. Ochratoxin A-producing species in grapes and sun-dried grapes and their relation to ecophysiological factors. Letters in Applied Microbiology 41: 196-201.

Varga, J., F. Kevei, C. Fekete, A. Coenen, Z. Kozakiewicz and J.H. Croft, 1993. Restriction fragment length polymorphisms in the mitochondrial DNAs of the *Aspergillus niger* aggregate. Mycological Research 97: 1207-1212.

Varga, J., F. Kevei, F. Debets, Z. Kozakiewicz and J.H. Croft, 1994. Mitochondrial DNA restriction fragment length polymorphisms in field isolates of the *Aspergillus niger* aggregate. *Canadian* Journal of Microbiology 40: 612-621.

Varga, J., E. Kevei, E. Rinyu, J. Téren and Z. Kozakiewicz, 1996. Ochratoxin production by *Aspergillus* species. Applied and Environmental Microbiology 62: 4461-4464.

Varga, J., F. Kevei, Z. Hamari, B. Tóth, J. Téren, J.H. Croft and Z. Kozakiewicz, 2000a. Genotypic and phenotypic variability among black Aspergilli. In: R.A. Samson and J.I. Pitt (eds.) *Integration of modern taxonomic methods for Penicillium and Aspergillus classification.* Harwood Academic Publishers, Amsterda, pp. 397-411.

Varga, J., E. Kevei, A. Palágyi, B. Tóth and Z. Kozakiewicz, 2000b. Analysis of genetic variability within the *Petromyces* genus. Antonie van Leeuwenhoek 77: 83-89.

Varga, J., K. Rigo, S. Kocsube, B. Farkas and K. Pal, 2003a. Diversity of polyketide synthase gene sequences in *Aspergillus* species. Research in Microbiology 154: 593-600.

Varga, J., K. Rigó, B. Tóth, J. Téren and Z. Kozakiewicz, 2003b. Evolutionary relationships among *Aspergillus* species producing economically important mycotoxins. Food Technology and Biotechnology 41: 29-36.

Varga, J., A. Juhász, F. Kevei and Z. Kozakiewicz, 2004. Molecular diversity of agriculturally important *Aspergillus* species. European Journal of Plant Pathology 110: 627-640.

Varga, J., R. Kiss, T. Mátrai, T. Mátrai and J. Téren, 2005. Detection of ochratoxin A in Hungarian wines and beers. Acta Alimentaria 34: 381-392.

Varga, J., S. Kocsubé, B. Tóth, J.C Frisvad, G. Perrone, A. Susca, M. Meijer and R.A. Samson, 2007a. *Aspergillus brasiliensis* sp. nov., a biseriate black Aspergillus species with world-wide distribution. International Journal of Systematic and Evolutionary Microbioliology 57: 1925-1932.

Varga, J., Z. Koncz, S. Kocsubé, T. Mátrai, J. Téren, V. Ostry, J. Skarkova, J. Ruprich, A. Kubatova and Z. Kozakiewicz, 2007b. Mycobiota of grapes collected in Hungarian and Czech vineyards in 2004. Acta Alimentaria 36: 329-341.

Visconti, A., G. Perrone, G. Cozzi and M. Solfrizzo, 2007. Managing ochratoxin A risk in the grape-wine food chain. Food Additives and Contaminants (in press).

Voetz, M. (2002). Development of specific DNA markers for characterization and quantification of ochratoxin forming mould fungi. Brau Forum 17: 7-9.

Wei, R-D., F.M. Strong and E.B. Smalley, 1971. Incorporation of chlorine-36 into ochratoxin A. Applied Microbiology 22: 276-277.

White, T.J., T. Bruns, S. Lee and J. Taylor, 1990. Amplification and direct sequencing of fungal ribosomal RNA genes for phylogenetics. In:D.H. Gelfand, J.J. Sninsky and T.J. White (Eds.) PCR protocols: A Guide to Methods and Applications. Academic Press, San Diego, USA, pp.315-322.

Xiao, H., R.R. Marquardt, A.A. Frohlich AND Y.Z. Ling, 1995. Synthesis and structural elucidation of analogs of ochratoxin A. Journal of Agriculture and Food Chemistry 43: 524-530.

Zimmerli, B. and R. Dick, 1996. Ochratoxin A in table wine and grape juice: occurrence and risk assessment. Food Additives and Contaminants 13: 655-668.

Aspergilli and ochratoxin A in coffee

Paramee Noonim[1,2], Warapa Mahakarnchanakul[2], János Varga[1,3] and Robert A. Samson[1]
[1]CBS Fungal Biodiversity Centre, Uppsalalaan 8, 3584 CT Utrecht, The Netherlands
[2]Department of Food Science and Technology, Agro-Industry Faculty, Kasetsart University, 10900 Bangkok, Thailand
[3]Department of Microbiology, Faculty of Sciences, University of Szeged, H-6701 Szeged, P.O. Box 533, Hungary

Abstract

Ochratoxin contamination of coffee can be regarded as a post-harvest problem. The diversity and ochratoxin producing potential of Aspergilli in coffee beans depends on a combination of various factors including coffee variety, geographic region, climate, susceptibility to insect pests and processing method. According to recent surveys, the most important ochratoxin producers are *A. westerdijkiae* and *A. niger* in Arabica coffee beans, and *A. carbonarius* in Robusta beans. Optimum water activities and temperatures for growth and ochratoxin production of *Aspergillus* species were found to be at a_w 0.95-0.99 at 24-31 °C. Some chemicals (caffeine, chlorogenic acid) have inhibiting effect on growth and toxin production of *Aspergillus* species in coffee. During storage, water activities above 0.80 pose a risk for fungal growth and ochratoxin contamination until roasting. In order prevent the risk of ochratoxin contamination, this critical period need to be minimised to appropriate levels to prevent growth and toxin production. The choice of coffee processing affects ochratoxin content in coffee, while roasting can reduce ochratoxin levels by 30-96% depending on temperature and time of roasting. Coffee contributes only to a small extent to ochratoxin intake of humans in Europe, even if it is prepared from relatively highly contaminated green beans.

Keywords: *Aspergillus*, coffee, ochratoxin A, roasting, processing

1. Introduction

Ochratoxin A (OTA), one of the most extensively studied mycotoxins, which contaminates different plant products (Varga *et al.*, 2001). Ochratoxins are cyclic pentaketids, dihydroisocoumarin derivatives linked to an L-phenylalanine moiety. The mycotoxin was first discovered in 1965 from an *Aspergillus ochraceus* isolate (Van der Merwe *et al.*, 1965). Since then, several *Aspergillus* and *Penicillium* species have been described as producers of this mycotoxin. OTA showed nephrotoxic, immunosuppressive, teratogenic and carcinogenic properties (Holzhauser *et al.*, 2003). Several nephropathies affecting animals as well as humans have been attributed to OTA; e.g. this mycotoxin is the aetiological agent of Danish porcine nephropathy, and renal disorders observed in other animals (Smith and Moss, 1985). In humans, OTA is frequently cited as the possible causative agent of Balkan endemic nephropathy, a syndrome characterised by contracted kidneys with tubular degeneration, interstitial fibrosis and hyalinisation of glomeruli (Krogh *et al.*, 1977). Recently, OTA has been suggested to play a role in chronic karyomegalic interstitial nephropathy and chronic interstitial nephropathy in Tunisia (Maaroufi *et al.*, 1999), urothelial tumours (end stage renal disease) in Egypt (Wafa *et al.*, 1998), and testicular cancer (Schwartz, 2002).

OTA has also been classified as 'possibly carcinogenic to humans' (group 2B) by the International Agency for Research in Cancer (IARC) because of the confirmed effect in animals but no concrete evidence in humans (IARC, 1993; JECFA, 2001). It is naturally found in various foods products, including cereals and cereal products (Molinie *et al.*, 2004; Araguás *et al.*, 2005; Zinedine *et al.*, 2006, 2007; Juan *et al.*, 2007), coffee beans and coffee products (Taniwaki *et al.*, 2003; Leong *et al.*, 2007), raisins and wine (Blesa *et al.*, 2004; Cabanes *et al.*, 2002), cocoa and cocoa products (Tafuri *et al.*, 2004), spices (Thirumala-Devi *et al.*, 2001), dried fruits (Iamanaka *et al.*, 2006) and milk (Breitholtz-Emanuelsson, 1993).

The presence of OTA in the coffee beverage was discovered by Tsubouchi *et al.* (1988). The perception that coffee may be an important source of OTA drew attention from many researchers worldwide and many countries have established regulatory limits for OTA. Moreover, pressure from European Union as they also established maximum limit of OTA in roasted coffee and soluble coffee led to much work on occurrence of OTA in coffee bean worldwide, its formation, destruction during roasting, and presence in coffee beans and products.

In this review we will summarise our knowledge on the occurrence of OTA in coffee, the producing organisms, factors affecting OTA contamination of coffee, and effect of coffee processing on OTA levels.

2. Coffee varieties and processing

Coffee is one of the world's most important primary commodities. With over 500 billion cups consumed every year, coffee is one of the world's most popular beverages, comprising about a third of tap water consumption. Worldwide, 25 million small producers rely on coffee for a living. For instance, in Brazil, where almost a third of the world's coffee is produced, over 5 million people are employed in the cultivation and harvesting of over 3 billion coffee plants (FAO, 2007). In 2006, 7.8 million metric tons of coffee have been produced, with Brazil alone producing 2.59 million tons of it (FAOSTAT, 2007). All other countries produced less than 1 million tons, including Vietnam (0.85 million tons), Colombia (0.70 million tons) and Indonesia (0.65 million tons), and several other countries located in Central and South America, Africa and South-East Asia.

Coffee belongs to the botanical family Rubiaceae, which has some 500 genera and over 6,000 species (Bremer *et al.,* 1999). There are 2 major coffee varieties grown and traded widely for consumption, *Coffea arabica* and *Coffea canephora* var. *robusta.* Key differences of these two varieties are listed in Table 1. A third species, *Coffea liberica* is also cultivated in some parts of the world including Malaysia and West Africa, but only very small quantities are traded (Clifford and Wilson, 1985). Proportions of each variety in world production are approximately 60-70% and 30-40% for Arabica and Robusta coffee, respectively (FAO, 2007). Coffee is normally harvested once a year. Methods for harvesting vary from farm to farm. There are 3 harvesting methods generally used: selective picking, stripping and mechanical harvesting. Quality, cost efficiency and prevention of OTA contamination are factors in determining to optimise and select the best method for each farm.

Table 1. Differences between the two most important coffee varieties, Arabica and Robusta (adapted from ICO, 2007).

Characters	**Arabica**	**Robusta**
Chromosomes (2n)	44	22
Yield (kg beans ha^{-1})	1500-3000	2300-4000
Soil requirements	Fertile soil	Poorer soil
Optimum temperature (yearly average)	15-24 °C	24-30 °C
Optimal rainfall	1500-2000 mm	2000-3000 mm
Optimum altitude for growth	1000-2000 m	0-700 m
Coffee berry disease	Susceptible	Resistant
Caffeine content of beans	0.8-1.4%	1.7-4.0%
Typical brew characteristics	Acid, fuller flavour	Bitter, weaker flavour
Body	Average = 1.2%	Average = 2.0%

The duration of invasion of ochratoxigenic fungi to coffee is significant, in order to follow the outcome of OTA throughout coffee production process and to set up control strategies. Mantle (2000) reported the possibility of trace OTA uptake of coffee tree from soil, but other researchers concluded that OTA contamination in coffee can be best regarded as a post-harvest problem. Consequently, OTA problem can be prevented and minimised.

For coffee processing, the wet process is applied for Arabica coffee while for Robusta coffee a simple-robust dry method is used. In each type of processing method, different possibilities of fungal contamination are involved. For the dry method, cherries are sun dried directly or after passing through water sorting. Bucheli *et al.* (2000) suggested that sun drying of Robusta coffee cherries led to OTA formation in the pulp and parchment of the cherries. The maturity of the cherries also influenced OTA contamination whereas green cherries being the least and overripe cherries the most susceptible.

Hulling or husk removal caused significant OTA reduction because of the removal of toxin-rich outer layer (Saurez-Quiroz *et al.*, 2005). For the wet method, 3 steps are involved: pulping, fermentation followed by washing and drying. Fermentation and removal of mucilage can either cause some extent of decontamination of OTA or cause penetration of OTA and/or OTA producing fungi into deeper zones of the bean. It was suggested that a physical wet method devoid of a fermentation step is the best way to prevent risk of OTA contamination (Saurez-Quiroz *et al.*, 2005).

3. Ochratoxin A contamination in coffee

Various methods can be used for OTA detection. For screening, ELISA is a good rapid and reliable method but with a limited level of detection. In order to confirm the quantitative results, immunoaffinity column-clean up followed by HPLC is recommended and generally used. The occurrence of OTA in coffee bean has been reported for many decades, first by Levi *et al.* (1974) and confirmed by several other reports from all over the world. For green coffee bean, OTA has been reported by several authors in wide concentration ranges from trace to 360 μg / kg (Table 2). High variability could be explained by the non-uniform distribution of OTA within samples. Lower concentration ranges were observed in roasted coffee or soluble coffee indicating that OTA contamination in coffee is a relatively minor source of OTA intake in humans. Regarding the correlation between origin and OTA content of coffee samples, relatively high OTA levels have been detected in some African coffee producing countries (ECF, 2005). Statistical analyses confirmed that some countries have more frequent problems than others and that these countries are spread around the globe.

Table 2. Occurrence of OTA in green coffee, roasted and soluble coffee.

Origin	Coffee type[1]	Range of OTA (µg / kg)	References
Angola	GC, NS	<20	Levi *et al.* (1974)
Brazil	"	Trace-360	"
Cameroon	"	<20	"
Columbia	"	Trace-50	"
Brazil	GC, Arabica	0.2-3.7	Micco *et al.* (1989)
Cameroon	GC, Robusta	Trace-2.2	"
Columbia	GC, Arabica	3.3	"
Zaire	GC, Robusta	8.4-15.0	"
Brazil	GC, NS	2.0-7.4	Studer-Rhor *et al.* (1995)
Columbia	"	1.2-9.8	"
Ivory Coast	"	9.9-56	"
Unknown	"	2.2-11.8	"
Yemen	GC, Arabica	0.7-17.4	Nakajima *et al.* (1997)
Tanzania	GC, Arabica	0.1-7.2	"
Indonesia	GC, Robusta	0.2-1.0	"
Central America	GC, Arabica	<0.1	"
South America	GC, Arabica	<0.1	"
East Africa	GC, NS	0.2-62.0	Heilmann *et al.* (1999)
West Africa	"	0.3-5.0	"
Asia	"	0.2-4.9	"
America	"	0.1-4.9	"
Africa	GC, Arabica	0.5-48.0	Romani *et al.* (2000)
Brazil	GC, Arabica	0.7-47.8	Leoni *et al.* (2000)
Brazil	GC, Arabica	0.47-4.82	Batista *et al.* (2003)
Brazil	GC, Arabica	0.2-7.3	Martins *et al.* (2003)
Brazil	GC, Arabica and Robusta	<0.16-6.24	Gollucke *et al.* (2004)
Africa	GC, Robusta	2.4-23.3	Pardo *et al.* (2004)
Vietnam	GC, Arabica	Trace-1.8	Leong *et al.* (2007)
Japan	RC, Arabica	3.2-17.0	Tsubouchi *et al.* (1988)
UK	RC, NS	0.2-2.1	Patel *et al.* (1997)
Europe	RC, NS	0.2-2.1	Van der Stegen *et al.* (1997)
Denmark	RC, NS	0.1-3.2	Jorgensen (1998)
Brazil	RC, NS	0.3-6.5	Leoni *et al.* (2000)
Brazil	RC, NS	0.99-5.87	Prado *et al.* (2000)
Germany	RC, NS	0.2-12.1	Ottender and Majerus (2001)

Table 2. Continued.

Origin	Coffee type[1]	Range of OTA (µg / kg)	References
Canada	RC, NS	0.1-2.3	Lombaert *et al.* (2002)
Australia	SC, NS	0.2-4.0	Pittet *et al.* (1996)
USA	"	1.52.1	"
Germany	"	0.3-2.2	"
UK	SC, NS	0.1-8.0	Patel *et al.* (1997)
Europe	SC, NS	<0.5-27.2	Van der Stegen *et al.* (1997)
Spain	SC, NS	0.31-0.1.78	Prado *et al.* (2000)
Brazil	SC, NS	0.5-5.1	Leoni *et al.* (2000)
Germany	SC, NS	0.28-4.8	Ottender and Majerus (2001)
Canada	SC, NS	0.1-3.1	Lombaert *et al.* (2002)
Brazil	SC, NS	0.17-6.29	De Almeida *et al.* (2007)

[1]NS = Not specified; GC = Green coffee; RC = Roasted coffee; SC = Soluble coffee.

4. Ochratoxin A producing fungi in coffee

Many studies concerning the occurrence OTA producing fungi in coffee beans have been published. However, conventional morphological methods for fungal identification were usually applied and consequently some of the responsible moulds were misidentified.

The sources of OTA contamination in coffee are species of *Aspergillus* section *Circumdati* and *Nigri*. One of the well-known OTA producers is *A. ochraceus*, a species belonging to section *Circumdati*, and which the toxin was named after (Van der Merwe *et al.*, 1965). In several studies, *A. ochraceus* was isolated and identified from coffee beans (Urbano *et al.*, 2001; Batista *et al.*, 2003; De Moraes *et al.*, 2003; Martins *et al.*, 2003; Saurez-Quiroz *et al.*, 2004; Taniwaki *et al.*, 2003). For many years, *A. ochraceus* was suggested as the only source of OTA in coffee (Frank, 1999), but since the taxonomy of the yellow Aspergilli was revised and *A. westerdijkiae* was recognised as a distinct species (Frisvad *et al.*, 2004), this conclusion should be revised. During a study of the mycobiota of Thai coffee beans using a polyphasic identification (P. Noonim, unpublished data), *A. ochraceus* was not detected in any samples. The most prevalent ochratoxigenic yellow *Aspergillus* species found in Arabica coffee was *A. westerdijkiae*. Recently, Morello *et al.* (2007) revealed that in former a study, most of the *A. westerdijkiae* (84%) isolates came from Brazilian coffee beans were misidentified as *A. ochraceus*. So it is likely that of the yellow Aspergilli *A. westerdijkiae* is the real source of OTA in coffee.

Other *Aspergillus* species from section *Circumdati* were also reported as OTA producers, for example, *A. melleus* (Batista *et al.*, 2003). Apart from *A. westerdijkiae*, Frisvad *et al.* (2004) also indicated that seven species in section *Circumdati* consistently produce large amounts of OTA including *A. cretensis*, *A. flocculosus*, *A. pseudoelegans*, *A. roseoglobulosus*, *A. sulphureus*, and *Neopetromyces muricatus*. Two other species, *A. ochraceus* and *A. sclerotiorum*, produce OTA, but less consistently. However, among the yellow *Aspergillus* isolated from Thai coffee beans, *A. steynii* was the best OTA producer compared to *A. westerdijkiae*, *A. sclerotiorum* and *A. melleus* (P. Noonim, unpublished data). Two other yellow-spored species, *Petromyces alliaceus* and *P. albertensis* are also able to produce OTA, but these species belong to *Aspergillus* section *Flavi* (Bayman *et al.*, 2002; Varga *et al.*, 1996; Rigó *et al.*, 2002) and are not found on coffee.

Aspergillus species assigned to section *Nigri* are also common in coffee beans. According to Taniwaki *et al.* (2003), *A. niger* was the most commonly occurring species in Brazilian green coffee, which was confirmed by Martins *et al.* (2003) and Leong *et al.* (2007). The last authors also reported *A. niger* as the predominant species with 83.3% and 89% incidence in Vietnamese coffee beans. *A. niger* has poor ochratoxigenic ability as confirmed by various studies. However there was inconsistency in the percent of toxigenicity reported. Heenan *et al.* (1998) and Taniwaki *et al.* (2003) indicated that only 2-3% of *A. niger* isolates isolated from coffee beans could produce ochratoxins, while 11-13% of *A. niger* isolates were toxigenic as found in a present study of Thai coffee beans (P. Noonim, unpublished data). Moreover, *A. niger* could produce both OTA and OTB.

A. carbonarius was also reported in coffee beans (Bucheli *et al.*, 2000; Joosten *et al.*, 2001; Taniwaki *et al.*, 2003; Leong *et al.*, 2007; Taniwaki, 2006). High percentage if not all isolates of *A. carbonarius* had abilities to produce OTA. Joosten *et al.* (2001) reported that all *A. carbonarius* isolates came from Robusta Thai coffee bean produced significant amounts of OTA, in agreement with a more recent study (P. Noonim, unpublished data). Pardo *et al.* (2004) also found that all *A. carbonarius* isolates came from coffee beans from various countries produced OTA, and Leong *et al.* (2007) observed that almost all (110/113) of the examined *A. carbonarius* isolates came from Vietnamese coffee beans could produce OTA. However, Taniwaki *et al.* (2003) observed that only 77% of the *A. carbonarius* isolates came from Brazilian coffee beans produced OTA. Differences in the ratio of *A. carbonarius* isolates able to produce OTA could be due to misidentification of the non-OTA producer *A. ibericus* as *A. carbonarius* in previous studies (Serra *et al.,* 2006). Due to its high ochratoxigenic potential, *A. carbonarius* can be considered as another important source of OTA contamination besides *A. westerdijkiae.* Considering the incidence of *A. carbonarius* in coffee, this species occurred less frequently in Arabica than in Robusta coffee beans (Leong *et al.,* 2007). During our survey of Thai coffee beans, *A. carbonarius* was not detected in Arabica coffee but was the most predominant species (35%) found in Robusta coffee beans which confirmed the data reported by Joosten *et al.* (2001). Leong *et al.* (2007) also detected *A. carbonarius* (12-14%) in Vietnamese Robusta coffee beans. Robusta coffee is normally grown at low altitudes with higher temperatures, which are favourable growth conditions for *A. carbonarius.* Besides, the use of the dry processing method, surface contact with dust and soil increase the chance of *A. carbonarius* contamination. In contrast, for Arabica coffee beans the incidence of *A. carbonarius* was 0-6% in Brazil (Batista *et al.,* 2003; Martins *et al.,* 2003; Taniwaki *et al.,* 2003). It can be concluded that *A. carbonarius* is an important source of OTA contamination in Robusta coffee, while the importance of *A. carbonarius* in OTA contamination of Arabica coffee beans is less.

Recently, four new species within the black aspergilli were found in coffee beans: *A. sclerotioniger, A. lacticoffeatus, A. sclerotiicarbonarius,* and *A. aculeatinus* (Samson *et al.,* 2004; Noonim *et al.,* 2008). *A. sclerotioniger* and *A. lacticoffeatus* able to produce OTA, but these species are rare and probably do not pose a serious health hazard to humans.

OTA producing Penicillia (*P. verrucosum* and *P. nordicum*) have not been isolated from coffee. Recently, OTA production has been claimed for some endophytes of coffee including one isolate each of *P. brevicompactum, P. crustosum, P. olsonii, and P. oxalicum* (Vega *et al.,* 2006a), and for some grape-derived isolates related to *P. radicum* and *P. rugulosum* (Torelli *et al.,* 2006). These studies need further experiments, because OTA have never been confirmed in these fungi.

Recently, several PCR assays have been developed in order to detect important ochratoxigenic species, for example, *A. carbonarius* (Patiňo *et al.,* 2005), *A. westerdijkiae* (Morello *et al.,* 2007), *A. ochraceus* (Patiňo *et al.,* 2005), by using different species specific primers designed based on ITS, calmodulin or β-tubulin sequences. These PCR assays are sensitive and specific, and can be used for rapid detection of OTA-producing *Aspergillus* species in coffee beans.

5. Factors affecting ochratoxin production by *Aspergillus* spp. in coffee beans

Environmental conditions such as moisture content, temperature, storage time and the nature of the substrate play an important role in fungal colonisation and the amount of OTA produced. Pardo *et al.* (2006) evaluated the significance of the origin of *A. ochraceus* isolates on OTA production. Results indicated no difference in OTA-production on different substrates. In contrast, abiotic assays revealed that water activity, temperature and substrate significantly affected OTA production. Kouadio *et al.* (2007) studied the effect of temperature and water activity (a_w) on growth and OTA production of three ochratoxigenic species, *A. niger*, *A. carbonarius* and *A. ochraceus*. These three species had optimal growth at 30 °C and a_w of 0.99. However, conditions for optimal OTA production varied, with a_w being a limiting factor. Suarez-Quiroz *et al.* (2004) also investigated effect of environmental factors on OTA production of *A. ochraceus*, with similar results; a_w was found to play an important role in growth and OTA production of *A. ochraceus*, and temperature affected the rate of OTA production at appropriate a_w. Optimum condition was 0.95 a_w and at 35 °C for OTA production. A critical stage in the process was drying. Similar results were also reported in other studies. In conclusion, ochratoxigenic species have optimal growth and OTA production in the range of a_w 0.95-0.99 at temperatures of 30-35 °C.

Water condensation that leads to remoistening of coffee beans can be an important cause of fungal growth and OTA contamination. Palacios-Cabrera *et al.* (2004) reported the effect of alternating temperatures during storage of coffee during a 2-month period. No OTA production was observed at 80% relative humidity. OTA production was higher under alternating temperatures than at constant temperature, possibly due to condensation that led to rapid increase of moisture content and a_w of coffee.

As water activity is the limiting factor for fungal development and OTA contamination, practical drying conditions should be tested in the field in order to reduce a_w to the level below the critical point. Mburu (1999) studied the variation in a_w during traditional sun drying of coffee, and found that a_w decreased slowly and coffee beans were in appropriate condition for fungal growth and OTA production for 5-7 days between a_w 0.85-0.99 and 2-4 days at a_w 0.95. These critical days need to be minimised.

Impact of storage on fungal growth on industrial Robusta green coffee beans was reported by Bucheli *et al.* (1998). Bag storage and silo storage under air-conditioning, aeration and non-aeration have been used for investigation. Overall fungal count decreased after storage and no fungal growth or OTA production were observed. OTA contamination appeared to have occurred before storage. From this research it can be concluded that storage under humid tropical conditions is safe enough if water activity of coffee bean is below the critical limit and there is no remoistening.

Saurez-Quiroz *et al.* (2004) also studied effects of some chemical compounds naturally found in coffee beans and the effect on development and OTA production of *A. ochraceus*. These compounds were caffeine (1,3,7-trimethylxanthine) and chlorogenic acids (compounds mainly formed by quinic esterification with either caffeic, ferulic or p-coumaric acids). The amount of these compounds vary in different coffee varieties, but the post-harvest process normally used did not significantly affected that. The chlorogenic acid content had little effect on the growth rate of *A. ochraceus*, but an increase in caffeine content did reduce the growth rate of this fungus. Both types of compounds acted in the same way by reducing OTA production. At 2.0 mg ml^{-1} of caffeine and 8.0 mg ml^{-1} of chlorogenic acids, OTA concentration became almost zero. However, OTA production also took place in coffee beans containing potentially inhibitory concentrations.

Another factor contributing to OTA contamination is insect damage. Vega and Mercadier (1998) found that the coffee berry borer *Hypothenemus hampei* serve as vector for *A. ochraceus*. Pérez *et al.* (2003) could also detect potentially ochratoxigenic fungi including *A. ochraceus*, *A. sclerotiorum* and *A. niger* associated with the coffee berry borer. Later, Vega *et al.* (2006b) detected *A. westerdijkiae* on the cuticle of an insect parasitoid, *Prorops nasuta,* which has been introduced from Africa to many coffee-producing countries in an attempt to control the coffee berry borer. The authors suggested that this insect parasitoid might be disseminating an ochratoxin-producing fungus in coffee plantations.

6. Reduction of ochratoxin A in coffee processing

The effect of cleaning and roasting on OTA concentration in coffee beans has extensively been studied, with variable results. One of the most detailed studies was carried out by Blanc *et al.* (1998) who studied the behaviour of OTA throughout the production chain, during green coffee roasting and soluble coffee manufacturer. In the initial cleaning step, only a small proportion of OTA was removed. Since the highest amount of OTA is usually concentrated in husks, the dehulling process used to remove the husks can reduce OTA content of coffee beans. The OTA concentration in the discarded fraction was high; the dust comprised less than 1% of the weight of the cleaned coffee. Sorting with colour sorters resulted in some reduction, and steaming caused a mean 25% reduction. Decaffeination with solvents is an effective process, resulting in 92% reduction (Heilmann *et al.*, 1999).

Coffee roasting is also one of the effective measures for OTA destruction. Several reports on the effect of roasting on the concentration of OTA are summarised in Table 3. Different temperature and time for roasting affected the effectiveness of OTA reduction. The highest OTA reduction was seen for the combination of long time and high temperature (dark) roasting.

A study of Micco *et al.* (1989) was performed to evaluate the effect of roasting on OTA contamination of coffee beans. Twenty-nine samples of green coffee showed a significantly high contamination percentage (58%) ranging from 0.2 to 15 µg / kg. Naturally and artificially contaminated samples were roasted at different operation times (5-6 min) to verify the percentage of destruction of the mycotoxin. The percentage ranged from 48% to 87% and from 90% to 100% in artificially and naturally contaminated samples, respectively. Urbano *et al.* (2004) found that roasting at the temperature of 200 °C for 10 minutes reduced OTA content only by 22%. However, increasing the temperature to 220 °C for 15 minutes time period, the OTA content was reduced by up to 94%.

Table 3. OTA reduction during coffee roasting.

Type/Origin	Toxin	Roasting condition	% reduction	References
Unknown/Unknown	artificially	5-6 min	48-87	Micco *et al.* (1989)
"	natural	5-6 min	90-100	"
Unknown/Unknown	natural	250 °C, 150 sec	14-62	Studer-Rohr *et al.*(1995)
"	artificially	250 °C, 150 sec	2-28	"
Robusta/Thailand	natural	223 °C, 4 min	84	Blanc *et al.* (1998)
Robusta/Cote d'Ivoire	natural	Light to dark	69-96	Van de Stegen *et al.* (2001)
Unknown/Italy	natural	175-204 °C, 7-9 min	>90	Romani *et al.* (2003)
Arabica/Brazil	artificially	200 °C, 10-15 min	22.5-48.1	Urbano *et al.* (2004)
"	artificially	210 °C, 10-15 min	39.2-65.6	"
"	artificially	220 °C, 10-15 min	88.4-93.9	"
Robusta/Vietnamese	natural	Industrial roasting	66.5	Pérez de Obanos *et al.* (2005)

Several of these studies mentioned physical removal of OTA with the husk or parchment as a possible mechanism for the observed reduction, but data show that this can only partially be the explanation. Partial isomerisation of OTA can be another possible explanation. However, thermo-instability of OTA under roasting conditions appears to be attributed to OTA reduction. The beverages prepared from artificially contaminated coffee using the most common types of coffee makers showed no residues of OTA. Several studies indicate that the most significant OTA reduction from roasting was observed in the most contaminated samples (Romani *et al.,* 2003; Van de Stegen *et al.,* 2001).

7. Risk of OTA from coffee consumption

The potential danger presented by OTA to humans and animals led many countries to take measures to restrain OTA contamination of foods and feeding stuff. About 100 countries were known to have mycotoxin regulations in 2003 (FAO, 2004). Among these, 37 had regulations for OTA levels in agricultural products. Maximum tolerable level of ochratoxins in roasted coffee beans in the EU is 5 μg / kg and 10 μg / kg for instant (soluble) coffee (Table 4). Furthermore, there are also national laws and regulations in the Member States covering other coffee products not regulated by European law, and some countries and buyers also carry out OTA controls and apply their own limits, especially for coffee. For example, although OTA levels are not regulated at present in green coffee beans by EU laws, several countries set national limits for this product (Table 4).

Following the evaluations for OTA carried out in the nineties, JECFA has established a provisional tolerable weekly intake (PTWI) of 100 ng / kg body weight (b.w.) per week. The mean daily dietary intake of OTA for humans in the European Union was found to be between 1-2 ng / kg b.w. (Hoehler, 1998). More recently, the Panel on Contaminants in the Food Chain of the European Food Safety Authority (EFSA) concluded that according to the estimated exposure levels (2 to 8 ng/kg b.w. per day), exposure to OTA varies between 15 to 60 ng / kg b.w. per week. Based on the evaluation of the toxicity of OTA, the EFSA established a Tolerable Weekly Intake (TWI) of 120 ng/kg b.w. for OTA (EFSA, 2006). The main contributors to OTA intake in the European Union are cereals and cereal products, and to a lesser degree wine, coffee, beer, pork, pulses and spices (Miraglia and Brera, 2002). Van der Stegen *et al.* (1997) screened European coffee final products for occurrence of OTA and also estimated the daily intake of OTA. It was suggested that OTA is fully extracted by the brewing methods used in Europe, and the consumption of four cups of coffee per day contributes to 10-19 ng of OTA intake. This level is far below PTWI as consumption of 28 cups/week contributes only up to 2% of PTWI. From this evidence it seems that

Table 4. Regulatory limits for OTA in different coffee products by various importing countries (adapted from FAO, 2004; ECF, 2005).

Country	Regulatory limit (μg / kg)		
	Green coffee	Roasted coffee	Instant coffee
European Union	-	5	10
Finland	5	5	5
Germany	-	3	6
Greece	20	5	10
Hungary	15	5	10
Italy	8	4	4
Portugal	8	4	4
Spain	8	4	4
Switzerland	5	5	5
Japan	5	-	-

-: no limits set.

coffee is only a relatively minor source of OTA intake in Europe even in Scandinavian countries including Finland where coffee consumption was over 10 kg/person/year in 2003 (ICO, 2007).

8. Conclusions

Ochratoxin contamination of coffee can be regarded as a post-harvest problem. The diversity and ochratoxin producing potential of Aspergilli in coffee beans depends on a combination of various factors including coffee variety, geographic region, climate, susceptibility to insect pathogens and processing method. The most important OTA producers are *A. westerdijkiae* and *A. niger* in Arabica coffee beans, and *A. carbonarius* in Robusta beans. During storage, water activities above 0.80 pose a risk for fungal growth and OTA contamination until roasting. In order prevent the risk of OTA contamination, this critical period need to be minimised to appropriate levels to prevent growth and toxin production. This could be carried out using Good Agricultural Practices (GAPs) by controlling cherry maturation, selection, drying, transportation and storage. Grading and roasting are also effective processes to reduce or eliminate OTA contamination.

References

Araguás, C., E. González-Peñas and A. López de Cerain, 2005. Study on ochratoxin A in cereal-derived products from Spain. Food Chemistry 92: 459-464.

Batista, L.R., S.M. Chalfoun, G. Prado, R.F. Schwan and A.E. Wheals, 2003. Toxigenic fungi associated with processed (green) coffee beans (*Coffea arabica* L.). International Journal of Food Microbiology 85: 293-300.

Bayman, P., J.L. Baker, M.A. Doster, T.J. Michailides and N. Mahoney, 2002. Ochratoxin production by the *Aspergillus ochraceus* group and *Aspergillus alliaceus*. Applied and Environmental Microbiology 68: 2326-2329.

Blanc, M., A. Pittet, R. Munoz-Box and R. Viani, 1998. Behavior of ochratoxin A during green coffee roasting and soluble coffee manufacture. Journal of Agricultural and Food Chemistry 46: 673-675.

Blesa, J., J.M. Soriano, J. Moltó and J. Mañes, 2004. Concentration of ochratoxin A in wines from supermarkets and stores of Valencian Community (Spain). Journal of Chromatography A 1054: 397-401.

Breitholtz-Emanuelsson, A., M. Olsen, A. Oskarsson, I. Palminger and K. Hult, 1993. Ochratoxin A in cow's milk and in human milk with corresponding human blood samples. Journal of AOAC International 76: 842-846.

Bremer, B., R.K. Jansen, B. Oxelman, M. Backlund, H. Lantz and K.J. Kim, 1999. More characters or more taxa for a robust phylogeny - case study from the coffee family (Rubiaceae). Systematic Biology 48: 413-435.

Bucheli, P., C. Kanchanomai, I. Meyer and A. Pittet, 2000. Development of ochratoxin A during robusta (*Coffea canephora*) coffee cherry drying. Journal of Agricultural and Food Chemistry 48: 1358-1362.

Bucheli, P., I. Meyer, A. Pittet, G. Vuataz and R. Viani, 1998. Industrial storage of green robusta coffee under tropical conditions and its impact on raw material quality and ochratoxin A content. Journal of Agricultural and Food Chemistry 46: 4507-4511.

Cabanes, F.J., F. Accensi, M.R. Bragulat, M.L. Abarca, G. Castella, S. Minguez and A. Pons, 2002. What is the source of ochratoxin A in wine? International Journal of Food Microbiology 79: 213-215.

Clifford M.N. and K.C. Willson (eds), 1985. Coffee; botany, biochemistry and production of beans and beverage. Croom Helm, London.

De Almeida, A.P., J. Alaburda, L. Shundo, V. Ruviera, S.A. Navas, L.C.A. Lamardo and M. Sabino, 2007. Ochratoxin A in Brazilian coffee. Brazilian Journal of Microbiology 38: 300-303.

De Moraes, M.H.P. and R.H. Luchese, 2003. Ochratoxin A in coffee: Influence of harvest and drying processing procedures. Journal of Agricultural and Food Chemistry 51: 5824-5828.

ECF (European Coffee Federation), 2005. OTA risk management: Guidelines for green coffee buying (http://www.ecf-coffee.org/files/GuidelinesGreenCoffeeBuyingDef110105.pdf).

EFSA (European Food Safety Authority), 2006. Opinion of the scientific panel on contaminants in the food chain on a request from the commission related to ochratoxin A in food. The EFSA Journal 365: 1-56.

FAO (Food and Agricultural Organization of the United Nations), 2007. Reducing ochratoxin A in coffee (http://www.coffee-ota.org/).

FAO, 2004. Worldwide regulations for mycotoxins in food and feed in 2003. FAO, Rome.

FAOSTAT, 2007. http://faostat.fao.org/site/567/default.aspx.

Frank, J.M, 1999. HACCP and its mycotoxin control potential: ochratoxin A (OTA) in coffee production. In: Proceedings of the 7th International Committee of Food Microbiology and Hygiene, The Netherlands, Veldhoven, pp. 1222-1225.

Frisvad, J.C., J.M. Frank J.A.M.P. Houbraken, A.F.A. Kuijpers and R.A. Samson, 2004. New ochratoxin A producing species of *Aspergillus* section *Circumdati*. Studies in Mycology 50: 23-43.

Gollücke, A.P.B., M.H. Taniwaki and D.Q. Tavares, 2004. Survey of ochratoxin A in Brazilian green coffee destined for exports. Ciência e Tecnologia de Alimentos Campinas 24: 641-645.

Heenan, C.N., K.J. Shaw and J.I. Pitt, 1998. Ochratoxin A production by *Aspergillus carbonarius* and *A. niger* isolates and detection using coconut cream agar. Journal of Food Mycology 1: 67-72.

Heilmann, W., A.G. Rehfeldt and F. Rotzoll, 1999. Behaviour and reduction of ochratoxin A in green coffee beans in response to various processing methods. European Food Research and Technology 209: 297-300.

Hoehler, D., 1998. Ochratoxin A in food and feed: occurrence, legislation and mode of action. Zeitschrift für Ernährungswissenschaft 37: 2-12.

Holzhauser, D., T. Delatour, M. Marin-Kuan, S. Junod, G. Guignard, D. Piguet, J. Richoz, C. Bezencon, B. Schilter and C. Cavin, 2003. Ochratoxin: A toxicity and carcinogenicity. Toxicology Letters 144: s65.

Iamanaka B.T., M.H. Taniwaki, E. Vincente and H.C. Menezes, 2006. Fungi producing ochratoxin in dried fruits. In: A.D. Hocking, J.I. Pitt, R.A. Samson and U. Thrane (eds.) Advances in Food Mycology. Springer, New York, pp. 181-188.

IARC (International Agency for Research on Cancer), 1993. Some naturally occurring substances: items and constituents, heterocyclic aromatic amines and mycotoxins. IARC, Lyon, France.

ICO (International Coffee Organisation), 2007. About coffee: Botanical aspects (http://www.ico.org/botanical.asp).

JECFA (Joint FAO/WHO Expert Committee on Food Additives), 2001. Safety evaluations of certain mycotoxins in food. WHO Food Additives Series 47. World Health Organization, Geneva.

Joosten, H.M.L.J., J. Goetz, A. Pittet, M. Schellenberg and P. Bucheli, 2001. Production of ochratoxin A by *Aspergillus carbonarius* on coffee cherries. International Journal of Food Microbiology 65: 39-44.

Jorgensen, K., 1998. Survey on pork, poultry, coffee, beer and pulses for ochratoxin A. Food Additives and Contaminants 15: 550-554.

Juan, C., J.C. Moltó, C.M. Lino and J. Mañes, 2007. Determination of ochratoxin A in organic and non-organic cereals and cereal products from Spain and Portugal. Food Chemistry (in press)

Kouadio, A. I., A. Lebrihi, G.N. Agbo, F. Mathieu, A. Pfohl-Leszkowicz and M.B. Dosso, 2007. Influence de l'interaction de la température et de l'activité de l'eau sur la croissance at la production de l'ochratoxine A par *Aspergillus niger, Aspergillus carbonarius* et *Aspergillus ochraceus* sur un milieu de base café. Cannadian Journal of Microbiology 53: 852-859.

Krogh, P., B. Hald, R. Plestina and S. Ceovic, 1977. Balkan (endemic) nephropathy and food-borne ochratoxin A: preliminary results of a survey of foodstuff. Acta Pathologica et Microbiologica Scandinavica Section B 85: 238-240.

Leong S.L., L.T. Hien, T.V. An, N.T. Trang, A.D. Hocking and E.S. Scott, 2007. Ochratoxin A-producing Aspergilli in Vietnamese green coffee beans. Letters in Applied Microbiology 45: 301-306.

Leoni, L.A.B., L.M. Valente Soares and P.L.C. Oliveira, 2000. Ochratoxin A in Brazilian roasted and instant coffee. Food Additives and Contaminants 17: 867-870.

Levi, C.P., H.L. Trenk and H.K. Mohr, 1974. Study of the occurrence of ochratoxin A in green coffee beans. Journal of Association of Official Analytical Chemists 57: 866-870.

Lombaert, G.A., P. Pellaers, M. Chettiar, D. Lavalee, P.M. Scott and B.P.Y. Lau, 2002. Survey of Canadian retail coffees for ochratoxin A. Food Additives and Contaminants 19. 869-877.

Maaroufi, K., A. Zakhama, I. Baudrimont, A. Achour, S. Abid, F. Ellouz, S. Dhouib, E.E. Creppy and H. Bacha, 1999. Karyomegaly of tubular cells as early stage marker of the nephrotoxicity induced by ochratoxin A in rats. Human and Experimental Toxicology 18: 410-415.

Mantle, P.G., 2000. Uptake of radiolabelled ochratoxin A from soil by coffee plants. Phytochemistry 53: 377-378.

Martins, M.L., H.M. Martins and A. Gimeno, 2003. Incidence of microflora and of ochratoxin A in green coffee beans (*Coffea arabica*). Food Additives and Contaminants 20: 1127-1131.

Mburu, J.K., 1999. Notes on coffee processing procedures and their influence on quality. Kenya Coffee 64: 2861-2867.

Micco, C., M. Grossi, M. Miraglia and C. Brera, 1989. A study of the contamination by ochratoxin A of green and roasted coffee beans. Food Additives and Contaminants 6: 333-339.

Miraglia, M. and C. Brera, 2002. Assessment of dietary intake of ochratoxin A by the population of EU member states. Reports on tasks for scientific cooperation. Reports of experts participating in SCOOP Task 3.2.7. Directorate-General Health and Consumer Protection, Rome, Italy.

Molinie, A., V. Faucet, M. Castegnaro and A. Pfohl-Leszkowicz, 2004. Analysis of some breakfast cereals on the French market for their contents of ochratoxin A, citrinin and fumonisin B1: development of a method for simultaneous extraction of ochratoxin A and citrinin. Food Chemistry 92: 391-400.

Morello, L.G., D. Sartori A.L. de Oliveira Martinez, M.L.C. Vieira, M.H. Taniwaki and M.H.P. Fungaro, 2007. Detection and quantification of *Aspergillus westerdijkiae* in coffee beans based on selective amplification of β-tubulin gene by using real-time PCR. International Journal of Food Microbiology (in press).

Nakajima, M., H. Tsubouchi, M. Miyabe and Y. Ueno, 1997. Survey of aflatoxin B1 and ochratoxin A in commercial green coffee beans by high-performance liquid chromatography linked with immunoaffinity chromatography. Food and Agricultural Immunology 9: 77-83.

Noonim, P., W. Mahakarnchanakul, J. Varga, J.C. Frisvad and R.A. Samson, 2008. Two new *Aspergillus* species of section *Nigri* isolated from Thai coffee bean. International Journal of Systematic and Evolutionary Microbiology (in press).

Ottender, H. and P. Majerus, 2001. Ochratoxin A (OTA) in coffee: nation-wide evaluation of data collected by German Food Control 1995-1999. Food Additives and Contaminants 18: 431-435.

Palacios-Cabrera, H., M.H. Taniwaki, H.C. Menezes and B.T. Iamanaka, 2004. The production of ochratoxin A by *Aspergillus ochraceus* in raw coffee at different equilibrium relative humidity and under alternating temperatures. Food Control 15: 531-535.

Pardo, E., S. Marin, A.J. Ramos and V. Sanchis, 2004. Occurrence of ochratoxigenic fungi and ochratoxin A in green coffee from different origins. Food Science and Technology International 10: 45-50.

Pardo, E., V. Sanchis, A.J. Ramos and S. Martin, 2006. Non-specificity of nutritional substrate for ochratoxin A production by isolates of *Aspergillus ochraceus*. Food Microbiology 23: 351-358.

Patel, S., C.M. Hazel, A.G. Winterton and A.E. Gleadle, 1997. Survey of ochratoxin A in UK retail coffees. Food Additives and Contaminants 14: 217-222.

Patiño, B., A. González-Salgado, M.T. González-Jaén and C. Vázquez, 2005. PCR detection assays for the ochratoxin-producing *Aspergillus carbonarius* and *Aspergillus ochraceus* species. International Journal of Food Microbiology 104: 207-214.

Pérez de Obanos, A., E. González-Peñas and A. López De Cerain, 2005. Influence of roasting and brew preparation on the ochratoxin A content in coffee infusion. Food Additives and Contaminants 22: 463-471.

Pérez, J., F. Infante, F.E. Vega, F. Holguín, J. Macías, J. Valle, G. Nieto, S.W. Peterson, C.P. Kurtzman and K. O'Donnell, 2003. Mycobiota associated with the coffee berry borer (*Hypothenemus hampei*) in Mexico. Mycological Research 107: 879-887.

Pittet, A., D. Tornare, A. Huggett and R. Viani, 1996. Liquid chromatography determination of ochratoxin A in pure and adulterated soluble coffee using an immunoaffinity column cleanup procedure. Journal of Agricultural and Food Chemistry 44: 3564-3569.

Prado, G., M.S. Oliveira and F.M. Abrantes, 2000. Incidência de ocratoxina A em café torrado e moído e em café solúvel consumido na cidade de Belo Horizonte, MG. Ciência e Tecnologia de Alimentos 20: 192-196.

Rigó, K., J. Varga, B. Tóth, J. Téren, Á. Mesterházy and Z. Kozakiewicz, 2002. Evolutionary relationships within *Aspergillus* section *Flavi* based on sequences of the intergenic spacer regions and the 5.8S rRNA gene. Journal of General and Applied Microbiology 48: 9-16.

Romani, S., G.G. Pinnavaia and M.D. Rosa, 2003. Influence of roasting levels on ochratoxin A content in coffee. Journal of Agricultural and Food Chemistry 51: 5168-5171.

Samson, R.A., J.A.M.P. Houbraken, A.F.A. Kuijpers, J.M. Frank J.M. and J.C. Frisvad, 2004. New ochratoxin or sclerotium producing species in *Aspergillus* section *Nigri*. Studies in Mycology 50: 45-61.

Schwartz, G.G., 2002. Hypothesis: does ochratoxin A cause testicular cancer? Cancer Causes Control 13: 91-100.

Smith, J.E. and M.O. Moss, 1985. Mycotoxins. Formation, analysis and significance. John Wiley & Sons, Chichester, UK.

Studer-Rohr, I., D.R. Dietrich, J. Schlatter and C. Schlatter, 1995. The occurrence of ochratoxin A in coffee. Food Chemistry and Toxicology 33: 341-355.

Suarez-Quiroz, M.L., O. Gonzalez-Rios, M. Barelb, B. Guyot, S. Schorr-Galindo and J.P. Guiraud, 2004. Effect of chemical and environmental factors on *Aspergillus ochraceus* growth and toxigenesis in green coffee. Food Microbiology 21: 629-634.

Suarez-Quiroz, M.L., O. Gonzalez-Rios, M. Barelb, B. Guyot, S. Schorr-Galindo and J.P. Guiraud, 2005. Effect of postharvest processing procedure on OTA occurrence in artificially contaminated coffee. International Journal of Food Microbiology 103: 339-345.

Tafuri, A., R. Ferracane and A. Ritieni, 2004. Ochratoxin A in Italian marketed cocoa products. Food Chemistry 88: 487-494.

Taniwaki, M.H., 2006. An update on ochratoxigenic fungi and ochratoxin A in coffee. In: A.D. Hocking, J.I. Pitt, R.A. Samson and U. Thrane (eds.) Advances in Food Mycology. Springer, New York, pp. 189-202.

Taniwaki, M.H., J.I. Pitt, A.A. Teixeira and B.T. Iamanaka, 2003. The source of ochratoxin A in Brazilian coffee and its formation in relation to processing methods. International Journal of Food Microbiology 82: 173-179.

Thirumala-Devi, K., M.A. Mayo, G. Reddy, E.K. Tangni, Y. Larondelle and D.V. Reddy, 2001. Occurrence of ochratoxin A in black pepper, coriander, ginger and turmeric in India. Food Additives and Contaminants 18: 830-835.

Torelli, E., G. Firrao, R. Locci and E. Gobbi, 2006. Ochratoxin A-producing strains of *Penicillium* spp. isolated from grapes used for the production of 'passito' wines. International Journal of Food Microbiology 106: 307-312.

Tsubouchi, H., H. Terada, K. Yamamoto, K. Hisada and Y. Sakabe, 1988. Ochratoxin A found in commercial roast coffee. Journal of Agricultural and Food Chemistry 36: 540-542.

Urbano, G.R., M.F. de Freitas Leitao, M.C. Vicentini and M.H. Taniwaki, 2004. Preliminary Studies on the destruction of Ochratoxin A in coffee during roasting. Food Science and Technolology International 10: 45-49.

Urbano, G.R., M.H. Taniwaki, M.F.D. Leitao and M.C. Vicentini, 2001. Occurrence of ochratoxin A-producing fungi in raw Brazilian coffee. Journal of Food Protection 64: 1226-1230.

Van der Merwe, K.J., P.S. Steyn, L. Fourie, D.B. Scott and J.J. Theron, 1965. Ochratoxin A, a toxic metabolite produced by *Aspergillus ochraceus* Wilh. Nature 205: 1112-1113.

Van der Stegen, G.H.D., P.J.M. Essens and J. van der Lijn, 2001. Effect of roasting conditions on reduction of ochratoxin A in coffee. Journal of Agricultural and Food Chemistry 49: 4713-4715.

Van der Stegen, G., U. Jorissen, A. Pittet, M. Saccon, W. Steiner, M. Vincenzi, M. Winkler, J. Zapp and C. Schlatter, 1997. Screening of European coffee final products for occurrence of ochratoxin A (OTA). Food Additives and Contaminants 14: 211-216.

Varga, J., É. Kevei, E. Rinyu, J. Téren and Z. Kozakiewicz, 1996. Ochratoxin production by *Aspergillus* species. Applied and Environmental Microbiology 62: 4461-4464.

Varga, J., K. Rigó, J. Téren and Á. Mesterházy, 2001. Recent advances in ochratoxin research I. Production, detection and occurrence of ochratoxins. Cereal Research Communications 29: 85-92.

Vega, F. and G. Mercadier, 1998. Insects, coffee and ochratoxin A. Florida Entomologist 81: 543-544.

Vega, F., F. Posada, S.W. Peterson, T. Gianfagna and F. Chaves, 2006a. *Penicillium* species endophytic in coffee plants and ochratoxin A production. Mycologia 98: 31-42.
Vega, F.E., F. Posada, T.J. Gianfagna, F.C. Chaves and S.W. Peterson, 2006b. An insect parasitoid carrying an ochratoxin producing fungus. Naturwissenschaften 93: 297-299.
Wafa, E.W., R.S. Yahya, M.A. Sobh, I. Eraky, H. El Baz, H.A.M. El Gayar, A.M. Betbeder and E.E. Creppy, 1998. Human ochratoxicosis and nephropathy in Egypt: a preliminary study, Human and Experimental Toxicology 17: 124-129.
Zinedine, A., C. Brera, S. Elakhdari, C. Catano, F. Debegnach, S. Angelini, B. De Santis, M. Faid, M. Benlemlih, V. Minardi and M. Miraglia, 2006. Natural occurrence of mycotoxins in cereals and spices commercialized in Morocco. Food Control 17: 868-874.
Zinedine, A., C. Juan, L. Idrissi and J. Mañes, 2007. Occurrence of ochratoxin A in bread consumed in Morocco. Microchemical Journal (in press).

Genetic diversity in *Aspergillus flavus* and its implications for agriculture

Kenneth C. Ehrlich
Southern Regional Research Center/ARS/USDA, New Orleans, LA 70124, USA

Abstract

Aspergillus flavus is the most common species associated with agricultural aflatoxin contamination. *A. flavus* soil populations contain isolates from two morphologically distinct sclerotial size variants, termed the L-strain for isolates with average sclerotial size greater than 400 μm and the S-strain for isolates with sclerotial size less that 400 μm. On typical laboratory growth media S-strain isolates produce higher levels of aflatoxins, more abundant sclerotia, and fewer conidia. Atoxigenic S-strain isolates are very rarely found in natural environments, whereas as much as 40% of the L-strain isolates can be atoxigenic. *A. oryzae*, used in some food fermentation processes, is a variant atoxigenic *A. flavus* with characteristics of either S or L strain isolates. *A. flavus* lacks the ability to produce G-aflatoxins due to a gap in the cluster that includes a required cytochrome P450-encoding gene, *cypA*. The size of the deletion that causes loss of a portion of *cypA* is 1.5 kb for S-strain isolates and 0.8 kb for L-strain isolates. Soil populations of *A. flavus* are typically composed of isolates from hundreds of different vegetative compatibility groups. Although frequent genetic exchange among these groups has not been observed, historical recombination in populations probably has occurred. Based on the phylogenetic relationship of *A. flavus* and *A. oryzae*, S-strain isolates and some *A. oryzae* isolates are most likely descended from an aflatoxin-producing L-strain common ancestor. We hypothesise that the frequent loss of aflatoxin-producing ability in *A. flavus* isolates from agricultural soil could be a consequence of adaptation to a carbon-rich environment that makes the aflatoxin cluster less genetically stable.

Keywords: fungal diversity, evolution, biosynthesis, sclerotia, vegetative compatibility

1. Introduction

Aspergillus flavus is the most common species associated with aflatoxin contamination of agricultural crops (Cotty *et al.*, 1994; Cotty, 1997). *A. flavus* is a haploid Plectomycetes found in temperate and tropical regions in soil and, in agricultural areas, on maize, cotton, tree and ground nuts. *A. flavus* populations are highly diverse and their stability in the soil and on the plant is not well understood. An atoxigenic relative of *A. flavus*, *A. oryzae*, is widely used in soybean and rice fermentation processes. It is now increasingly clear that *A. oryzae* is not a separate species, but actually is only one of many examples of atoxigenic variants of *A. flavus* (Chang *et al.*, 2006; Geiser *et al.*, 2000). Other aflatoxin-producing fungi have been implicated in contamination of agricultural commodities. *A. parasiticus* has been associated with contaminations of peanuts in the United States (Horn, 2005), Argentina (Vaamonde *et al.*, 2003), and West Africa (Ismail, 2001), but generally, the predominant contaminating organism is *A. flavus* (Cotty *et al.*, 1994). *A. flavus* appears to be more invasive and out-competes *A. parasiticus* when both species are together in the soil. *A. nomius* is more rarely found in the soil, and usually is not associated with agricultural contamination episodes (Bhatnagar *et al.*, 2001; Cardwell and Cotty, 2002; Cotty *et al.*, 1994). Mis-identification of the contaminating organism, in some cases is possible. For example, in Thailand, some aflatoxin B-and G-producing organisms, found to be common in the soil resemble *A. flavus*, but have been conclusively identified as a new clade of *A. nomius* (Ehrlich *et al.*, 2007a).

2. Biosynthesis

Biosynthesis of aflatoxins in *A. flavus* involves 25 genes in a 70 kb gene cluster. The genes encode 23 enzymes and two regulatory proteins (Ehrlich *et al.*, 2005). *A. flavus* makes B-type aflatoxins while its nearest relatives, *A. parasiticus* and an *A. parasiticus*-like unnamed species from West Africa, Argentina and Australia (also called *A. flavus* Group II (Geiser *et al.*, 2000)) produce both B-and G-type aflatoxins. *A. parasiticus* requires three extra genes for the production of G-aflatoxins (Yu *et al.*, 2004). The sequences of most of the genes in the *A. parasiticus* cluster are >95% identical to those in *A. flavus* (Ehrlich *et al.*, 2005).

Formation of the aflatoxin precursor polyketide, noranthrone, requires transfer of hexanoyl CoA to the polyketide synthase. Two non-primary metabolism fatty acid synthases are required to form the hexanoyl CoA precursor. These are encoded by the genes *hexA* and *hexB* in the cluster (Watanabe and Townsend, 2002). The fatty acid synthases form a complex with the polyketide synthase that allows an N-terminal, 'starter' acyl transferase (SAT) domain in this protein to receive the hexanoyl CoA, a step required before further iterative addition of malonyl CoA units can occur. SAT domains have now been implicated in the formation of many fungal polyketides, but only some require hexanoyl CoA as the starter unit (Crawford *et al.*, 2006). Iterative addition concludes when the polyketide chain fills the cavity of the polyketide synthase and is excised by a C-terminal portion of the protein that acts as both a thioesterase and a Claisen-like-cyclase (Fujii *et al.*, 2001).

The loss of the ability to produce G-aflatoxins in *A. flavus* results from a deletion in the promoter and N-terminal amino acid-encoding regions (Figure 1) of the gene for a cytochrome P450 enzyme, CypA. CypA is required for conversion of an as yet uncharacterised aflatoxin precursor (common to formation of both aflatoxins) to an epoxide intermediate (Ehrlich *et al.*, 2004). The neighboring gene (*norB*), a homologue of another cluster gene, *norA*, is predicted to encode an aryl alcohol dehydrogenase and is also not expected to be functional in *A. flavus* due to loss of its promoter region and translational start site. This enzyme and the predicted enzyme encoded by the gene, *nadA*, at the opposite end of the gene cluster may be involved in further enzymatic conversion of the product resulting from the CypA-catalysed oxidisation of the presumed aflatoxin precursor.

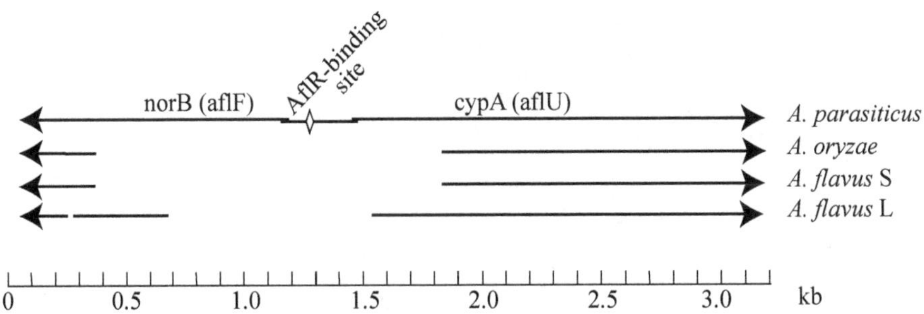

Figure 1. Schematic diagram of the norB-cypA *sequences of different aflatoxin biosynthesis gene cluster homologues. Thick arrows indicate coding regions and direction of transcription of* norB *and* cypA. *Gaps represent deletions of 32 and 854 bp in* A. flavus *L and 1516 bp in* A. flavus *S and* A. oryzae.

3. Sclerotial morphotypes

A. flavus soil populations contain isolates from two morphologically distinct sclerotial size variants, termed the L-strain (also called *A. flavus* Group IB (Geiser *et al.*, 2000)) for isolates with average sclerotial size greater than 400 μm and the S-strain (Group IA) for isolates with sclerotial size less that 400 μm (Cotty, 1997) On typical laboratory growth media, when grown in the dark, S-strain isolates produce higher levels of aflatoxins, more abundant sclerotia, and fewer conidia. Atoxigenic S-strain isolates are very rarely found in natural environments (Orum *et al.*, 1997). Differences in sclerotial morphology correlate with the differences between the S- and L-strain *A. flavus* in the size of the deletion in the *norB-cypA* gene (Ehrlich *et al.*, 2004). Because the 0.8 kb deletion in S-strain isolates is identical to the deletion in those *A. oryzae* isolates that possess most of the aflatoxin cluster, such isolates probably descended from a common ancestor that had the S-strain-type deletion (Chang *et al.*, 2005). Many L-strain isolates that lack the ability to produce aflatoxins also have an identical 0.8 kb *norB-cypA* deletion. When this gap size is included in a phylogenetic dataset that also includes *omtA* gene aflatoxin cluster region sequence (used by Geiser *et al.* (2000) to compare *A. flavus* isolates), separate clades were revealed that included members of aflatoxin-producing S-strain isolates and L-strain isolates incapable of aflatoxin production and *A. oryzae* and L-strain isolates incapable of AF production (Figure 2). From this data we conclude that the L-strain is ancestral to the S-strain and to both *A. oryzae* and atoxigenic L-strain isolates (Chang *et al.*, 2006).

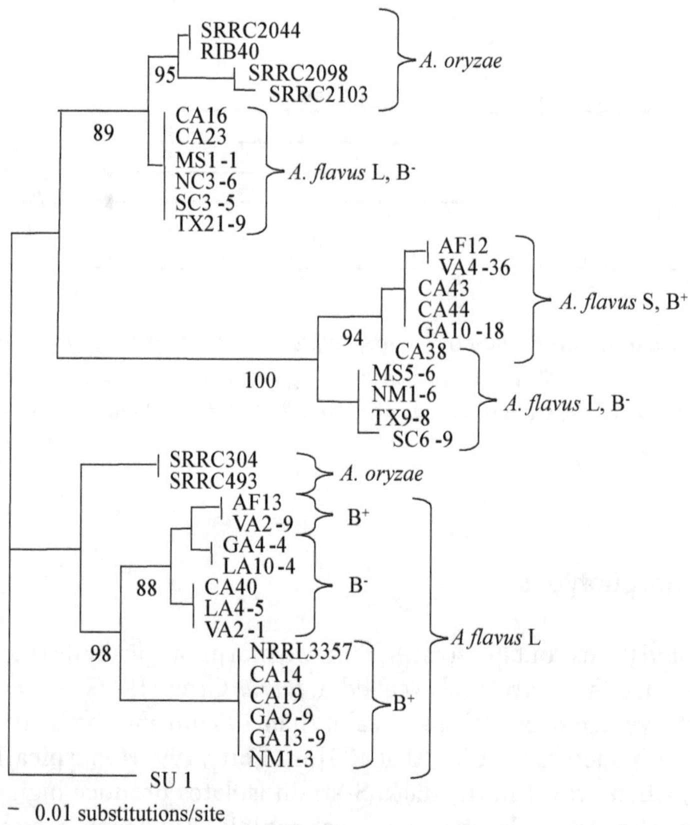

Figure 2. Phylogenetic analysis of A. flavus *toxin-producing and non-toxin producing isolates. Analysis was performed in Paup* (Swofford, 1998) and includes polymorphisms in the aflatoxin biosynthesis gene,* omtA, *and* norB-cypA *sequence gap size as well as sclerotial type, if applicable.*

On average 30% of the *A. flavus* soil isolates in Arizona were identified as belonging to the S-strain (Cotty, 1997; Orum *et al.*, 1997). Because S-strain isolates consistently produce more aflatoxin than L-strain isolates and aflatoxin production in this strain is more strongly affected by nitrogen source (Table 1), the concentration of S-strain isolates in the soil appears to be better correlated with major outbreaks of aflatoxin contamination in cotton-growing areas in Arizona and Texas (Jaime-Garcia and Cotty, 2006; Orum *et al.*, 1997). Furthermore, up to 40% of the L-strain soil isolates of *A. flavus* found in Arizona and other regions of the United States (Horn and Dorner, 1999) were incapable of producing aflatoxins while S-strain isolates rarely were atoxigenic (Cotty *et al.*, 1994). One strategy currently being used to remediate aflatoxin contamination of cotton and peanuts involves addition of a selected atoxigenic L-strain isolate of *A. flavus* to the soil of susceptible crops to dilute out toxin-producing strains (Cotty and Bayman, 1993; Horn *et al.*, 2000; Horn and Dorner, 2002). Interestingly, for cottonseed, the isolate chosen because of its proven ability to compete well with naturally occurring soil isolates, has the S-strain type *norB-cypA* deletion (Ehrlich, unpublished data).

Table 1. Aflatoxin production as a function of nitrogen source. Aflatoxin yield is given as µg / g mycelia after 4 days of shaking at 31 °C in the dark. AM, Aflatoxin Minimal Medium with 5 mM ammonium sulfate as nitrogen source, NO-10 mM sodium nitrate replacing ammonium in AM; UR, urea replacing ammonium in AM. The medium was buffered with 50 mM morpholinoethylsulfonate (MES) at pH 5.8. Data are averages of yields from 6 different isolates. AFL, AFS: L and S-strain A. flavus, *AP,* A. parasiticus.

Fungus	AM	NO	UR	Ratio AM to NO
AFL	240	190	102	1.3
AFS	2230	670	1130	3.3
AP	3260	330	5200	9.8

4. Vegetative compatibility and diversity

Isolates of *Aspergillus* are said to be vegetatively compatible if they have the ability to undergo hyphal fusion to form stable heterokaryons (Leslie, 1993). Such isolates are assumed to belong to the same vegetative compatibility group (VCG). In *Aspergillus* as well as other fungal genera, such as *Neurospora, Podospora,* and *Fusarium*, vegetative compatibility is controlled by multiple genes that occur on unlinked loci (Anwar *et al.*, 1993; Glass *et al.*, 2000; Kohn, 2005; Saupe, 2000). A single soil sample from agricultural fields usually contains isolates from many different *A. flavus* VCGs (Cotty *et al.*, 1994; Horn *et al.*, 2000). In over 10 years of study, natural genetic complementation among isolates belonging to different VCGs has not been observed in the laboratory using tester cultures containing marker mutations. How strong a barrier vegetative compatibility is to recombination within a fungal species that appears to lack the ability to sexually recombine is unknown. The high degree of variability in *A. flavus* isolates in morphology, ability to produce aflatoxins and other secondary metabolites, and in their ability to infect and decay plants may reflect phenotypic diversity among VCGs (Cotty *et al.*, 1994). *A. flavus* VCGs consisting of either aflatoxin-producing or non-producing isolates can be obtained from the same soil, but much flux has been found in these populations in soil samples collected from the same place in different years (Bayman and Cotty, 1993; Chang *et al.*, 2006). One atoxigenic isolate from VCG YV36 is currently being used as a competitor to remediate aflatoxin contamination in cotton fields in Arizona (Antilla and Cotty, 2002). The long-term success of this strategy, in part, depends on whether or not such populations can be stably maintained over the course of the growing season.

5. Recombination as a source of diversity

In an effort to determine if geographically separated populations of *A. flavus* isolates from the same VCG could be recombining, we examined sets of polymorphisms in three genes from different chromosomes. We found that isolates of the same VCG collected from soil samples obtained from regions of the United States separated by over 2000 km contained the same sets of polymorphisms in the three different genes (Figure 3) (Ehrlich *et al.*, 2007b). Therefore, at least over the time-span of the collection of these isolates, recombination could not be demonstrated. The study was complicated because some of the VCGs shared the same set of polymorphisms and therefore it was impossible to prove that these are not a recombining population. That recombination in these otherwise asexual fungi has occurred in the more distant past has been demonstrated by phylogenetic tests and sequencing (Berbee *et al.*, 2003; Burt *et al.*, 1996; Geiser *et al.*, 1998; Paoletti *et al.*, 2005). In the phylogenetic tree shown in Figure 2, two of the aflatoxin-producing L-strain isolates (AF13 and VA2-9) fit into a subclade containing mostly atoxigenic isolates. These isolates have a very different set of *omtA* gene polymorphisms than do most of the aflatoxin-producing isolates which fall into a separate subclade in the tree (Chang *et al.*, 2006). This divergence could best be explained by assuming a recombination event occurred that caused the joining of genes from the opposite ends of the cluster.

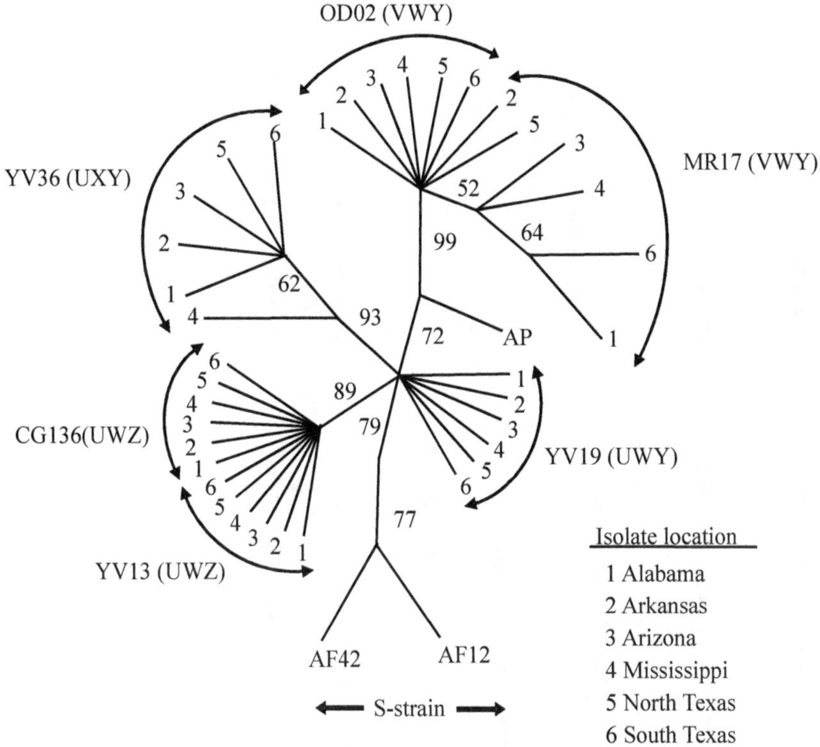

Figure 3. Strict consensus phylogenetic tree of different strain L A. flavus *VCGs. Analysis used Paup* and was done on the combined dataset consisting of partial sequences of the genes,* taa *and* xynF3, *and the* aflJ-aflR *intergenic region. Numbers next to the branches are the bootstrap values based on 1000 replicates; numbers at the branch termini refer to* A. flavus *isolates from the geographic regions surveyed (see inset).*

6. Evolution of the aflatoxin cluster

As many as 30 species and several genera (*Aspergillus, Bipolaris, Chaetomium, Dothistroma*) are capable of producing either aflatoxin or known aflatoxin precursors (Barnes *et al.*, 1994; Bradshaw and Zhang, 2006; Maskey *et al.*, 2003). The gene clusters responsible for sterigmatocystin production in *A. nidulans* (Brown *et al.*, 1996) and aflatoxin production in *A. ochraceoroseus* (Cary *et al.*, unpublished results) have similar genes to those in the AF cluster, but the gene order is somewhat different (Figure 4). The structure of the gene cluster in *Dothistroma septosporum* (which produces a versicolorin A metabolite) is quite different from that in *A. nidulans* and *A. flavus* (Figure 4).

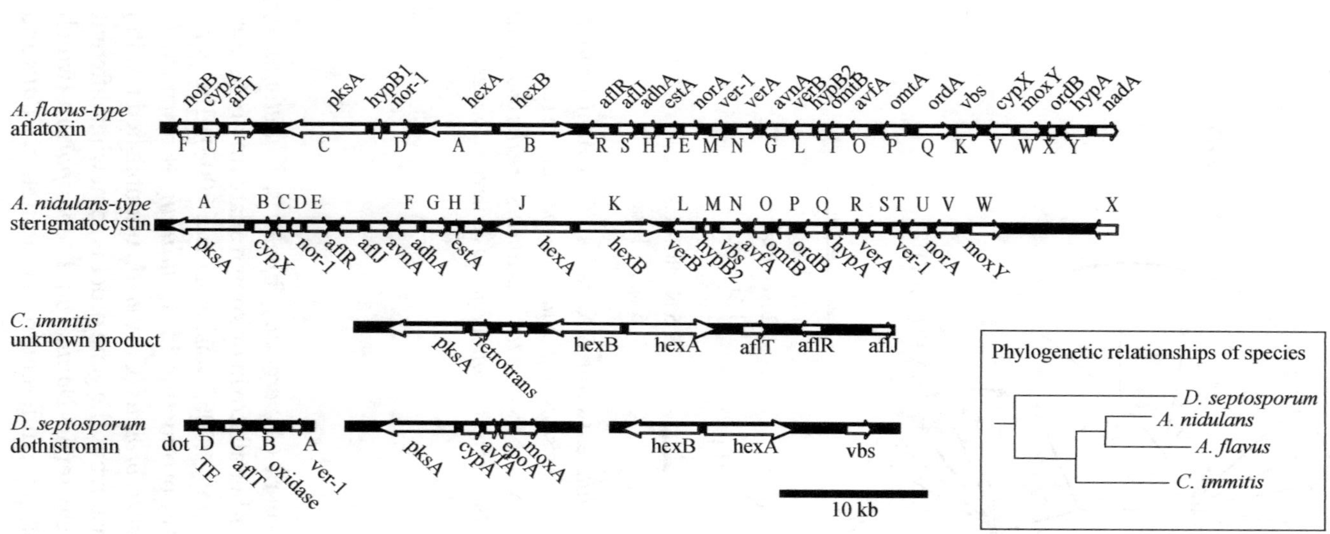

Figure 4. Comparison of some polyketide gene clusters. All of the clusters probably use hexA *and* hexB *for generation of the polyketide starter unit. The polyketide cluster in C. immitis is based on genomic sequence, only, and the polyketide product is unknown. For the C. immitis and D. septosporum gene clusters not all of the biosynthetic genes have been identified. The phylogenetic relationship of these species is shown in the inset.*

The product of the *Coccidioides immitis* cluster is unknown, but clearly some of the genes are similar to those in the other clusters shown in Figure 4 suggesting that the formation of the polyketide occurs via a starter fatty acid produced by its *hexA* and *hexB* orthologs. The gene clusters appear to share many similar regulatory and modifying enzyme genes. Recent evidence from whole genome comparisons of fungal taxa suggests that evolution of the gene clusters occurs by combinations of gene duplications, gene movement, and adaptive selection (Carbone *et al.*, 2007; Ward *et al.*, 2002; Wong and Wolfe, 2005). Because gene clusters often reside in or near the subtelomeric arms of chromosomes they are subject to a high degree of genetic flux. In spite of this, the evolutionary rate of change of most genes in the clusters occurs at a comparable rate to that of essential genes in the organism (Ehrlich *et al.*, 2003; Ward *et al.*, 2002).

Early studies suggested that secondary metabolites serve no survival function for the fungus (Demain and Fang, 2000). If they are not required for fungal adaptation, than explanation has to given for why some of the genes encoding secondary metabolite-producing polyketides have been maintained in almost all fungi for over 1 billion years. One possible explanation is that production of metabolites in the aflatoxin cluster correlates with increased conidiation, a marker of fungal fitness (Wilkinson *et al.*, 2004). During the course of evolution, new blocks of genes have been added to gene clusters to increase the ability of the fungus to modify its preformed polyketide (Carbone *et al.*, 2007). Two such blocks contain the genes required for G-toxin *(cypA* and *norB*) and B-toxin formation (*ordA* and *omtA*). They are at or near the ends of the cluster, are divergently transcribed, and share a single binding site for the pathway-specific transcription factor, AflR (Ehrlich *et al.*, 2005). Genes common to biosynthesis of the related metabolites, aflatoxin, sterigmatocystin, and dothistromin, appear to be in Pezizomycotina, before the divergence of Eurotiomycetes from Sordariomycetes (ES split), an event that has been estimated to have occurred at least 300 Mya. Fungi and insects evolved over the same time scale. I previously hypothesized that polyketide aflatoxin precursor anthraquinones and xanthones served as insect attractants that fostered spore dispersal (Ehrlich, 2006). Certainly, if this was the 'original' function that these colourful polyketide metabolites served, it has been a highly successful strategy for dispersal of the Aspergilli!

A time-scale for divergence of the main species of aflatoxin-producing fungi is shown in the tree in Figure 5. To put a time-scale on this tree, another tree (based on ITS sequence) was obtained that was rooted with a Sordariomycete as outgroup. This allowed an estimate of 100 million years to be made for how long ago *A. nidulans* diverged from *A. nomius*, assuming the ES split to have occurred 300 Mya (Berbee and Taylor, 2001; Heckman *et al.*, 2001). Based on these admittedly crude estimates,

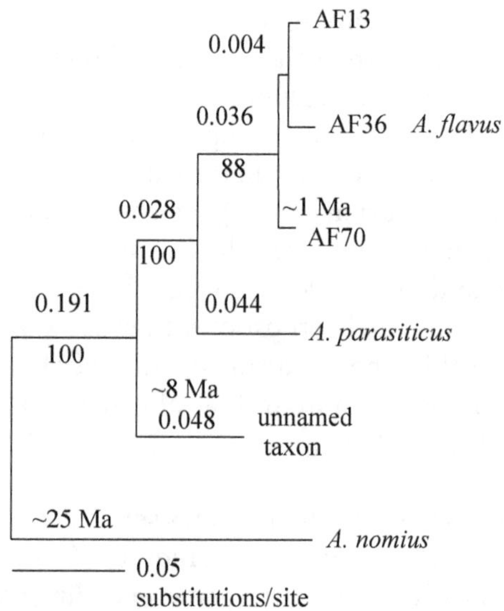

Figure 5. Phylogenetic trees based on alignments of the combined intergenic regions of genes in the 65 kb AF cluster. The tree was generated using the maximum likelihood (ML) option in PAUP. The tree is rooted with* A. nomius. *The bootstrap support values based on 1000 replicates are shown below the branches.*

the divergence of *A. nomius* and *A. parasiticus* would have occurred about 25 Mya and *A. parasiticus* and *A. flavus*, about 5 Mya. Therefore, the S- and L-strains of *A. flavus* would have become reproductively separated at least one million years ago, well before the emergence of agriculture (Ehrlich *et al.*, 2005).

7. Agriculture and *A. flavus* diversity

As the most successful species in its ability to colonise plants of the aflatoxin-producing Aspergilli, *A. flavus* populations also show unusual diversity in their ability to produce aflatoxins. Could the adaptation of *A. flavus* to the carbon-rich environment of certain agricultural communities be conducive to gene loss? Many of the isolates incapable of aflatoxin production have multiple mutations in their aflatoxin cluster genes and some have lost the entire cluster (Chang *et al.*, 2005; Lee *et al.*, 2006). In some of these isolates the remaining AF genes neighbored the telomere. Proximity to the telomere may make the cluster more unstable. In *A. parasiticus* when normal development is thwarted, by forced repeated mycelial transfer, the resulting isolate permanently loses some of its normal developmental functions, including the ability to produce aflatoxins (Kale *et al.*, 2003). The defects in chromosomes of these isolates remain to be determined.

We offer the hypothesis that the intersection of *A. flavus* with agricultural commodities has partially offset the selective pressure that prevented loss of function of the genes required for aflatoxin formation, whereas the harsher and more competitive soil environment has stimulated conservation of the gene cluster. As support for this hypothesis, *A. oryzae* isolates that have been maintained in culture for 100s of years have lost their ability to produce many secondary metabolites, but have increased their genome size and their ability to utilise carbohydrate. It is well understood that stress conditions stimulate aflatoxin biosynthesis in strains that have retained the ability to produce these metabolites. It is possible that, in the absence of normal stress, *A. flavus*, growing on agricultural soils and on plants, will eventually lose the ability to produce aflatoxins. In the laboratory common practice is to maintain fungal cultures on sub-optimal media in order to foster conidial development and to prevent unexpected loss of some of the genes involved in toxin formation.

Unlike the protective role of melanin on reproductive structures, no evidence has been found that the conidia or sclerotia are protected by the aflatoxin cluster metabolites. Even though the red-pigmented dothistromin may be a virulence factor for *D. septosporum* responsible for its pathogenicity to pine (Bradshaw *et al.*, 2002), there is no evidence that any of the aflatoxin pathway metabolites act as similar types of virulence factors. Like dothistromin, most of the aflatoxin precursor metabolites are red or orange. Because of their colour, such metabolites could have helped to foster fungal dispersal. In addition, since section *Flavi* isolates are normally saprophytic, polyketide metabolites may increase fungal survival in soil, whereas in a carbon- and nitrogen-rich agricultural environment, polyketide production may be a disadvantage. When *A. flavus* isolates were collected from non-agricultural soils, almost all of the isolates were capable of producing aflatoxins (Ehrlich *et al.*, 2007a). Furthermore, in some soils, *A. flavus* was not the most prominent species. Understanding the role of aflatoxin production and in general secondary metabolite production may only be possible if attempts are made to duplicate, in the laboratory, the conditions of the natural environment for these Aspergilli.

References

Antilla, L. and P.J. Cotty, 2002. The ARS-ACRPC partnership to control aflatoxin in Arizona cotton: current status. *Mycopathologia* 155: 64.

Anwar, M.M., J.H. Croft and R.B. Dales, 1993. Analysis of heterokaryon incompatibility between heterokaryon- compatibility (h-c) groups R and GL provides evidence that at least eight het loci control somatic incompatibility in *Aspergillus nidulans*. Journal of General Microbiology 139: 1599-1603.

Barnes, S.E., T. P. Dola, J.W. Bennett and D. Bhatnagar, 1994. Synthesis of sterigmatocystin on a chemically defined medium by species of *Aspergillus* and *Chaetomium*. Mycopathologia 125: 173-178.

Bayman, P. and P.J. Cotty, 1993. Genetic diversity in *Aspergillus flavus*: association with aflatoxin production and morphology. Canadian Journal of Botany 71: 23-31.

Berbee, M.L. and J.W. Taylor, 2001. Fungal molecular evolution: gene trees and geologic time. In: D. J. McLaughlin, E. McLaughlin and P. A. Lemke (eds). *The Mycota*. Springer, Berlin, pp. 229-246.

Berbee, M.L., B.P. Payne, G. Zhang, R.G. Roberts and B.G. Turgeon, 2003. Shared ITS DNA substitutions in isolates of opposite mating type reveal a recombining history for three presumed asexual species in the filamentous ascomycete genus Alternaria. Mycological Research 107: 169-182.

Bhatnagar, D., P.J. Cotty and T.E. Cleveland, 2001. Genetic and biological control of aflatoxigenic fungi. In: C. L. Wilson and S. Droby (eds). *Microbial Food Contamination*. CRC Press, Boca Raton, pp. 208-240.

Bradshaw, R.E., D. Bhatnagar, R.J. Ganley, C.J. Gillman, B.J. Monahan and J.M. Seconi, 2002. *Dothistroma pini*, a forest pathogen, contains homologs of aflatoxin biosynthetic pathway genes. Applied and Environmental Microbiology 68: 2885-2892.

Brown, D.W., J.H. Yu, H.S. Kelkar, M. Fernandes, T.C. Nesbitt, N.P. Keller, T.H. Adams and T.L. Leonard, 1996. Twenty-five coregulated transcripts define a sterigmatocystin gene cluster in *Aspergillus nidulans*. Proceedings of the National Academy of Sciences USA 93: 1418-1422.

Burt, A., D.A. Carter, G.L. Koenig, T.J. White and J.W. Taylor, 1996. Molecular markers reveal cryptic sex in the human pathogen *Coccidioides immitis*. Proceedings of the National Academy of Sciences USA 93: 770-773.

Carbone, I., J.H. Ramirez-Prado, J.L. Jakobek and B.W. Horn, 2007. Gene duplication, modularity and adaptation in the evolution of the aflatoxin gene cluster. BMC Evolutionary Biology 7: 111. doi:10.1186/1471-2148-7-111.

Cardwell, K. and P.J. Cotty, 2002. Distribution of *Aspergillus* Section *Flavi* among field soils from the four agroecological zone of the Republic of Benin, West Africa. Plant Disease 86: 434-439.

Chang, P.K., B.W. Horn and J.W. Dorner, 2005. Sequence breakpoints in the aflatoxin biosynthesis gene cluster and flanking regions in nonaflatoxigenic *Aspergillus flavus* isolates. Fungal Genetics and Biology 42: 914-923.

Chang, P.-K., K.C. Ehrlich, S.-S.H. Hua, 2006. Cladal relatedness among *Aspergillus oryzae* isolates and *Aspergillus flavus* S and L morphotypes. International Journal of Food Microbiology 108: 172-177.

Cotty, P.J. and P. Bayman, 1993. Competitive exclusion of a toxigenic strain of *Aspergillus flavus* by an atoxigenic strain. Phytopathology 83: 1283-1287.

Cotty, P.J., D.S. Bayman, D.S. Egel and K.S. Elias, 1994. Agriculture, aflatoxins and *Aspergillus.* In: K. Powell (ed). *The Genus Aspergillus.* Plenum Press, New York, pp. 1-27.

Cotty, P.J., 1997. Aflatoxin-producing potential of communities of *Aspergillus* section *Flavi* from cotton producing areas in the United States. Mycological Research 101: 698-704.

Crawford, J. M., B.C. Dancy, E.A. Hill, D.W. Udwary and C.A. Townsend, 2006. Identification of a starter unit acyl-carrier protein transacylase domain in an iterative type I polyketide synthase. Proceedings of the National Academy of Sciences USA 103: 16728-16733.

Demain, A.L. and A. Fang, 2000. The natural functions of secondary metabolites. Advances in Biochemical Engineering and Biotechnology 69: 1-39.

Ehrlich, K.C., 2006. Evolution of the aflatoxin gene cluster. Mycotoxin Research 22: 9-15.

Ehrlich, K.C., B.G. Montalbano and P.J. Cotty, 2003. Sequence comparison of *aflR* from different *Aspergillus* species provides evidence for variability in regulation of aflatoxin production. Fungal Genetics and Biology 38: 63-74.

Ehrlich, K.C., B.G. Montalbano and P.J. Cotty, 2007b. Analysis of single nucleotide polymorphisms in three genes shows evidence for genetic isolation of certain *Aspergillus flavus* vegetative compatibility groups. FEMS Microbiology Letters 268: 231-236.

Ehrlich, K.C., J. Yu and P.J. Cotty, 2005. Aflatoxin biosynthesis gene clusters and flanking regions. Journal of Applied Microbiology 99: 518-527.

Ehrlich, K.C., P.K. Chang, J. Yu and P.J. Cotty. 2004. Aflatoxin biosynthesis cluster gene *cypA* is required for G aflatoxin formation. Applied and Environmental Microbiology 70: 6518-6524.

Ehrlich, K.C., K. Kobbeman, B.G. Montalbano and P.J. Cotty, 2007a. Aflatoxin-producing *Aspergillus* species from Thailand. International Journal of Food Microbiology 114: 153-159.

Fujii, I., A. Watanabe, U. Sankawa and Y. Ebizuka, 2001. Identification of Claisen cyclase domain in fungal polyketide synthase WA, a naphthopyrone synthase of *Aspergillus nidulans.* Chemistry and Biology 8: 189-197.

Geiser, D.M., J.I. Pitt and J.W. Taylor, 1998. Cryptic speciation and recombination in the aflatoxin-producing fungus, *Aspergillus flavus.* Proceedings of the National Academy of Sciences USA 95: 388-393.

Geiser, D.M., J.W. Dorner, B.W. Horn and J.W. Taylor, 2000. The phylogenetics of mycotoxin and sclerotium production in *Aspergillus flavus* and *Aspergillus oryzae.* Fungal Genetics and Biology 31: 169-179.

Glass, N.L., D.J. Jacobson and P.K. Shiu, 2000. The genetics of hyphal fusion and vegetative incompatibility in filamentous Ascomycete fungi. Annual Reviews in Genetics 34: 165-186.

Heckman, D.S., D.M. Geiser, B.R. Eidell, R.L. Stauffer, N.L. Kardos and S.B. Hedges, 2001. Molecular evidence for the early colonization of land by fungi and plants. Science 293: 1129-1133.

Horn, B.W. and J.W. Dorner. 1999. Regional differences in production of aflatoxin B1 and cyclopiazonic acid by soil isolates of *Aspergillus flavus* along a transect within the United States. Applied and Environmental Microbiology 65: 1444-1449.

Horn, B.W., R.L. Greene and J.W. Dorner, 2000. Inhibition of aflatoxin B1 production by *Aspergillus parasiticus* using nonaflatoxigenic strains: role of vegetative compatibility. Biological Control 17: 147-154.

Horn, B.W. and J.W., Dorner. 2002. Effect of competition and adverse culture conditions on aflatoxin production by *Aspergillus flavus* through successive generations. Mycologia 94: 741-751.

Horn, B.W. 2005. Colonization of wounded peanut seeds by soil fungi: selectivity for species from *Aspergillus* section *Flavi*. Mycologia. 97: 202-217.

Ismail, M.A., 2001. Deterioration and spoilage of peanuts and desiccated coconuts from two sub-Saharan tropical East African countries due to the associated mycobiota and their degradative enzymes. Mycopathologia 150: 67-84.

Jaime-Garcia, R. and P.J. Cotty, 2006. Spatial distribution of *Aspergillus flavus* and its toxigenic strains on commercial cottonseed from south Texas and its relationship to aflatoxin contamination. Plant Pathology 55: 358-366.

Kale, S.P., J.W. Cary, C. Baker, D. Walker, D. Bhatnagar and J.W. Bennett. 2003. Genetic analysis of morphological variants of *Aspergillus parasiticus* deficient in secondary metabolite production. Mycological Research 107: 831-840.

Kohn, L.M., 2005. Mechanisms of fungal speciation. Annual Reviews in Phytopathology 43: 279-308.

Lee, Y.H., M. Tominaga, R. Hayashi, K. Sakamoto, O. Yamada and O. Akita, 2006. *Aspergillus oryzae* strains with a large deletion of the aflatoxin biosynthetic homologous gene cluster differentiated by chromosomal breakage. Applied Microbiology and Biotechnology 72: 339-45.

Leslie, J.F., 1993. Fungal vegetative compatibility. Annual Reviews in Phytopathology 31: 127-150.

Maskey, R.P., I. Grun-Wollny and H. Laatsch, 2003. Isolation, structure elucidation and biological activity of 8-O-methylavrufin and 1,8-O-dimethylaverantin as new antifungal agents from *Penicillium chrysogenum*. Journal of Antibiotics 56: 459-463.

Orum, T.V., D.M. Bigelow, M.R. Nelson, D.R. Howell and P.J. Cotty, 1997. Spatial and temporal patterns of *Aspergillus flavus* strain composition and propagule density in Yuma County, arizona, soils. Plant Disease 81: 911-916.

Paoletti, M., C. Rydholm, E.U. Schwier, M.J. Anderson, G. Szakacs, F. Lutzoni, J.P. Debeaupuis, J. P. Latge, D. W. Denning and P.S. Dyer, 2005. Evidence for sexuality in the opportunistic fungal pathogen *Aspergillus fumigatus*. Current Biology 15: 1242-1248.

Saupe, S.J., 2000. Molecular genetics of heterokaryon incompatibility in filamentous ascomycetes. Microbiology and Molecular Biology Reviews 64: 489-502.

Swofford, D.L. 1998. *PAUP*. Phylogenetic analysis using parsimony (*and other methods).* Sinauer Associates, Sunderland MA.

Vaamonde, G., A. Patriarca, V. Fernandez Pinto, R. Comerio and C. Degrossi, 2003. Variability of aflatoxin and cyclopiazonic acid production by *Aspergillus* section *flavi* from different substrates in Argentina. International Journal of Food Microbiology 88: 79-84.

Ward, T.J., J.P. Bielawski, H.C. Kistler, E. Sullivan and K. O'Donnell, 2002. Ancestral polymorphism and adaptive evolution in the trichothecene mycotoxin gene cluster of phytopathogenic *Fusarium*. Proceedings of the National Academy of Sciences USA 99: 9278-9283.

Watanabe, C.M. and C.A. Townsend, 2002. Initial characterization of a type I fatty acid synthase and polyketide synthase multienzyme complex NorS in the biosynthesis of aflatoxin B1. Chemistry and Biology 9: 981-988.

Wilkinson, H., A. Ramaswamy, S.C. Sim and N.P. Keller, 2004. Increased conidiation associated with progression along the sterigmatocystin biosynthetic pathway. Mycologia 96: 1190-1198.

Wong, S. and K.H. Wolfe, 2005. Birth of a metabolic gene cluster in yeast by adaptive gene relocation. Natrue Genetics 37: 777-782.

Yu, J., P.-K. Chang, K.C. Ehrlich, J.W. Cary, D. Bhatnagar, T.E. Cleveland, G.A. Payne, J.E. Linz, C.P. Woloshuk and J.W. Bennett, 2004. Clustered pathway genes in aflatoxin biosynthesis. Applied and Environmental Microbiology 70: 1253-1262.

The importance of Aspergilli and regulatory aspects of *Aspergillus* nomenclature in biotechnology

Piet W. M. van Dijck
DSM Nutritional Products, Dept. Regulatory Affairs, P.O. Box 1, 2600 MA Delft, The Netherlands

Abstract

This contribution deals with the development of large scale production processes for enzymes and organic acids with Aspergilli in Europe and the United States starting in the late 19th century and based on old fermentation practices from the Orient. These products have an established history of safe use in the food and food processing industry. The importance of the Aspergilli, and in particular *Aspergillus niger* and *Aspergillus oryzae* for the large scale fermentation industry and some regulatory and taxonomic aspects are discussed.

Keywords: *Aspergillus niger, Aspergillus oryzae,* industrial/large-scale production, enzymes, citric acid, regulatory aspects

1. History of *Aspergillus* in biotechnology

The use of *Aspergillus* species for fermentation purposes goes back millennia. Old Chinese recipes have been found that describe the use of fungal cultures in the preparation of foodstuffs. In Japan mold mixtures, the so-called koji, are used in the production of certain foodstuffs and flavor additives based on soya protein (shoyu, miso) and rice (sake). Koji is prepared from steamed rice as substrate for a mixture of fungi, the composition of which is passed on from generation to generation as family secrets. This use of strains from the *Aspergillus oryzae / sojae* group is one of the finest examples of sold-state fermentation. And although these processes in the orient were known for centuries it took till the 1890's before fermentation processes with Aspergilli were developed.

The Japanese scientist and businessman Jokichi Takamine was the first to transfer this oriental knowledge to the west and in 1890 he established the Takamine Ferment Company in Peoria, Illinois to manufacture commercial products with diastatic / amylolytic activities using the koji process and *Aspergillus oryzae* as the production organism. Although the enzyme products were intended for use in alcohol fermentations as a faster and cheaper alternative for malt, his enterprice became successful when he in 1894 patented the product as a medicinal diastatic preparation called Takadiastase (Bennett, 2001).

At about the same time 1895 Albert Boidin in France developed a process for the production of alcohol called the Amyloprocess. It comprised of cooking grains, inoculated with *Aspergillus* species which formed starch degrading enzymes such as α-amylase and glucoamylase, with subsequent fermentation with yeast.

Other pioneers in the fermentative production of enzymes using Aspergilli are Leo Wallerstein (USA) and Otto Röhm (Germany). They laid the foundation of today's major producers of enzymes originating from *Aspergilli.* Boidin's and Wallerstein's legacy became part of DSM, Takamine's company now is part of Genencor-Danisco, while Röhm now is AB Enzymes. The origin of Novozymes interests in fungal enzymes is their facility in Switzerland (Novo Ferment), which started in 1915 as the Schweizerische Ferment AG in Basel with the production of malt amylase by *A. oryzae.* In 1945 they introduced their first pectinase products made by surface fermentation.

Many of the *Aspergillus* enzyme processes started either as sold-state fermentations or as surface fermentations. Although solid-state production processes still are in use, in particular in Japan, India and China, surface fermentations no longer are cost-effective and most production processes now are submerged fermentations in large stain-less steel vessels, which can be up to 200 m^3 in size. The three major fermentation industries, DSM, Novozymes and Genencor-Danisco all employ *Aspergillus* as their most important production organism. For DSM this is *Aspergillus niger,* for Novozymes it is *Aspergillus oryzae,* and Genencor uses both organisms.

Besides for the production of enzymes, *Aspergillus* species, and in particular *A. niger* is used for the production of citric acid. Its development was based on the studies of Wehmer in 1891 of the production of oxalic acid by *A. niger.* He showed that *A. niger* is able to decompose sugars with an abundant production of oxalic acid. However, this fermentation was not developed to a commercial process because oxalic acid could be produced cheaper by other means. In 1917 Currie demonstrated that the production of oxalic acid could be completely inhibited by the sugar concentration in the process and by depleting the medium for metal ions, in particular manganese; in result citric acid accumulated in high levels. Commercialisation of this process in the USA was done by Pfizer. More or less simultaneously the commercial production of citric acid in Europe started in 1919 at Citrique Belge (Tienen) by surface fermentation of *A. niger* on beet molasses.

Similar to the situation with enzymes, surface fermentation processes still are in use, but citric acid and other organic acids today are also produced in submerged fermentations.

2. Regulatory approval and nomenclature

In all countries regulatory approvals for microbial products are given for the combination of the product with the production organism. For Instance, the US FDA in 1958, in response to public concern about the increased use of chemicals in foods and food processing and with the support of the food industry, implemented the GRAS (Generally Recognised as Safe) system. This was done by the US Congress with an amendment to the Food Additives Amendment. The basic change with this amendment was the requirement that, before a new additive could be used in food, its producer demonstrate the safety of the additive to FDA by submitting a dossier: the food additive petition. For new products the producer had to submit a file with a science based- risk assessment of the safety of the product and for microbial products this included a complete description and analysis of potential risks based on a taxonomic identification of the producing organism and including safety studies of the product in animal and in-vitro safety tests. For microbial products which already were on the market before the GRAS system was implemented the petition could be based on a history of safe use. The petition for microbial enzyme products already on the market prior to 1958 was submitted to FDA by the US branch organisation of enzyme producers and became known as the GRAS16 petition. For *Aspergillus niger* it were products based on the enzymes pectinase, protease, carbohydrase, catalase and glucose oxidase. So if now a producer wants to introduce a new pectinase product in the US market and he can demonstrate that it is produced with a safe strain of *A. niger* he can legally do this without having to submit a new dossier. If he does the same but with another non-approved production organism he has to file a new safety dossier. Currently the system of GRAS petitions has been replaced with GRAS notifications: the FDA no longer does formally approves the dossier, but only states that after careful examination of the data provided the FDA has no further questions, but all responsibility for the product is with the producing company.

Many other countries in the world that approve microbe derived products use positive lists of permitted combinations of product and producing micro-organism. Therefore when due to progress in taxonomy name changes are proposed regulatory authorities (mostly layman with no feeling for taxonomy) will become concerned because the combination of product and producing organism is not present on the positive lists of approved combinations.

This raises the question of what constitutes a proper identification of a strain. Many regulatory bodies have a strong preference for applying genetic tools in a proper identification. This may be appropriate for bacterial strains, but for filamentous fungi this is not the case. Here an expert evaluation of morphological and physiological data, combined with information on the extrolite composition and genetic data may yield the best identification but it is for the fungal expert to decide which data are the most appropriate and this will be species-dependent.

Another problem which may arise with identification is that as a consequence of industrial strain improvement programs species characteristics may be lost to such an extent that proper identification of a production strain becomes difficult. This can be solved by identification of the original isolate from nature supported by a confirmation of the identification on the final production strain.

And let's not forget that strain identification only is a tool in the risk analysis process of identification of possible hazards for the safety of the final product. When the strain has been identified properly all available literature data on the species can be used to see which safety parameters need to be checked to ensure the safety of the final product. When taxonomists decide to adopt another name for a given species this does not change the safety of a strain with a documented history of safe use nor the safety of the derived products.

The safety of *A. niger* and *A. oryzae* as production organisms has been documented in the literature (Barbesgaard *et al.*, 1992; Schuster *et al.*, 2002). The main concern, as with many filamentous fungi, is the potential production of extrolites with toxic properties. Blumenthal (2004) has given an overview of the toxic metabolites produced by *A. niger* and *A. oryzae*. For *A. oryzae* the metabolites which are of toxicological concern are 3-nitropropionic acid and cyclopiazonic acid. *A. niger* isolates may produce ochratoxin A. In the development of industrial production strains the producers have to secure that their production strains do not produce such compounds, at least not in levels which might be of toxicological concern. For the DSM *A. niger* enzyme production strains this is documented in various publications (Van Dijck *et al.*, 2002, 2003).

3. Commercially produced enzymes and pharmaceutical proteins

With the development for recombinant technology both *A. niger* and *A. oryzae* have been developed into important hosts to overproduce enzymes and pharmaceutical proteins. An overview of safe enzyme products made with *A. niger* and *A. oryzae*, from both traditionally improved and recombinant sources has been published by

Pariza and Johnson (2001) and Olempska-Beer *et al.* (2006). The enzyme producers, organised in AMFEP (Association of manufacturers and formulators of enzyme products, Brussels) and ETA (Enzyme Technical Association, Washington, D.C.) maintain up-to-date overviews on their websites: (http://www.amfep.org/list.html and http://www.enzymetechnicalassoc.org/fcclist.pdf). A list of commercial enzymes produced by various *Aspergillus* species is presented in Table 1.

Table 1. Commercial enzyme preparations produced by Aspergillus *species for food and feed processing [adapted from AMFEP (Association of Manufacturers and Formulators of Enzyme Products, Brussels)].*

Enzyme	Host organism	Donor organism	IUB number	Application		
				Food	Feed	Techn.
Aminoacylase	*Aspergillus melleus*	none	3.5.1.14	Y	N	Y
Aminopeptidase	*Aspergillus niger*	none	3.4.11.x	Y	N	N
Aminopeptidase	*Aspergillus oryzae*	none	3.4.11.x	Y	N	Y
AMP deaminase	*Aspergillus melleus*	none	3.5.4.6	Y	N	N
Amylase (alpha)	*Aspergillus niger*	none	3.2.1.1	Y	N	N
Amylase (alpha)	*Aspergillus niger* var. *awamori*	none	3.2.1.1	Y	N	N
Amylase (alpha)	*Aspergillus oryzae*	none	3.2.1.1	Y	Y	Y
Arabinanase	*Aspergillus níger*	none	3.2.1.99	Y	Y	N
Arabinofuranosidase	*Aspergillus niger*	*Aspergillus* sp.	3.2.1.55	Y	N	N
Arabinofuranosidase	*Aspergillus niger*	none	3.2.1.55	Y	N	N
Asparaginase	*Aspergillus niger*	*Aspergillus* sp.	3.5.1.1	Y	N	N
Carboxypeptidase (serine-type)	*Aspergillus niger*	*Aspergillus* sp.	3.4.16.x	Y	N	N
Catalase	*Aspergillus niger*	*Aspergillus* sp.	1.11.1.6	Y	N	Y
Catalase	*Aspergillus niger*	none	1.11.1.6	Y	N	Y
Catalase	*Aspergillus oryzae*	*Scytalidium* sp.	1.11.1.6	N	N	Y
Cellulase	*Aspergillus niger*	none	3.2.1.4	Y	Y	N
Cellulase	*Aspergillus oryzae*	*Humicola* sp.	3.2.1.4	N	N	Y
Cellulase	*Aspergillus oryzae*	*Myceliopthora* sp.	3.2.1.4	N	N	Y
Cellulase	*Aspergillus oryzae*	*Thielavia* sp.	3.2.1.4	N	N	Y
Galactosidase (alpha)	*Aspergillus niger*	none	3.2.1.22	Y	Y	N
Galactosidase (alpha)	*Aspergillus oryzae*	*Aspergillus* sp.	3.2.1.22	N	Y	N
Glucanase (beta)	*Aspergillus aculeatus*	none	3.2.1.6	Y	Y	N

Table 1. Continued.

Enzyme	Host organism	Donor organism	IUB number	Application		
				Food	Feed	Techn.
Glucanase (beta)	*Aspergillus niger*	none	3.2.1.6	Y	Y	N
Glucanase (beta)	*Aspergillus oryzae*	*Thermoascus* sp.	3.2.1.6	Y	N	N
Glucoamylase or Amyloglucosidase	*Aspergillus niger*	*Aspergillus* sp.	3.2.1.3	Y	N	N
Glucoamylase or Amyloglucosidase	*Aspergillus niger*	none	3.2.1.3	Y	N	Y
Glucoamylase or Amyloglucosidase	*Aspergillus niger*	*Talaromyces* sp.	3.2.1.3	Y	N	Y
Glucose oxidase	*Aspergillus niger*	*Aspergillus* sp.	1.1.3.4	Y	N	N
Glucose oxidase	*Aspergillus niger*	none	1.1.3.4	Y	N	Y
Glucose oxidase	*Aspergillus oryzae*	*Aspergillus* sp.	1.1.3.4	Y	N	N
Glucosidase (alpha)	*Aspergillus niger*	none	3.2.1.20	Y	N	N
Glucosyltransferase or Transglucosidase	*Aspergillus foetidus*	none	2.4.1.24	Y	N	Y
Hemicellulase	*Aspergillus foetidus*	none	-	Y	Y	Y
Hemicellulase	*Aspergillus niger*	none	-	Y	Y	N
Inulase	*Aspergillus niger*	none	3.2.1.7	Y	N	N
Laccase	*Aspergillus oryzae*	*Myceliopthora* sp.	1.10.3.2	Y	N	Y
Laccase	*Aspergillus oryzae*	*Polyporus* sp.	1.10.3.2	Y	N	Y
Lactase or Galactosidase (beta)	*Aspergillus oryzae*	*Aspergillus* sp.	3.2.1.23	Y	N	N
Lactase or Galactosidase (beta)	*Aspergillus oryzae*	none	3.2.1.23	Y	N	N
Lipase triacylglycerol	*Aspergillus niger*	*Candida* sp.	3.1.1.3	Y	N	N
Lipase triacylglycerol	*Aspergillus niger*	none	3.1.1.3	Y	N	N
Lipase triacylglycerol	*Aspergillus oryzae*	*Aspergillus* sp.	3.1.1.3	N	N	Y
Lipase triacylglycerol	*Aspergillus oryzae*	*Candida* sp.	3.1.1.3	Y	N	Y
Lipase triacylglycerol	*Aspergillus oryzae*	*Fusarium* sp.	3.1.1.3	Y	N	N
Lipase triacylglycerol	*Aspergillus oryzae*	*Humicola* sp.	3.1.1.3	N	N	Y
Lipase triacylglycerol	*Aspergillus oryzae*	*Rhizomucor* sp.	3.1.1.3	Y	N	Y
Lipase triacylglycerol	*Aspergillus oryzae*	*Thermomyces* sp.	3.1.1.3	Y	N	Y
Mannanase (endo-1.4-beta)	*Aspergillus niger*	none	3.2.1.78	Y	Y	N
Pectin lyase	*Aspergillus japonicus*	none	4.2.2.10	Y	N	N

Table 1. Continued.

Enzyme	Host organism	Donor organism	IUB number	Application		
				Food	Feed	Techn.
Pectin lyase	*Aspergillus niger*	*Aspergillus* sp.	4.2.2.10	Y	N	N
Pectin lyase	*Aspergillus niger*	none	4.2.2.10	Y	Y	N
Pectin lyase	*Aspergillus niger* var. *awamori*	*Aspergillus* sp.	4.2.2.10	Y	Y	Y
Pectin lyase	*Aspergillus sojae*	none	4.2.2.10	Y	N	N
Pectin methylesterase or Pectinesterase	*Aspergillus sojae*	none	3.1.1.11	Y	N	N
Pectin methylesterase or Pectinesterase	*Aspergillus japonicus*	none	3.1.1.11	Y	N	N
Pectin methylesterase or Pectinesterase	*Aspergillus niger*	*Aspergillus* sp.	3.1.1.11	Y	Y	N
Pectin methylesterase or Pectinesterase	*Aspergillus niger*	none	3.1.1.11	Y	Y	N
Pectin methylesterase or Pectinesterase	*Aspergillus oryzae*	*Aspergillus* sp.	3.1.1.11	Y	N	Y
Pentosanase	*Aspergillus niger*	none	-	Y	N	N
Peroxidase	*Aspergillus oryzae*	*Coprinus* sp.	1.11.1.7	N	N	Y
Phosphatase	*Aspergillus niger*	none	3.1.3.2	Y	N	N
Phospholipase A	*Aspergillus niger*	*Aspergillus* sp.	3.1.1.4	Y	N	N
Phospholipase A	*Aspergillus oryzae*	*Fusarium* sp.	3.1.1.4	Y	N	N
Phospholipase B	*Aspergillus niger*	*Aspergillus* sp.	3.1.1.5	Y	N	N
Phospholipase B	*Aspergillus niger*	none	3.1.1.5	Y	N	N
Phospholipase B	*Aspergillus niger var. awamori*	none	3.1.1.5	Y	Y	N
Phytase	*Aspergillus niger*	*Aspergillus* sp.	3.1.3.8	Y	N	N
Phytase	*Aspergillus niger*	none	3.1.3.8	Y	N	N
Phytase	*Aspergillus oryzae*	*Peniophora* sp.	3.1.3.8	Y	Y	N
Polygalacturonase or Pectinase	*Aspergillus aculeatus*	none	3.2.1.15	Y	N	N
Polygalacturonase or Pectinase	*Aspergillus niger*	*Aspergillus* sp.	3.2.1.15	Y	N	N
Polygalacturonase or Pectinase	*Aspergillus niger*	none	3.2.1.15	Y	Y	Y
Polygalacturonase or Pectinase	*Aspergillus pulverulentus*	none	3.2.1.15	Y	N	N

Table 1. Continued.

Enzyme	Host organism	Donor organism	IUB number	Application		
				Food	Feed	Techn.
Protease (incl. milkclotting enzymes)	*Aspergillus melleus*	none	3.4.2x.x	Y	N	N
Protease (incl. milkclotting enzymes)	*Aspergillus niger*	*Aspergillus* sp.	3.4.2x.x	Y	N	N
Protease (incl. milkclotting enzymes)	*Aspergillus niger*	none	3.4.2x.x	Y	Y	N
Protease (incl. milkclotting enzymes)	*Aspergillus niger* var. *awamori*	Calf stomach	3.4.2x.x	Y	N	N
Protease (incl. milkclotting enzymes)	*Aspergillus oryzae*	none	3.4.2x.x	Y	Y	Y
Protease (incl. milkclotting enzymes)	*Aspergillus oryzae*	*Rhizomucor* sp.	3.4.2x.x_	Y	N	Y
Protease (incl. milkclotting enzymes)	*Aspergillus sojae*	none	3.4.2x.x	Y	Y	Y
Tannase	*Aspergillus niger*	none	3.1.1.20	Y	Y	N
Transglucosidase	*Aspergillus niger*	none	2.4.1.24	N	N	Y
Xylanase	*Aspergillus foetidus*	none	3.2.1.8	Y	Y	Y
Xylanase	*Aspergillus niger*	*Aspergillus* sp.	3.2.1.8	Y	N	N
Xylanase	*Aspergillus niger*	none	3.2.1.8	Y	Y	N
Xylanase	*Aspergillus níger* var. *awamori*	*Aspergillus* sp.	3.2.1.8	Y	Y	N
Xylanase	*Aspergillus níger* var. *awamori*	none	3.2.1.8	Y	Y	N
Xylanase	*Aspergillus oryzae*	*Aspergillus* sp.	3.2.1.8	Y	N	N
Xylanase	*Aspergillus oryzae*	*Thermomyces* sp.	3.2.1.8	Y	Y	N

References

Barbesgaard, P., H.P. Heldt-Hansen and B. Diderichsen, 1992. On the safety of *Aspergillus oryzae*: a review. Applied Microbiology and Biotechnology 36: 569-572.

Bennett, J.W., 2001. *Aspergillus* and koji: history, practice and molecular biology. Society for Industrial Microbiology News 51: 65-71.

Blumenthal, C.Z., 2004. Production of toxic metabolites in *Aspergillus niger*, *Aspergillus oryzae*, and *Trichoderma reesei*: justification of mycotoxin testing in food grade enzyme preparations derived from the three fungi. Regulatory Toxicology and Pharmacology 39: 214-228.

Olempska-Beer, Z.S., R.I. Merker, M.D. Ditto and M.J. DiNovi, 2006. Feed-processing enzymes from recombinant microorganisms – a review. Regulatory Toxicology and Pharmacology 45: 144-158.

Pariza, M.W. and E.A. Johnson, 2001. Evaluating the safety of microbial enzyme preparations used in food processing: update for a new century. Regulatory Toxicology and Pharmacology 33: 173-186.

Schuster, E, N. Dunn-Coleman, J.C. Frisvad and P.W.M. van Dijck, 2002. On the safety of *Aspergillus niger* – a review. Applied Microbiology and Biotechnology 59: 426-435.

Van Dijck, P.W.M, G.C.M. Selten and R.A. Hempenius, 2003. On the safety of a new generation of DSM *Aspergillus niger* production strains. Regulatory Toxicology and Pharmacology 38: 27-35.

Van Dijck, P.W.M., H. Spierenburg, J.C. Frisvad and R.A. Samson, 2002. The influence of strain improvement and genetic modification on the pattern of secondary metabolites in industrial strains of *Penicillium chrysogenum* and *Aspergillus niger*. Presentation M-60 (Book of abstracts p. 23) 10th IUMS (International Union of Mycological Societies), Paris, France.

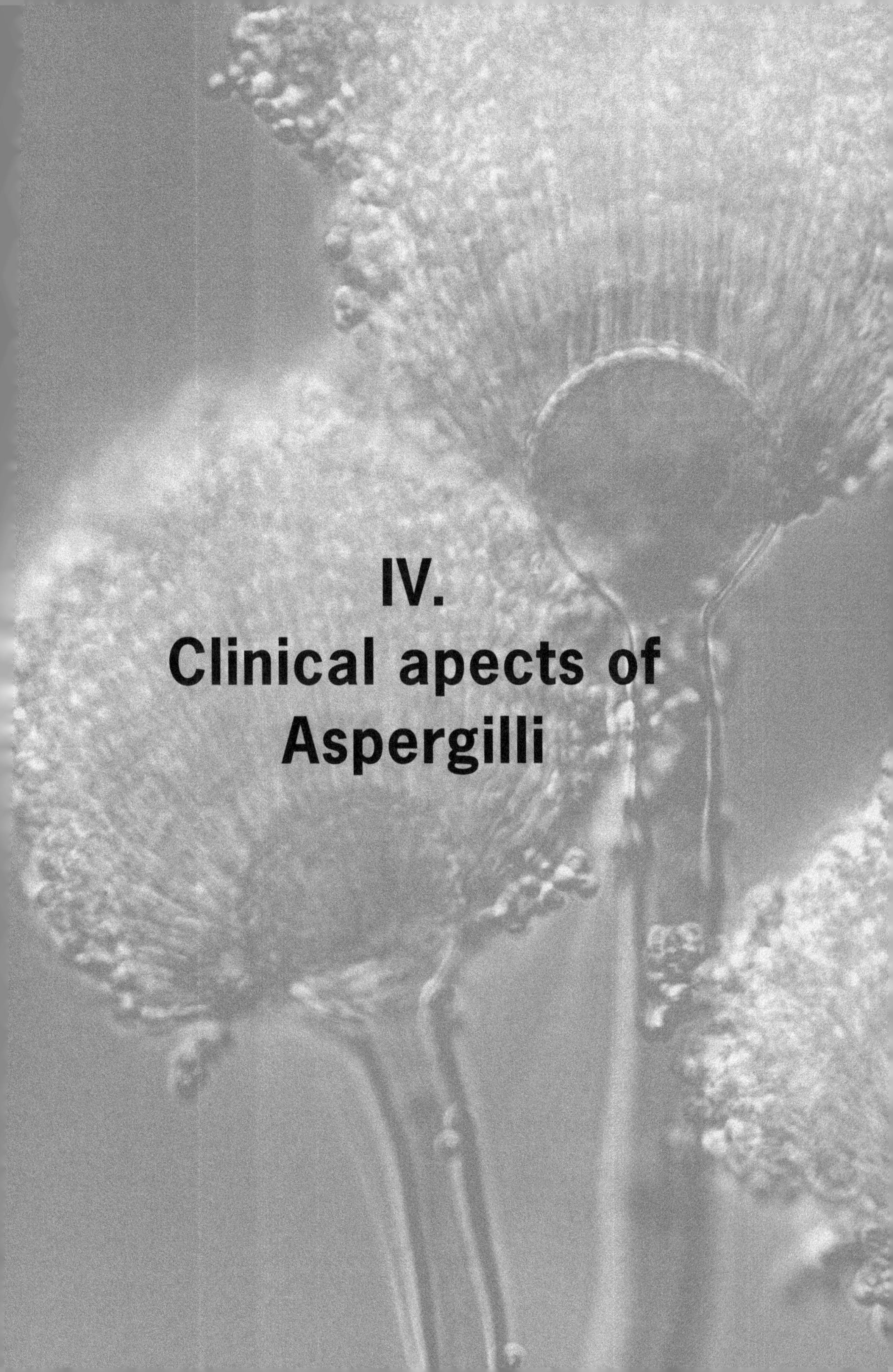

IV.
Clinical apects of Aspergilli

DNA sequence based methods for species identification in the genus *Aspergillus*

S. Arunmozhi Balajee
Mycotic Diseases Branch, Centers for Disease Control and Prevention, Mail stop - G 11, 1600 Clifton Road, Atlanta, GA 30333, USA

Abstract

Identification of species within the genus *Aspergillus* has entered an exciting era where molecular methods such as DNA sequence based schemes facilitate more rapid and objective species diagnosis. Currently available sequencing schemes for the genus *Aspergillus* are discussed in this chapter along with potential targets that can be used for genus and species level identification.

Keywords: identification, genotyping, comparative sequence methods, *Aspergillus* species

1. Introduction

Aspergilli are truly diverse organisms and a species concept should reflect this tremendous diversity appropriately as well as serve as a framework that is functional and relevant. Identification of aspergilli remains a vital task in the clinical microbiology laboratory and can be helpful in predicting drug resistant *Aspergillus* species, detecting novel agents of infections and identifying clusters of nosocomial isolates. Historically, methods based on the morphological species concept have been employed for identification of *Aspergillus* isolates, but these phenotype based identification formats have been plagued by many factors including (1) presence of overlapping morphological features among closely related species; (2) lack of / or poor sporulation of some clinical aspergilli; (3) presence of aberrant phenotypes in clinical isolates that are not consistent with those of 'type' isolates; (4) inexperience of trained professionals largely because of increasing recovery of less common and therefore unfamiliar aspergilli from clinical specimens; (5) requirement of several days to weeks to accomplish identification. In contrast, molecular methods of *Aspergillus* species identification largely using comparative sequence based methodologies ensure that identification formats are more rapid and objective and not subject to inter-observer variations. This chapter will outline the current DNA sequence based identification strategy that is available for species identification for the genus *Aspergillus* with particular reference to clinically relevant aspergilli.

2. Comparative DNA sequencing strategy

2.1. Steps in generating a DNA sequence

The first step in utilising DNA sequence based methodologies for *Aspergillus* identification is to generate appropriate sequence data from an unknown organism. Determination of a DNA sequence entails several steps and will be briefly outlined here: (1) the 'unidentified organism' is grown in broth or solid media and genomic DNA extracted using standard methods; (2) this DNA is used as the template for PCR to amplify a segment of the desired gene; (3) the resultant PCR amplicons are purified to remove excess primers and nucleotides by one of several commercially available methods (enzyme based such as Exo-SapIT (Cleveland, OH) or column based purification methods); (4) the purified amplicon is then subjected to cycle sequencing utilising (usually) the same primer pairs as the PCR reaction; (5) the resulting products are purified to remove unincorporated dye terminators and the sequence is then determined using capillary electrophoresis (6) the sequence is then imported into any commercially available software such as Sequencher^{TM} (Ann Arbor, MI) or DNAStar (Madison, WI) and edited manually. The DNA sequences are usually assembled by aligning the forward and reverse sequences and a consensus sequence is generated. The resultant sequence from an unknown organism can then be analysed in one of several ways to obtain a species identification.

2.2. Obtaining a species identification from the DNA sequence

The consensus sequence obtained from the unknown organism can be queried against a database library and sequence comparison and evaluation for species identification can then be performed by generating dendograms, examining percent similarity/ percent dissimilarity, or executing more sophisticated phylogenetic analyses. Sequence-comparing software packages include PAUP, BLAST and Phylip; among these the BLAST or Basic alignment tool is simple to use and freely available through the publicly available database such as the GenBank database (http://www.ncbi.nlm.nih.gov/BLAST/). Sequence comparisons can be shown as dendograms, using methods such as the Neighbor-joining (NJ method) or the unweighted pair group method with arithmetic averages (UPGMA). Here, closely related species are clustered together based on the sequence information and can therefore be used to assess the identity of an unknown sequence.

Another way of interpreting sequence comparison results is the use of percent identity scores where the percent identity value is a single numeric score determined for each pair of aligned sequences. Thus, it measures the number of identical nucleotide matches in relation to the length of the alignment when the 'query sequence' (sequence from the unknown isolate) is aligned with another sequence from a sequence database. The percent identity values can then be used to assign a species identity and has been previously investigated in bacterial populations (Clarridge, 2004). It must be remembered that assignment of such cut-off values for species identification is entirely arbitrary and limitations include considerable variability of the percent identity values depending on the type of database used, the length of the sequence fragment and the software program employed. In addition, a single value for the definition of a species or genus may not be appropriate for all genera.

Recently, the genealogical concordance of phylogenetic species recognition concept (GCPSR) has been proposed for fungal species recognition by Taylor and colleagues (Taylor *et al.*, 2000). In this multigene phylogenetic approach, species boundaries are recognised at the transition point between concordance and non-concordance of gene genealogies (sequences of the multiple loci investigated) given that there is a lack of genetic exchange between species. Thus, for asexual fungi, GCPSR would identify any two isolates as separate species, where a species is defined based on evidence of genetic isolation by analysis of the congruence of genealogies. Given that almost 40% of described fungi are asexual and morphology appears to be a poor marker for species for several of these fungi, the GCPSR concept is finding wide spread use in mycology for species recognition and for determining population structure of fungi (Taylor and Fisher, 2003; Taylor *et al.*, 2000). A similar strategy, the Multilocus Sequence Typing method (MLST) uses nucleotide sequences from multiple protein coding genes to characterise genetic diversity of the organism in question by generating sequence types and allelic profiles.

One limitation of any comparative sequence identification scheme is the lack of quality controlled, and often incomplete sequences available in public databases such as GenBank. In addition, sequences from non-type strains that may not be a representative of 'true' species further complicates correct species identification. Data from one elegant study assessing the utility of internal transcribed spacer regions (ITS) for *Aspergillus* identification revealed that a GenBank query of some sequences gave best match results not only with reference sequences of relevant type strains but also gave excellent scores to other completely unrelated taxa (Hinrikson *et al.*, 2005). Closer analyses of these results revealed that several of the GenBank entries displayed ambiguous sequence annotations, thus presenting as 'false molecular siblings'. Caution must therefore be exercised when performing sequence searches where the results should be scrutinised carefully before a conclusive identification is reached. Alternatively, several commercially available curated databases are becoming available for *Aspergillus* identification such as the Microseq®D2 (Applied

Biosystems, Foster City, Calif.) and the Smartgene IDNS (SmartGene, Inc., Raleigh, NC) identification systems. However, these databases are expensive and as yet not comprehensive, and need to be enlarged to include more sequences to ensure correct identification of diverse species.

3. The genus *Aspergillus*

Aspergilli remain the most common filamentous fungi that cause invasive infections in immunocompromised populations such as hematopoietic stem cell transplant and solid organ transplant patients. According to a recent, large multicenter study monitoring trends in the incidence of invasive fungal infections (IFI) in recipients of organ transplants [Transplant Associated Infection Surveillance Network Study 2001-2006], almost half of the IFIs were attributed to the genus *Aspergillus* (43.6%, unpublished data). *Aspergillus fumigatus* is the predominant etiological agent of IFI, but other Aspergilli such as *A. terreus, A. flavus* and *A. ustus* are also emerging as significant opportunistic pathogens. In addition, *in vitro* and *in vivo* resistance of Aspergilli to several classes of antifungals is being reported. For instance, *A. terreus* appears to be intrinsically resistant to amphotericin B (AMB), while *A. nidulans* and *A. flavus* have a tendency to have lower susceptibilities to AMB and the antifungal drug itraconazole (ITZ). Although antifungal drug resistance in *A. fumigatus* is rare, several recent reports have demonstrated the presence of multi-drug resistant isolates of *A. fumigatus. Aspergillus lentulus,* a species within the section *Fumigati,* has decreased *in vitro* susceptibilities to multiple antifungals while *A. udagawae* and *N. pseudofischeri* (isolates that are rarely recovered from clinical samples) have lower *in vitro* susceptibility to ITZ (Balajee *et al.*, 2004, 2005a,b, 2006).

4. Molecular identification of the infrageneric species

The genus *Aspergillus,* of which there is approximately 250 members (and predicted to increase) is divided into 7 subgenera which in turn are subdivided into sections (Figures 1 and 2). Each section can comprise a few to several closely related species. This classification is illustrated in Figure 1.

Aspergilli that have been recovered from clinical samples are found scattered throughout the genus with the most commonly recovered clinical fungi belonging to sections *Fumigati, Flavus, Terrei, Nigri* and *Usti.* For identification of species at the inter-section level such as *A. fumigatus, A. terreus* and *A. flavus,* sequence comparison of the various ribosomal gene regions appears useful. The most prevalent of these rRNA targets to date have been the D1-D2 regions of the large-subunit rRNA and the internal transcribed spacer 1 and 2 (ITS1 and ITS2 with 5.8S rRNA)

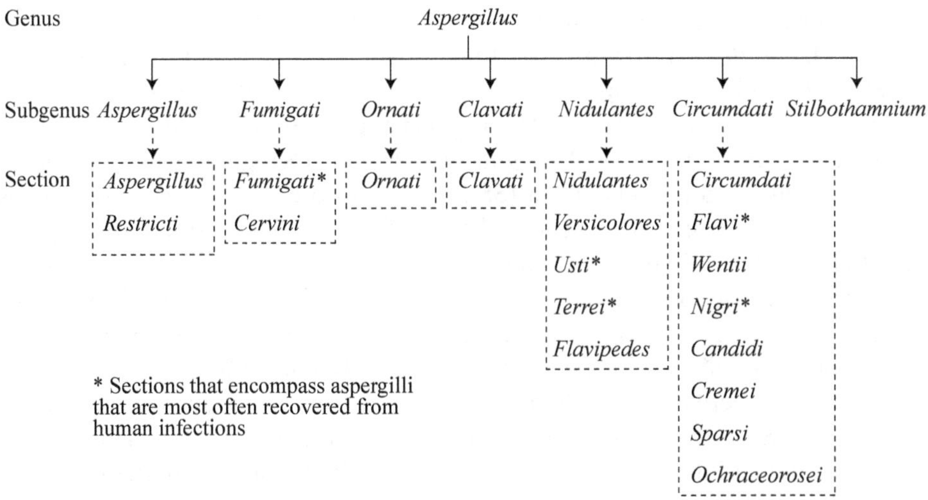

Figure 1. The genus Aspergillus *is divided into 7 subgenus; subgenus is subdivided into sections. Sections that encompass Aspergilli that are most often recovered from human infections are denoted with an asterisk. This figure has been modified from Table 1 (Klich, 2006) with permission from Maren A. Klich.*

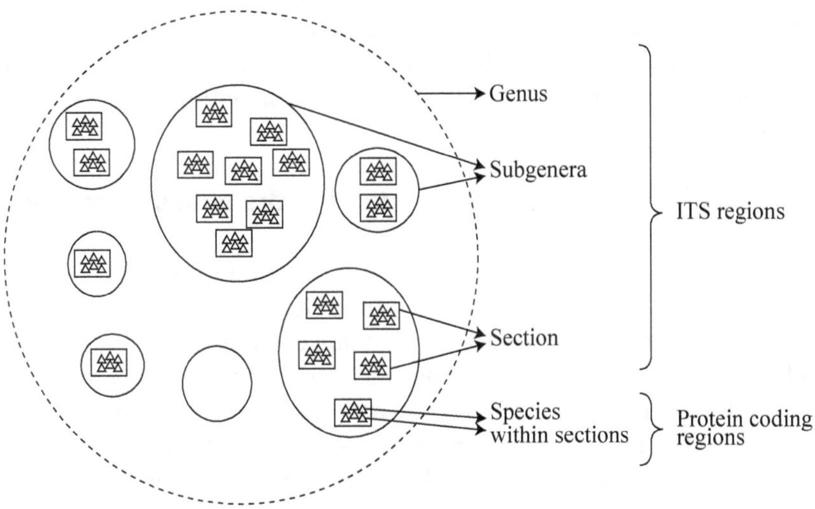

Figure 2. The members of the genus Aspergillus *and target loci are shown in this schematic representation. Species within the sections are represented as triangles; the different sections are denoted as rectangles within the individual subgenus (each subgenus shown as solid circles). The two tiers of identification using comparative sequence analyses of the ITS and protein coding regions are specified.*

regions. Recent data show convincingly that the D1-D2 regions may be useful to distinguish *Aspergillus* sp. from other fungal genera but have limited utility for intraspecies identification (Hinrikson *et al.*, 2005) because of negligible sequence diversity. In contrast, the ITS1 and ITS2 regions appear to discriminate the several species that constitute the genus *Aspergillus* where the ITS region primers (Table 1) make use of conserved regions of the 18S (ITS 1) and the 28S (ITS 4) rRNA genes to amplify the intervening 5.8S gene and the ITS 1 and ITS 2 noncoding regions (Henry *et al.*, 2000). PCR amplification using these primer sets yields a 550 – 600 bp amplicon size; the resultant sequence can be used to query the GenBank database for species identification. Although at present there are no consensus cut-off values for *Aspergillus* species delineation, a percent identity match of 100% with a type isolate should assure respective species identity in most cases. An additional advantage of employing ITS region for species identification include the universality of the primer sets, amenability of the regions to PCR amplification and sequencing and the availability of a large and ever expanding ITS sequence database in the publicly available GenBank database.

Thus, the ITS regions have several advantages and can therefore be a useful workhorse for sub genus and section level species identification (inter-section level) within the genus *Aspergillus* such as *A. fumigatus* and *A. terreus*. However, the ITS regions have limited utility for finer levels of discrimination such as delineation of species within the individual sections (intra-section level), for example *A. lentulus* and *N. udagawae* within the section *Fumigati*. The following paragraphs will outline the sequence based strategies that are available for species recognition within the sections in the genus *Aspergillus*.

5. Identification of species within the sections in the genus *Aspergillus*

Comparative sequence analyses of the ITS regions is useful for intersection species level discrimination but has limited utility in intrasection level species identification. Recent studies suggest that comparative sequencing of multiple protein coding loci may provide intra-section level resolution and can delineate species within the sections. The subsequent parts of the review will discuss the relevant loci that can be useful targets within the sections *Fumigati*, *Terrei*, *Usti* and *Flavi*. Discussions are limited to the identification of clinically relevant species within these four sections.

5.1. *Aspergillus fumigatus*, section *Fumigati*

Aspergillus fumigatus is notoriously difficult to identify by morphological characteristics alone because of significant overlap in phenotypic features among

Table 1. Primer sequences useful for Aspergillus *molecular identification.*

Purpose	Useful targets	Primer sequence	Reference
Genus level identification and Intra-species identification	ITS1-5.8S-ITS2	**ITS 1 F**: TCC GTA GGT GAA CCT GCG G	(Henry *et al.*, 2000)
		ITS 4 R: TCC TCC GCT TAT TGA TAT	
Identification of species within the section			
Section *Fumigati*	*benA*	***benA* F**: ATTGGTGCCGCTTTCTGG	(Balajee *et al.*, 2005)
		***benA* R**: AGTTGTCGGGACGGAATAG	
Section *Terrei*	*benA*	***benA* F:** GGGGATAGGATGTTTTGTGACA	(Balajee *et. al.*, unpublished results)
		***benA* R:** GGTCAACGAGGACGGCACGA	
Section *Flavi*	*Omt12*	***omt1* F**: GGAGTATCAGAGGATTTA	(Geiser *et al.*, 1998)
		***omt2* R**: AGTGCTGTAATAGTCAAA	
Section *Usti*	*benA*	**Bt2a F**: 5'-GGTAACCAAATCGGTGCTGCTTTC-3'	(Varga *et al.*, unpublished results)
		Bt2b R: 5'-ACCCTCAGTGTAGTGACCCTTGGC-3'	

ITS - the internal transcribed spacer 1 and 2 (ITS1 and ITS2 with 5.8S rRNA), *benA* - β tubulin, *Omt12* - *O*-methyltransferase.

organisms constituting section *Fumigati* (Balajee and Marr, 2006). Thus, recent molecular studies have unveiled several new species and previously recognised species amongst isolates identified phenotypically as *A. fumigatus* (Balajee *et al.*, 2005a,b; Hong *et al.*, 2005, 2006; Pringle *et al.*, 2005; Yaguchi *et al.*, 2007). Two studies will be illustrated to demonstrate the role of molecular based methods in species recognition within the section *Fumigati*. The first study involved the discovery of a new species, *A. lentulus* recovered from isolates identified as *A. fumigatus*. Interestingly, this newly discovered species was first described as a slow sporulating variant of *A. fumigatus* using molecular methods such as random amplified polymorphic DNA (Balajee *et al.*, 2004). Detailed phylogenetic analyses of the ITS and protein coding locus such as the salt responsive gene regions, *benA* and *rodA* regions revealed that these 'variants' of *A. fumigatus* represented a new species within section *Fumigati*. *Post hoc* morphological analyses of the *A. lentulus* type isolate revealed minor differences between *A lentulus* and *A. fumigatus* and major differences in secondary metabolite profiles (Larsen *et al.*, 2007). Notably, none of the *A. lentulus* isolates produced the purported 'virulence' factor gliotoxin, but produced other metabolites such as cyclopiazonic acid and neosartorin (Larsen *et al.*, 2007). Subsequent studies have demonstrated a global distribution of *A. lentulus*, with members of this species being recovered from environmental sources such as soil (Hong *et al.*, 2005) as well as clinical samples. Significantly, all *A. lentulus* isolated so far have low susceptibilities *in vitro* to multiple antifungals that are available for therapy, thereby underlining the clinical relevance of identification of this species.

The second study revealed the risk of relying on morphological features alone for species identification within the section *Fumigati*. Several members in the section *Fumigati* are sexual and the teleomorphic (sexual) states of these isolates within section *Fumigati* are assigned to the genus *Neosartorya*. Thus, on artificial media in the laboratory, the *Neosartorya* species within the section *Fumigati* produce sexual fruiting bodies called cleistothecia which serve as diagnostic markers for these taxa. Unfortunately, clinical isolates often do not produce these sexual structures on laboratory media and are thus refractory to traditional methods of identification. Comparative DNA sequence based analyses appears to be useful in species delimitation of such isolates. For instance, a recent molecular investigation of several clinical *A. fumigatus* isolates revealed that these isolates were not *A. fumigatus* but *N. pseudofischeri*. The anamorphic state of *N. pseudofischeri*, *A. thermomutatus*, is indistinguishable from *A. fumigatus* and since initially none of the isolates produced diagnostic fruiting bodies all of these isolates were misidentified as *A. fumigatus* (Balajee *et al.*, 2005a). Sequence comparison of the *benA* and *rodA* regions revealed that all these isolates were 100% homologous to *N. pseudofischeri* and were eventually identified as *A. thermomutatus* (anamorphic state).

Likewise, a previously recognised *Aspergillus* species, *A. udagawae* was also identified by sequence based methods from clinical specimens (Balajee *et al.*, 2006). Two new species, *A. fumigatiaffinis* and *A. novofumigatus*, within the section *Fumigati* have also been described using sequence based methods in combination with phenotype and secondary metabolite analyses (Hong *et al.*, 2005). Both of these species were isolated from soil and have not been implicated in invasive infections. Taken together, molecular methods appear to be more useful in species identification than traditional phenotype based methods for species recognition within the section *Fumigati*. Because the different species within this section have differential *in vitro* susceptibilities to various antifungals, species delineation appears crucial and may influence therapeutic decision making.

Several loci have been evaluated and include protein coding gene regions such as β tubulin (*benA*), calmodulin (*cal*), rodlet A (*rodA*) and other variable loci that encompass microsatellite repeat motifs (Pringle *et al.*, 2005), and intergenic regions between genes including *inter1* (found between *sec61* and *ecm40* on chromosome 5), *inter2* (found between *yll034C* and *erb1* on chromosome 1), and *inter3* (found between *glc3* and *atp2* on chromosome 5) (Rydholm *et al.*, 2006). Additionally, a MLST scheme specific for *A. fumigatus* has also been evaluated (Bain *et al.*, 2007) with the following gene regions: annexin (*ANX4*), β-1,3-glucanosyl transferase (*BGT1*), catalase (*CAT*), lipase (*LIP*), mating type protein (*MAT1-2*), superoxide dismutase (*SOD*) and zinc transporter (*ZRF2*). Most of the evaluated genes have been polymorphic and suitable for species delimitation, recombination studies and population structure analyses. Numerous studies have shown that the *benA* gene can be a useful marker for species delimitation within the section *Fumigati* (Table 1).

5.2. Aspergillus flavus, *section* Flavi

Raper and Fennell (1965) accepted 9 species and 2 varieties within the section *Flavi* using morphological methods. However, subsequent research has shown that this section is more heterogeneous and thus far includes 23 species or varieties based on morphological characteristics alone (Hedayati *et al.*, 2007). Morphological methods of species identification for section *Flavi* remain challenging due to the absence of clear phenotypic differences between the many species within this complex. Numerous molecular studies have resulted in varying numbers of species to be included within this section. Using comparative sequence based methods with the partial regions of the mitochondrial cytochrome b, seven genotypes were distinguished within this section (Wang *et al.*, 2001). Another study using sequences from the rDNA regions delineated 17 type strains within this section and excluded some isolates from this section (Rigó *et al.*, 2002). The ITS regions have also found use in species identification within this section with a recent study demonstrating that *A. flavus* isolates recovered from ocular isolates may be distinct from those recovered from environmental sources (Bagyalakshmi *et al.*, 2007).

Aspergillus flavus is a clinically important member within this section and cause a wide spectrum of disease ranging from sinus infections to invasive infections, and is the second most common cause of invasive and non-invasive aspergillosis. This fungus also infects insects and economically important crops such as maize and peanuts by the production of potent mycotoxins such as aflatoxin. The most severe episode of human aflatoxin poisoning occurred in 2004 in Kenya where aflatoxicoses by *A. flavus* resulted in 125 human deaths with a case fatality rate of 39% (Probst *et al.*, 2007). DNA sequencing and phylogenetic analyses have demonstrated considerable heterogeneity within *A. flavus* isolates. Geiser *et al.* (Geiser *et al.*, 1998) analysed 11 different protein-encoding loci that included acetamidase (*amdS12*), polygalacturonase (*pecA12*), *O*-methyltransferase (*omt12*), nitrate reductase (*niaD12*), β tubulin (*benA*), phosphoglycerate kinase (*pgkA12*), acetate regulation (*facB34*), glucoamylase (*glaA12*), tryptophan synthetase (*trpC13*), calmodulin (*cmdA78*), and glucose-6-phosphate dehydrogenase (*gsdA12*) by exploiting single restriction site polymorphisms in portions of these different protein-encoding loci. Subsequent sequence analyses of 5 of these 11 loci revealed the existence of 16 different genotypes forming two genetically distinct clades I and II within the *A. flavus* isolates; *A. oryzae* isolates grouped in Clade I, while some *A. flavus* isolates producing small sclerotia clustered together in Clade II. Although all genes used in this study were phylogenetically useful, the *omt12* locus appeared to be a hypervariable region useful for *A. flavus* species identification (Table 1). This study also revealed the presence of cryptic species within morphologically indistinguishable isolates and apparent recombinant events in these supposedly asexual organisms.

Although considerable molecular studies have been performed with *A. flavus* isolates recovered from the environment, little is known of the population structure and molecular characteristics of *A. flavus* isolates recovered from clinical samples. Such research data will be of great significance in terms of modes of infection, pathogenesis and drug susceptibilities of the different 'genotypes' that may constitute *A. flavus*. *Aspergillus flavus* appears to have decreased susceptibilities to the antifungal drug AMB and ITZ and curiously, data appear to suggest that not all *A. flavus* isolates have similar susceptibility profiles (Hedayati *et al.*, 2007). The difference in drug susceptibilities between various *A. flavus* isolates may be due to presence of cryptic species in this yet to be tested clinical population. This speculation needs to be tested with a robust set of clinical isolates and comparative sequence analyses of multiple gene regions.

5.3. Aspergillus terreus, *section* Terrei

Aspergillus terreus is a cosmopolitan fungus with a global distribution and has been recovered from soils of deserts, grasslands, from compost heaps, and as contaminants of stored corn, barley and peanuts. This organism produces a range of secondary metabolites, some of which are economically valuable (Varga *et al.*, 2005). Members of this species are also clinically significant as they can cause superficial infections such as onychomycoses, as well as invasive infections in severely immunocompromised hosts (Symoens *et al.*, 2000). Infections due to *A. terreus* are of growing concern since this organism is often associated with disseminated infection and higher mortality in comparison with other *Aspergillus* spp. (Baddley *et al.*, 2003). Further, *A. terreus* appears to have decreased susceptibility to the antifungal drug AMB *in vitro* and this correlates with poor *in vivo* response to therapy (Baddley *et al.*, 2003; Steinbach *et al.*, 2004).

Aspergillus terreus isolates may be variable in their cultural characteristics, and to a limited extent have variable micromorphology. *Aspergillus terreus* grows as yellowish-brown colonies with a yellow reverse with dense conidiophores and smooth walled stipes giving rise to vesicles bearing globose, smooth conidia. Based on these morphological criteria, *A. terreus* is recognised as a single morphospecies within the Section *Terrei* along with 2 other atypical variants, *A. terreus* var. *africanus* and *A. terreus* var. *aureus* (Raper and Fennell, 1965). Phylogenetic analyses of the sequences derived from the ITS regions revealed three main clades that included *A. terreus, A. carneus* and *A. niveus* respectively within the section *Terrei* (Varga *et al.*, 2005). This study concluded that section *Terrei* warranted further detailed molecular analyses; anecdotally, it has been known for sometime now that the single morphospecies *A. terreus* may harbor several cryptic species that may be revealed by DNA sequence based studies. Studies in our laboratory have demonstrated that the β tubulin and calmodulin regions may be useful in species delimitation (Balajee *et al.*, unpublished data, Table 1).

5.4. Aspergillus ustus, *section* Usti

Aspergillus ustus is a soil saprophyte that belongs to section *Usti*, subgenus *Nidulantes.* This species is rarely encountered from clinical specimens. Less than 20 cases of systemic infection have been reported so far with more than half of these infections occurring in the past decade (Panackal *et al.*, 2006). Thus, there is concern that *A. ustus* may be an emerging fungal pathogen that causes fatal human infections. The spectrum of infections due to *A. ustus* includes onychomycosis, otitis media, eye infections, primary cutaneous infection, endocarditis, pneumonia, and disseminated infection, the latter cases occurring largely among immunocompromised hosts such as hematopoietic stem cell transplant recipients. *Aspergillus ustus* may be of particular concern, as they exhibit decreased susceptibility to multiple antifungal drugs and outcome of infections has been generally poor.

Most of the reported studies have used morphological methods for *A. ustus* identification. *Aspergillus ustus* grows as drab olive to dull brown or gray and woolly colonies with occasional dark purple or yellow exudates. Microscopically, the conidia are large (3.0–4.5 μm) and are rough-walled; elongate and irregular-shaped Hülle cells that are resistant to desiccation may also be produced (Klich, 2002). Very few studies have used DNA sequence based identification schemes to study *A. ustus* isolates. Peterson (Peterson, 2000) analysed the phylogenetic relationships based on the D1 and D2 regions of the large subunit ribosomal RNA (lsu-DNA) and demonstrated that the type isolate of *A. ustus* fell within subgenus *Nidulantes*. A recent MLST study using the partial regions of β tubulin, calmodulin and actin genes and the ITS region revealed the presence of a new species *A. calidoustus* (sp. nov.) from predominantly clinical samples (Houbraken *et. al.*, unpublished data). The D1-D2 region appears to have limited utility in delimiting species within this section and protein coding genes such as the *benA* region (Table 1) appear to be an excellent marker for species identification within this section (Jos Houbraken, personal communication).

5.5. Other Aspergilli

Apart from the above Aspergilli, other organisms that have been recovered from human infections include *A. niger*, *A. versicolor* and *A. nidulans*. Recently, a rare isolate of *Petromyces alliaceus* (anamorph *A. alliaceus*) was recovered from a clinical specimen (Balajee *et al.*, 2007). In this study, DNA sequence analysis of the 28SrRNA and the ITS regions secured identification of this fungus. *Petromyces alliaceus* rarely causes human infections and this case report is an example of how previously unrecognised aspergilli that can be refractory to classical methods of species identification may be identified by molecular methods.

6. Conclusion

The decision to place Aspergilli within a subgenus/section or to further identify individuals to species within a section would ultimately depend on the needs of the individual laboratories. If the goal is to identify every individual to species within sections primarily for taxonomic purposes then it would be appropriate to analyse gene genealogies using several protein coding loci outlined in this chapter such as the *benA* and the *cal* region. On the other hand, such a fine level of resolution may not be necessary to a routine clinical microbiology laboratory. If, for instance, the intra-section species identity would indicate a drug resistant phenotype thereby influencing therapeutic decisions and / or the isolate was recovered from a sterile body site of a high risk patient (such as a hematopoietic stem cell transplant recipient), the isolate in question may warrant detailed identification to the species level. Here, a two tier identification strategy may be necessary: (1) an inter-section level identity using the ITS regions to establish the section/complex to which the isolate belongs, (2) a second tier of molecular analyses using one or more of the protein coding regions (as outlined in the preceding sections) for intra-section level identification. This two level identification schema is depicted in Figure 2 where the species within the sections (represented as triangles) can be identified by sequence analysis of protein coding regions while sequence analyses of the ITS regions would be adequate for intersection and subgenus level species identification. To summarise, as a frontline strategy, sequence information from the ITS locus can be used to quickly flag atypical isolates that may need more detailed taxonomical analyses using multiple protein coding loci.

Comparative DNA sequence based methods are powerful schemes that allow for rapid and objective species delineation, enabling communication of information between laboratories. In addition, with the availability of public sequence databases it is now possible to quickly recognise unusual and rare *Aspergilli* that cause disease. Significant progress made in sequencing methodologies and the availability of several fully sequenced *Aspergillus* genomes will propel these DNA sequence identification strategies into fully successful endeavors.

Disclaimer

The findings and conclusions in this article are those of the author and do not necessarily represent the views of the CDC.

References

Baddley, J.W., P.G. Pappas, A.C. Smith and S.A. Moser, 2003. Epidemiology of *Aspergillus terreus* at a university hospital. Journal of Clinical Microbiology 41: 5525-5529.

Bagyalakshmi, R., K.L. Therese and H.N. Madhavan, 2007. Nucleotide polymorphisms associated with Internal Transcribed Spacer (ITS) regions of ocular isolates of *Aspergillus flavus*. Journal of Microbiological Methods 68: 1-10.

Bain, J.M., A. Tavanti, A.D. Davidson, M.D. Jacobsen, D. Shaw, N.A. Gow and F.C. Odds, 2007. Multilocus sequence typing of the pathogenic fungus *Aspergillus fumigatus*. Journal of Clinical Microbiology 45: 1469-1477.

Balajee, S.A. and K.A. Marr, 2006. Phenotypic and genotypic identification of human pathogenic Aspergilli. Future Microbiology 1: 435-445.

Balajee, S.A., D. Nickle, J. Varga and K.A. Marr. 2006. Molecular studies reveal frequent misidentification of *Aspergillus fumigatus* by morphotyping. Eukaryotic Cell 5: 1705-1712.

Balajee, S.A., J.L. Gribskov, E. Hanley, D. Nickle and K.A. Marr, 2005b. *Aspergillus lentulus* sp nov., a new sibling species of *A. fumigatus*. Eukaryotic Cell 4: 625-632.

Balajee, S.A., J. Gribskov, M. Brandt, J. Ito, A. Fothergill and K.A. Marr, 2005a. Mistaken identity: *Neosartorya pseudofischeri* and its anamorph masquerading as *Aspergillus fumigatus*. Journal of Clinical Microbiology 43: 5996-5999.

Balajee, S.A., M.D. Lindsley, N. Iqbal, J. Ito, P.G. Pappas and M.E. Brandt, 2007. A non-sporulating clinical isolate identified as *Petromyces alliaceus* (anamorph *Aspergillus alliaceus*) by morphological and sequence based methods. Journal of Clinical Microbiology 45: 2701-2703.

Balajee, S.A., M. Weaver, A. Imhof, J. Gribskov and K.A. Marr, 2004. *Aspergillus fumigatus* variant with decreased susceptibility to multiple antifungals. Antimicrobial Agents and Chemotherapy 48:1197-203.

Clarridge, J.E., 2004. Impact of 16S rRNA gene sequence analysis for identification of bacteria on clinical microbiology and infectious diseases. Clinical Microbiology Reviews 17: 840-862.

Geiser, D.M., J.I. Pitt and J.W. Taylor, 1998. Cryptic speciation and recombination in the aflatoxin-producing fungus *Aspergillus flavus*. Proceedings of the National Academy of Sciences USA 95: 388-393.

Hedayati, M.T., A.C. Pasqualotto, P.A. Warn, P. Bowyer and D.W. Denning, 2007. *Aspergillus flavus*: human pathogen, allergen and mycotoxin producer. Microbiology 153: 1677-1692.

Henry, T., P.C. Iwen and S.H. Hinrichs, 2000. Identification of *Aspergillus* species using internal transcribed spacer regions 1 and 2. Journal of Clinical Microbiology 38: 1510-1515.

Hinrikson, H.P., S.F. Hurst, T.J. Lott, D.W. Warnock and C.J. Morrison, 2005. Assessment of ribosomal large-subunit D1-D2, internal transcribed spacer 1, and internal transcribed spacer 2 regions as targets for molecular identification of medically important *Aspergillus* species. Journal of Clinical Microbiology 43: 2092-2103.

Hong, S.B., H.S. Cho, H.D. Shin, J.C. Frisvad and R.A. Samson, 2006. Novel *Neosartorya* species isolated from soil in Korea. International Journal of Systematic and Evolutionary Microbiology 56: 477-486.

Hong, S.B., S.J. Go, H.D. Shin, J.C. Frisvad and R.A. Samson, 2005. Polyphasic taxonomy of *Aspergillus fumigatus* and related species. Mycologia 97: 1316-1329.

Klich, M.A., 2002. Identification of common *Aspergillus* species. Centraalbureau voor Schimmelcultures, Utrecht, The Netherlands.

Klich, M.A., 2006. Identification of clinically relevant Aspergilli. Medical Mycology 44 (Suppl): 127-131.

Larsen, T.O., J. Smedsgaard, K.F. Nielsen, M.A. Hansen, R.A. Samson and J.C. Frisvad, 2007. Production of mycotoxins by *Aspergillus lentulus* and other medically important and closely related species in section *Fumigati*. Medical Mycology 45: 225-232.

Panackal, A.A., A. Imhof, E.W. Hanley and K.A. Marr, 2006. *Aspergillus ustus* infections among transplant recipients. Emerging Infectious Diseases 12: 403-408.

Peterson, S.W., 2000. Phylogenetic relationships in *Aspergillus* based upon rDNA sequence analysis. In R.A. Samson and J.I. Pitt (eds.), Integration of modern taxonomic methods for *Penicillium* and *Aspergillus* classification. Harwood Academic Publishers, Amsterdam, pp. 323-355.

Pringle, A., D.M. Baker, J.L. Platt, J.P. Wares, J.P. Latge and J.W. Taylor, 2005. Cryptic speciation in the cosmopolitan and clonal human pathogenic fungus *Aspergillus fumigatus*. Evolution 59: 1886-1899.

Probst, C., H. Njapau and P.J. Cotty, 2007. Outbreak of an acute aflatoxicosis in Kenya in 2004: identification of the causal agent. Applied and Environmental Microbiology 73: 2762-2764.

Raper, B.K. and D.I. Fennell, 1965. The genus *Aspergillus*. The Williams and Wilkins Company, Baltimore, Md.

Rigó, K., J. Varga, B. Tóth, J. Téren, Á. Mesterházy and Z. Kozakiewicz, 2002. Evolutionary relationships within *Aspergillus* section *Flavi* based on sequences of the intergenic spacer regions and the 5.8S rRNA gene. Journal of General and Applied Microbiology 48: 9-16.

Rydholm, C., G. Szakacs and F. Lutzoni, 2006. Low genetic variation and no detectable population structure in *Aspergillus fumigatus* compared to closely related *Neosartorya* species. Eukaryotic Cell 5: 650-657.

Steinbach, W.J., D.K. Benjamin, Jr., D.P. Kontoyiannis, J.R. Perfect, I. Lutsar, K.A. Marr, M.S. Lionakis, H.A. Torres, H. Jafri and T.J. Walsh, 2004. Infections due to *Aspergillus terreus*: a multicenter retrospective analysis of 83 cases. Clinical Infectious Diseases 39: 192-198.

Symoens, F., J.P. Bouchara, S. Heinemann and N. Nolard, 2000. Molecular typing of *Aspergillus terreus* isolates by random amplification of polymorphic DNA. Journal of Hospital Infection 44: 273-280.

Taylor, J.W. and M.C. Fisher, 2003. Fungal multilocus sequence typing--it's not just for bacteria. Current Opinion in Microbiology 6: 351-6.

Taylor, J.W., D.J. Jacobson, S. Kroken, T. Kasuga, D.M. Geiser, D.S. Hibbett and M.C. Fisher, 2000. Phylogenetic species recognition and species concepts in fungi. Fungal Genetics and Biology, 31: 21-32.

Varga, J., B. Toth, S. Kocsube, B. Farkas, G. Szakacs, J. Teren and Z. Kozakiewicz, 2005. Evolutionary relationships among *Aspergillus terreus* isolates and their relatives. Antonie van Leeuwenhoek 88: 141-150.

Wang, L., K. Yokoyama, H. Takahasi, N. Kase, Y. Hanya, K. Yashiro, M. Miyaji and K. Nishimura, 2001. Identification of species in *Aspergillus* section *Flavi* based on sequencing of the mitochondrial cytochrome b gene. International Journal of Food Microbiology 71: 75-86.

Yaguchi, T., Y. Horie, R. Tanaka, T. Matsuzawa, J. Ito and K. Nishimura, 2007. Molecular phylogenetics of multiple genes on *Aspergillus* section *Fumigati* isolated from clinical specimens in Japan. Nippon Ishinkin Gakkai Zasshi 48: 37-46.

Azole resistance in *Aspergillus fumigatus*

Paul E. Verweij, Eveline Snelders and Willem J.G. Melchers
Department of Medical Microbiology, Nijmegen University Center for Infectious Diseases, P.O. box 9101, 6500 HB Nijmegen, The Netherlands

Abstract

Invasive aspergillosis remains an important complication of immunosuppressive treatment regimens in humans. Triazoles are increasingly used in the prevention and management of invasive aspergillosis. Voriconazole is first choice for the primary therapy of invasive aspergillosis, and posaconazole was shown to be highly effective in preventing invasive aspergillus disease when given to high-risk groups prophylactically. Resistance in moulds is uncommon in clinical medicine, but might occur in specific patient groups. It appears that azole resistance might evolve in patients that harbor a high number of reproducing *Aspergillus* and are treated with azoles for long period of time. These conditions are present in patients with aspergilloma or other cavitary lung diseases caused by *Aspergillus* species. Azole resistance has been described to emerge in this patient group. Recently azole resistance was also reported in patients with acute invasive aspergillosis, where the conditions for resistance development appear to be less that optimal: fungal proliferation through hyphal elongation and relative short treatment episodes. This might suggest that azole resistance is caused by exposure outside the patient, for instance in the environment. Molecular studies have shown that triazole resistance in *A. fumigatus* is associated with amino acid substitutions in the cyp51A protein. Alterations might result in different patterns of azole resistance, primarily of itraconazole, with varying reduced activity of other triazoles such as voriconazole and posaconazole. Azole resistance in *A. fumigatus* has been associated with treatment failure. At present clinical microbiology laboratories do not routinely perform in vitro susceptibility testing, but it appears to be appropriate to test *A. fumigatus* isolates at least in those isolates recovered from patients that fail to azole therapy.

Keywords: triazole, *Aspergillus fumigatus*, resistance, molecular mechanism

1. Introduction

Aspergillus species cause a spectrum of diseases in humans including allergic syndromes, mycotoxicosis, non-invasive infection and invasive aspergillosis. Infections by aspergillus are associated with a significant morbidity and mortality, with *Aspergillus fumigatus* being the principle etiological agent. Patients with aspergilloma and other cavitary diseases are difficult to manage and in most patients curation is not achieved (Denning *et al.*, 2003). Invasive aspergillosis is a feared complication of immunosuppressive treatment, most notably in patients with hematological malignancies such as acute myeloid leukemia or myelodysplastic syndrome. Also patients that receive a hematopoietic stem cell transplant or a solid organ transplant have an increased risk for invasive aspergillosis, although the risk of infection varies for the different transplant categories (Maschmeyer *et al.*, 2007). Depending on host factors, less acute invasive infections might occur in those less immunosuppressed, such as in critically ill patients. The fatality rate of invasive aspergillosis is high in most patients (Table 1).

Although the lung is the most commonly affected site, the mould can primarily affect other sites including the sinus. Invasive infections might disseminate to other sites including the cerebrum. The prognosis of disseminated diseases is dismal. The management of *Aspergillus* infection relies on surgery, antifungal therapy and reduction of immunosuppressive therapy if applicable.

Table 1. Fatality rates in invasive aspergillosis for different patient groups (Maschmeyer et al., *2007).*

Patients	Fatality rate (%)
Acute leukemia	30 – 40
Allogeneic hematopoietic stem cell transplantation	60
Solid organ transplantation	50 – 60
Other causes of immunosuppression (intensive care unit, burns, AIDS)	70 – 85

2. Antifungal agents

Several classes of antifungal agents are available for treatment of patients with mycotic diseases (Table 2).

Amphotericin B is a polyene antifungal drug, often used intravenously for systemic fungal infections. It was originally extracted from *Streptomyces nodosus*, a filamentous bacterium, in 1955. Amphotericin B associates with ergosterol, a membrane chemical of fungi, forming a pore that leads to K^+ leakage and fungal cell death. For

Table 2. Antifungal drugs with activity against Aspergillus *spp., spectrum and target.*

Class	Target	Compounds	Spectrum
Polyenes	Binding to ergosterol (fungal cell membrane)	Amphotericin B deoxycholate, liposomal amphotericin B, amphotericin B lipid complex, amphotericin B colloid dispersion	*Candida, Cryptococcus, Trichosporon, Aspergillus, Fusarium*, Zygomycetes, fungi causing endemic mycoses. Not active against most *A. terreus* and *Pseudallescheria* sp.
Azoles	Ergosterol biosynthesis, inhibition of 14-α demethylase	Itraconazole	*Candida, Cryptococcus*, and *Aspergillus*. Variably active in vitro against members of the zygomycetes and no activity against *A. calidoustus*.
		Voriconazole	*Candida, Cryptococcus, Aspergillus, Fusarium, Scedosporium*. No activity against zygomycetes, *A. calidoustus* and *S. prolificans*
		Posaconazole	*Candida, Cryptococcus, Aspergillus, Fusarium, Scedosporium* and zygomycetes. No activity against *A. calidoustus*.
Echinocandines	1,3-β-D-glucan (glucan synthase in fungal cell wall)	Caspofungin, micafungin, anidulafungin.	*Candida* and *Aspergillus*. Reduced activity against *C. parapsilosis* and *C. guillermondii*. No significant activity against *Fusarium, Scedosporium, Trichosporon*, zygomycetes and *Cryptococcus*.

many years conventional amphotericin B, amphotericin B-desoxycholate, was the standard treatment for invasive aspergillosis. Due to its toxicity new formulations have been developed that retain their efficacy, but have less side effects compared to conventional amphotericin B. The lipid-formulations of amphotericin B include liposomal amphotericin B (Ambisome), amphotericin B lipid complex (ABLC, Abelcet) and amphotericin B colloid dispersion (ABCD, Amphocil). Liposomal amphotericin B was shown to be effective for the primary treatment of invasive aspergillosis (Cornely *et al.,* 2007a).

The triazoles are an important group of antifungal agents that have an increasing role in the treatment of aspergillus disease. The triazoles are potent inhibitors of fungal cytochrome P450 rate-limiting enzyme 14α-sterol demethylase. Mammalian demethylase activity is much less sensitive to triazoles than fungal demethylase. This inhibition prevents the conversion of lanosterol to ergosterol, an essential component of the fungal cytoplasmic membrane, and subsequent accumulation of 14α-methyl sterols. The representative preparations include fluconazole, which has no activity against *Aspergillus,* itraconazole, voriconazole and posaconazole. The drugs are relatively well tolerated and can be administered intravenously (itraconazole and voriconazole) or orally (itraconazole, voriconazole and posaconazole). Itraconazole was registered for use in 1991 and is primarily used for the treatment of dermatomycoses. Its role in the treatment of invasive aspergillosis has been modest, primarily due to the limited scientific evidence that supports its use for this indication. Voriconazole is an extended spectrum, synthetic triazole derivative of fluconazole (Scott and Simpson, 2007) and was shown to be superior to amphotericin B and other licensed antifungal therapy for the primary treatment of invasive aspergillosis (Herbrecht *et al.,* 2002). This drug is used as first line therapy for invasive aspergillosis in many centers throughout the world. Posaconazole is a new broad-spectrum triazole, with a molecule structure that is similar to that of itraconazole. Posaconazole is available as an oral solution and was evaluated for its efficacy to prevent invasive aspergillosis in high-risk patients. The drug was shown to be highly effective in preventing invasive aspergillosis in patients with acute myeloid leukemia and myelodysplastic syndrome when it was administered during episodes of neutropenia (Cornely *et al.,* 2007b). Also patients receiving therapy for acute-graft-versus-host disease were found to benefit from posaconazole prophylaxis (Ullmann *et al.,* 2007).

The echinocandins are lipopeptides that have been synthetically modified from the fermentation broths of various fungi. There are three drugs of this group that are available for treatment or will become available shortly. These include caspofungin, micafingin and anidulafungin. The echinocandins work at the level of the cell wall of fungi by inhibiting β-1,3-glucan synthesis. Specifically, they bind to β-1,3-D-glucan synthase, blocking the synthesis of β-1,3-D glucan, which along with chitin provides the integrity and shape of the cell wall (Cappelletty and Eiselstein-McKitrick, 2007). The echinocandins are active against *Aspergillus* species, but they cause aberrant growth of hyphae at the apical tips, and killing does not occur. Caspofungin has been shown to be effective for salvage therapy of patients who failed to primary therapy, with a favorable response to caspofungin therapy in 37 (45%) of 83 patients, including 32 (50%) of 64 with pulmonary aspergillosis and 3 (23%) of 13 with disseminated aspergillosis (Maertens, 2004). The efficacy of caspofungin was evaluated for the primary therapy of invasive aspergillosis in a phase II study, in predominantly neutropenic patients with a proven or probable diagnosis. Caspofungin had a response rate of 33% in the overall population (Viscoli *et al.*, 2007).

Treatment guidelines have been developed that recommend the use of drugs for treatment or prevention of invasive fungal infections in different host groups, based on the scientific evidence available in the medical literature. The different guidelines, i.e. from the first European Conference on Infections in Leukemia (ECIL1)(Herbrecht *et al.*, 2007) and the Infectious Disease Society of America (IDSA)(Patterson, 2007), recommend an important role for azoles in the treatment and prevention of invasive aspergillosis. It can be expected that the volume of use of triazoles will increase due to the increased use of these drugs for the management of invasive fungal disease.

3. Intrinsic or primary resistance

Primary resistance among *Aspergillus* species appears to be uncommon. Infections caused by *A. terreus* are commonly refractory to treatment with amphotericin B. In one study *A. terreus* infections were associated with a lower response rate to amphotericin B therapy, compared with patients with non-*A. terreus* infections (20% versus 47%, respectively; $P < 0.05$)(Lass-Flörl *et al.*, 2005). In comparison to *A. fumigatus, A. terreus* was resistant to the in vitro fungicidal effects of safely achievable concentrations of amphotericin B. These in vitro findings correlated directly with resistance of *A. terreus* to amphotericin B in experimental invasive pulmonary aspergillosis (Walsh *et al.*, 2003). Other species that have reduced susceptibility to antifungal agents include *A. calidoustus* (formerly classified as *A. ustus)* to the triazoles (J. Varga, personal communication) and *Emericella nidulans* to amphotericin B (Kontoyiannis *et al.*, 2002).

Primary resistance of *A. fumigatus* to antifungal agents has not been described, although reduced susceptibility has been described in a distinct variant of *A. fumigatus* (Balajee *et al.*, 2004). However, the taxonomy of the *Aspergillus* section *Fumigati* has recently been reclassified, based on micro- and macro morphology, extrolite profiles, and gene sequence analysis of the β-tubulin, calmodulin and actin genes, resulting in the description of many new species that are closely related to *A. fumigatus* (Hong *et al.*, 2007). These new species are morphologically very similar to *A. fumigatus*, and will not be discriminated in the clinical microbiology laboratory. However, some of the new species were found to cause infections in humans and exhibit a more resistant phenotype in comparison with *A. fumigatus*. This was described for *Neosartorya pseudofischeri* (anamorph *A. thermomutatus*), *A. lentulus*, and *A. udagawae* (Balajee *et al.*, 2005a,b). These new species might be misidentified in the clinical microbiology laboratory, which rely on identification by morphological features only (Balajee *et al.*, 2006). Most published *A. fumigatus* isolates with a resistant phenotype have been found following exposure to antifungal drugs, most notably the triazoles.

4. Evolution of resistance to antifungal drugs

Exposure of microorganisms to antimicrobial agents possesses the risk of development of resistance to that drug. Numerous resistance mechanisms have been described in bacteria, but in fungi resistance appears to be less prominent. Factors that increase the likelihood of resistance occurring are the fungal population size and the mutation rate. A high population size will increase the probability, that following exposure to an antifungal compound, survivors will be present. The mutation rate depends on features of the drug, the drug target and the pathogen (Anderson, 2005). The fate of resistant determinants is important for the spread of resistant fungi. In bacteria numerous factors are present that facilitate the transmission of resistance. These include plasmids, integrons and gene islands that contain the antibiotic resistance that facilitate transmission within the same species but also across taxa (horizontal transfer)(Anderson, 2005). In fungi, however, drug-resistance and other genes are not spread horizontally, but through vertical transfer (Anderson, 2005). Genetic exchange and recombination might give rise to recombinant genotypes that contain multiple mechanisms of resistance that have been acquired by the fungus over time from different origins.

Although the above-mentioned process occurs in the environment, also in patient's resistance mechanisms might be acquired especially in those patients with a high population of reproducing fungi and long periods of treatment with fungistatic agents (Table 3). Secondary resistance has been described in fungi that are exposed to azoles, most well known in patients with oropharyngeal candidiasis who received repeated courses of treatment with fluconazole (Johnson *et al.*, 1995). The yeast *Candida albicans* was shown to rapidly evolve to a highly resistant phenotype by expression of efflux pumps that reduce drug accumulation, alteration of the structure

Table 3. Probability of emergence of resistance in fungal infection.

Fungus	Disease	Fungal population	Reproduction	Duration of drug exposure	Resistance expected	Frequency observed
Candida	Oropharyngeal candidiasis	High	Yes	Long or frequently repeated courses	Yes	Fluconazole resistance 7-12% (Barchiesi *et al.* 2002; Sobel *et al.* 2001)
	Candidemia	Low	Yes	Short, usually two weeks	No	Fluconazole resistance was ≤ 3% for all species except *C. glabrata* (9%) and *C. krusei* (40%)(Pfaller and Diekema 2007)
Aspergillus	Aspergilloma/ cavitary disease	High	Yes	Long, months to years	Yes	Sporadic, those cases reported are primarily in this patient group
	Invasive aspergillosis	Low	No	Short, weeks to months	No	Unknown, series of cases was reported from the Netherlands

or concentration of antifungal target proteins, and alteration of membrane sterol composition (Sanglard and Odds, 2002).

5. Azole resistance in *Aspergillus fumigatus*

Azole resistance has been reported in *A. fumigatus*. Sporadic cases have been reported in patients with aspergilloma or other cavitary pulmonary aspergillus lesions (Table 4). In a cavity the fungal burden is high with reproducing and sporulating *Aspergillus* mould, and patients with this condition are commonly treated with itraconazole for longer periods of time. These factors increase the likelihood of emergence of resistance and this has been shown over time in several patients (Chen *et al.,* 2005; Dannaoui *et al.,* 2001).

Molecular studies have shown that triazole resistance in *A. fumigatus* is associated with amino acid substitutions in the Cyp51A protein (Table 4). In *A. fumigatus* two distinct but closely related *cyp51* genes were found (*cyp51A* and *cyp51B*)(Mellado *et al.,* 2001). Functional analyses of these genes indicated that the *cyp51A* gene is not essential for viability (Mellado, 2005). In another study, however, it was shown that the ERG11 gene family (ERG11A and ERG11B) was essential in *A. fumigatus* despite neither member being essential individually (Hu *et al.,* 2007).

Mutant strains with only an active *cyp51B* gene showed in vitro susceptibility to ketoconazole and fluconazole, which suggests that Cyp51B could be more susceptible to these agents (Mellado *et al.,* 2005). In itraconazole-resistant *A. fumigatus* strains, two molecular mechanisms of resistance to azole drugs have been described: first, azole drug resistance in *A. fumigatus* seems to be mostly related to point mutations in Cyp51A which indicates that this homologue is the target of antifungal azoles (Table 4). A second proposed mechanism is reduced intracellular accumulation, due to either increased expression of efflux pumps or reduced penetration of the drug (Mellado *et al.,* 2007). Regarding the modification of *A. fumigatus* Cyp51, specific mutations in *cyp51A* have been associated with different susceptibility profiles. A phenotype characterised by cross-resistance to itraconazole and posaconazole has been associated with amino acid substitutions at glycine 54 (G54) (Table 4), and a pattern of itraconazole resistance, characterised by different patterns of elevated minimum inhibitory concentrations (MICs) for the other triazole drugs, has been linked to different amino acid substitutions at methionine 220 (M220). A multiple-triazole-resistant (MTR) phenotype was also associated with a substitution at codon 98 (leucine substitution for histidine) in combination with a tandem repeat in the promoter region of the *cyp51A* gene (Mellado *et al.,* 2007). It was shown that both alterations were required for the MTR-phenotype and that the 34 basepair tandem

Table 4. Mutations in the Cyp51A gene and expression of transporter genes of A. fumigatus *resulting in an azole-resistant phenotype.*

Origin of isolate	Underlying condition	Target	Mutations/expression	Resistant phenotype	Azole exposure	Reference
Clinical	HIV, unknown	*cyp51A* gene	G54E	ITZ	Unknown	Diaz-Guerra *et al.*, 2003
Clinical	Unknown	*cyp51A* gene	G54Q, -R or -W	ITZ	Unknown	Mann *et al.*, 2003
Laboratory	-	*cyp51A* gene	G54W	ITZ + PCZ	Exposed to PCZ	Mann *et al.*, 2003
Laboratory	-	*cyp51A* gene	G54E, -K, or –R	ITZ	Exposed to ITZ	Nascimento *et al.*, 2003
Laboratory	-	*Afu*MDR3, *Afu*MDR4	Overexpression	ITZ	Exposed to ITZ	Nascimento *et al.*, 2003
Laboratory	-	*Afu*MDR3, *Afu*MDR4, *Atr*F	Overexpression	ITZ	Exposed to ITZ	Da Silva Ferreira *et al.*, 2004
Clinical	Unknown	*cyp51A* gene	M220 V, -K or –T	Multi-azole	Unknown	Mellado *et al.*, 2004
Clinical	Aspergilloma	*cyp51A* gene	M220I and G54R	Multi-azole	Exposed	Chen *et al.*, 2005
Clinical	Aspergilloma	*cyp51A* gene	M220L, increased expression of AtrG and AtrF	ITZ	Exposed to ITZ	Dannauoi *et al.*, 2006
Clinical	Invasive aspergillosis	*cyp51A* gene	L98H and tandem repeat	Multi-azole	Naïve and exposed to ITZ or VCZ	Mellado *et al.*, 2007; Verweij *et al.*, 2007a
Clinical	Aspergilloma	*cyp51A* gene	G138C	Multi-azole	Exposed to ITZ and VCZ (and intracavity AmB)	Howard *et al.*, 2006

repeat increased the expression levels of the *cyp51A* gene (Mellado *et al.,* 2007). The MTR phenotype consisted of resistance to itraconazole and elevated MICs of voriconazole, ravuconazole and posaconazole (Verweij *et al.,* 2007a,b).

The latter change was found in *A. fumigatus* isolates of nine patients with invasive aspergillosis in the Netherlands (Verweij *et al.,* 2007a), which was not in accordance with previous research that failed to find azole resistance in patients with invasive aspergillosis (Dannaoui *et al.,* 2004). However, a recently presented survey of itraconazole susceptibility of 1,753 clinical *A. fumigatus* isolates from 1,119 patients admitted to a University Medical Center in Nijmegen, showed that resistant isolates were encountered from the year 2000 and onwards, and not in those collected between 1994 and 1999 (Verweij *et al.,* 2007b). This might indicate the recent emergence of azole resistance in *A. fumigatus,* although confirmative international surveys are required to confirm this observation.

Five of nine patients had presented with breakthrough invasive aspergillosis while on prophylaxis or treatment with itraconazole or voriconazole (Verweij *et al.,* 2007a). In these patients, the duration of exposure had varied significantly from 4 weeks to over 10 years. However, it is difficult to understand that azole resistance had evolved in the patient. In invasive aspergillosis the population of fungi is low with hyphal elongation as mode of growth as opposed to reproduction through sporulation. It is also unlikely that two genomic changes are induced in the setting of this acute invasive disease. However, patients might be colonised with aspergillus before invasive disease develops, although even in that scenario, the population of fungi would be low. Among the nine patients, four were present without a record of azole use indicating that azole resistance had emerged out-side the patient, in the environment. It has been suggested that the use of triazole compounds in the environment for plant protection or material protection might cause resistance in medically relevant fungi. However, most studies that have investigated this relationship have failed to show an association. In one study, clinical and environmental *A. fumigatus* isolates were not susceptible to non-azole agricultural agents and little impact of azole resistance was observed in both clinical and environmental isolates. When detected, azole resistance was compound-specific (Meneau and Sanglard, 2005). In another study, however, cross-resistance was found for *Candida* isolates between triazoles used in clinical medicine and those use as fungicides (Müller *et al.,* 2007). There is at present, however, no evidence that proves that the use of azole fungicides causes clinically relevant resistance in *A. fumigatus.*

6. Clinical implications of azole resistance

Although the frequency of azole-resistance in *A. fumigatus* is probably low, the finding of resistance in patients with invasive aspergillosis might have impact on the management of these patients. Triazoles are the primary agents used for the prevention and treatment of invasive aspergillosis. Also evidence is accumulating that indicates that the timing of treatment is essential with respect to favorable outcome. Patients with early pulmonary lesions on the CT scan, the halo sign, had a better response to treatment as well as a better survival compared with those patients with other lesions (Greene *et al.*, 2007). Improvement of diagnostic tests and procedures together with potent antifungal agents, such as voriconazole, have contributed to a significant decrease in mortality in patients with a diagnosis of invasive aspergillosis following hematopoietic stem cell transplantation (Upton *et al.*, 2007). However, although an increasing number of centers employ rapid diagnostic tests and high resolution CT scans for early diagnosis of invasive aspergillosis, in vitro susceptibility testing of clinical relevant *A. fumigatus* isolates is not routinely performed in clinical microbiology laboratories (Hassan *et al.*, 2006). As opposed to amphotericin B, a good correlation is found between in vitro susceptibility testing of itraconazole and in vivo efficacy in a murine model of invasive aspergillosis (Denning *et al.*, 1997a,b). Therefore, itraconazole resistance can be detected using in vitro susceptibility tests, such as the CLSI M38-A reference method. For other agents, such as amphotericin B, the relation between in vitro susceptibility test and clinical response is less clear. However, in the majority of patients an isolate is not obtained, which precludes the opportunity to test the susceptibility of the isolate. Furthermore, patients might be infected by multiple *A. fumigatus* strains that differ in susceptibility to triazoles. One patient was reported in whom at diagnosis an itraconazole susceptible isolate was cultured, but at autopsy a MTR *A. fumigatus* isolate (Van Leer-Buter *et al.*, 2007). This patient was shown to be infected by two genotypes. It could be the case that the resistant strains continue to grow during azole therapy. This observation and the prevalence of azole resistance in *A. fumigatus* will determine if the practice and treatment guidelines require reconsideration. However, it appears to be appropriate for clinical microbiology laboratories to test for azole susceptibility in *A. fumigatus* at least in those isolates that are cultured from patients that are failing to azole therapy.

References

Anderson, J.B., 2005. Evolution of antifungal-drug resistance: mechanisms and pathogen fitness. Nature Reviews Microbiology 3: 547-556.

Balajee, S.A., D. Nickle, J. Varga and K.A. Marr, 2006. Molecular studies reveal frequent misidentification of *Aspergillus fumigatus* by morphotyping. Eukaryotic Cell 5: 1705-1712.

Balajee, S.A., J.L. Gribskov, E. Hanley, D. Nickle and K.A. Marr, 2005a. *Aspergillus lentulus* sp. nov., a new sibling species of A. fumigatus. Eukaryotic Cell 4: 625-632.

Balajee, S.A., J. Gribskov, M. Brandt, J. Ito, A. Fothergill and K.A. Marr, 2005b. Mistaken identity: *Neosartorya pseudofischeri* and its anamorph masquerading as *Aspergillus fumigatus*. Journal of Clinical Microbiology 43: 5996-5999.

Balajee, S.A., M. Weaver, A. Imhof, J. Gribskov and K.A.Marr, 2004. *Aspergillus fumigatus* variant with decreased susceptibility to multiple antifungals. Antimicrobial Agents and Chemotherapy 48: 1197-1203.

Barchiesi, F., M. Maracci, B. Radi, D. Arzeni, I. Baldassarri, A. Giacometti and G. Scalise, 2002. Point prevalence, microbiology and fluconazole susceptibility patterns of yeast isolates colonizing the oral cavities of HIV-infected patients in the era of highly active antiretroviral therapy. Journal of Antimicrobial Chemotherapy 50: 999-1002.

Cappelletty, D. and K. Eiselstein-McKitrick, 2007. The echinocandins. Pharmacotherapy 27: 369-388.

Chen, J., H. Li, R. Li, D. Bu and Z. Wan, 2005. Mutations in the cyp51A gene and susceptibility to itraconazole in *Aspergillus fumigatus* serially isolated from a patient with lung aspergilloma. Journal of Antimicrobial Chemotherapy 55: 31-37.

Cornely, O.A., J. Maertens, D.J. Winston, J. Perfect, A.J. Ullmann, T.J. Walsh, D. Helfgott, J. Holowiecki, D. Stockelberg, Y.T. Goh, M. Petrini, C. Hardalo, R. Suresh R and D. Angulo-Gonzalez, 2007b. Posaconazole vs. fluconazole or itraconazole prophylaxis in patients with neutropenia. New England Journal of Medicine 356: 348-359.

Cornely, O.A., J. Maertens, M. Bresnik, R. Ebrahimi, A.J. Ullmann, E. Bouza, C.P. Heussel, O. Lortholary, C. Rieger, A. Boehme, M. Aoun, H.A. Horst, A. Thiebaut, M. Ruhnke, D. Reichert, N. Vianelli, S.W. Krause, E. Olavarria and R. Herbrecht, 2007a. Liposomal amphotericin B as initial therapy for invasive mold infection: a randomized trial comparing a high-loading dose regimen with standard dosing (AmBiLoad trial). Clinical Infectious Diseases 44: 1289-1297.

Da Silva Ferreira, M.E., J.L. Capellaro, E. dos Reis Marques, I. Malavazi, D. Perlin, S. Park, J.B. Anderson, A.L. Colombo, B.A. Arthington-Skaggs, M.H. Goldman and G.H. Goldman, 2004. In vitro evolution of itraconazole resistance in *Aspergillus fumigatus* involves multiple mechanisms of resistance. Antimicrobial Agents and Chemotherapy 48: 4405-4413.

Dannaoui, E., D. Garcia-Hermoso, J.M. Naccache, I. Meneau, D. Sanglard, C. Bouges-Michel, D. Valeyre and O. Lortholary, 2006. Use of voriconazole in a patient with aspergilloma caused by an itraconazole-resistant strain of *Aspergillus fumigatus*. Journal of Medical Microbiology 55: 1457-1459.

Dannaoui, E., E. Borel, M.F. Monier, M.A. Piens, S. Picot and F. Persat, 2001. Acquired itraconazole resistance in *Aspergillus fumigatus*. Journal of Antimicrobial Chemotherapy 47: 333-340.

Dannaoui, E., J. Meletiadis, A.M. Tortorano, F. Symoens, N. Nolard, M.A. Viviani, M.A. Piens, B. Lebeau, P.E. Verweij and R. Grillot, 2004. Susceptibility testing of sequential isolates of *Aspergillus fumigatus* recovered from treated patients. Journal of Medical Microbiology 53: 129-134.

Denning, D.W., K. Riniotis, R. Dobrashian and H. Sambatakou, 2003. Chronic cavitary and fibrosing pulmonary and pleural aspergillosis: case series, proposed nomenclature change, and review. Clinical Infectious Diseases 37 (Suppl 3): S265-S280.

Denning, D.W., K. Venkateswarlu, K.L. Oakley, M.J. Anderson, N.J. Manning, D.A. Stevens, D.W. Warnock and S.L. Kelly, 1997a. Itraconazole resistance in *Aspergillus fumigatus*. Antimicrobial Agents and Chemotherapy 41: 1364-1368.

Denning, D.W., S.A. Radford, K.L. Oakley, L. Hall, E.M. Johnson and D.W. Warnock, 1997b. Correlation between in-vitro susceptibility testing to itraconazole and in-vivo outcome of *Aspergillus fumigatus* infection. Journal of Antimicrobial Chemotherapy 40: 401-414.

Diaz-Guerra, T.M., E. Mellado, M. Cuenca-Estrella and J.L. Rodriguez-Tudela, 2004. A point mutation in the 14alpha-sterol demethylase gene cyp51A contributes to itraconazole resistance in *Aspergillus fumigatus*. Antimicrobial Agents and Chemotherapy 47: 1120-1124.

Greene, R.E., H.T. Schlamm, J.W. Oestmann, P. Stark, C. Durand, O. Lortholary, J.R. Wingard, R. Herbrecht, P. Ribaud, T.F. Patterson, P.F. Troke, D.W. Denning, J.E. Bennett, B.E. de Pauw and R.H. Rubin, 2007. Imaging findings in acute invasive pulmonary aspergillosis: clinical significance of the halo sign. Clinical Infectious Diseases 44: 373-379.

Hassan, I.A., P. Critten, B. Isalska and D.W. Denning, 2006. Audit of laboratory mycology services for the management of patients with fungal infections in the northwest of England. Journal of Clinical Pathology 59: 759-763.

Herbrecht, R., D.W. Denning, T.F. Patterson, J.E. Bennett, R.E. Greene, J.W. Oestmann, W.V. Kern, K.A. Marr, P. Ribaud, O. Lortholary, R. Sylvester, R.H. Rubin, J.R. Wingard, P. Stark, C. Durand, D. Caillot, E. Thiel, P.H. Chandrasekar, M.R. Hodges, H.T. Schlamm, P.F. Troke and B. de Pauw, 2002. Voriconazole versus amphotericin B for primary therapy of invasive aspergillosis. New England Journal of Medicine 347: 408-415.

Herbrecht, R., U. Flückiger, B. Gachot, P. Ribaud, A. Thiebaut and C. Cordonnier, 2007. Treatment of invasive *Candida* and *Aspergillus* infections in adult haematological patients. European Journal of Cancer Supplements 5: 49-59.

Hong, S.B., H.D. Shin, J.B. Hong, J.C. Frisvad, P.V. Nielsen, J. Varga and R.A. Samson, 2007. New taxa of *Neosartorya* and *Aspergillus* in *Aspergillus* section *Fumigati*. Antonie van Leeuwenhoek (in press)

Howard, S.J., I. Webster, C.B. Moore, R.E. Gardiner, S. Park, D.S. Perlin and D.W. Denning, 2006. Multi-azole resistance in Aspergillus fumigatus. International Jornal of Antimicrobial Agents 28: 450-453.

Hu, W., S. Sillaots, S. Lemieux, J. Davison, S. Kauffman, A. Breton, A. Linteau, C. Xin, J. Bowman, J. Becker, B. Jiang and T. Roemer, 2007. Essential gene identification and drug target prioritization in *Aspergillus fumigatus*. PLoS Pathogenesis 3: e24.

Johnson, E.M., D.W. Warnock, J. Luker, S.R. Porter and C. Scully, 1995. Emergence of azole drug resistance in *Candida* species from HIV-infected patients receiving prolonged fluconazole therapy for oral candidosis. Journal of Antimicrobial Chemotherapy 35: 103-114.

Kontoyiannis, D.P., R.E. Lewis, G.S. May, N. Osherov and M.G. Rinaldi, 2002. *Aspergillus nidulans* is frequently resistant to amphotericin B. Mycoses 45: 406-407.

Lass-Flörl, C., K. Griff, A. Mayr, A. Petzer, G. Gastl, H. Bonatti, M. Freund, G. Kropshofer, M.P. Dierich and D. Nachbaur, 2005. Epidemiology and outcome of infections due to *Aspergillus terreus*: 10-year single centre experience. British Journal of Haematology 131: 201-207.

Maertens, J., I. Raad, G. Petrikkos, M. Boogaerts, D. Selleslag, F.B. Petersen, C.A. Sable, N.A. Kartsonis, A. Ngai, A. Taylor, T.F. Patterson, D.W. Denning and T.J. Walsh TJ, 2004. Efficacy and safety of caspofungin for treatment of invasive aspergillosis in patients refractory to or intolerant of conventional antifungal therapy. Clinical Infectious Diseases 39: 1563-1571.

Mann, P.A., R.M. Parmegiani, S.Q. Wei, C.A. Mendrick, X. Li, D. Loebenberg, B. DiDomenico, R.S. Hare, S.S. Walker and P.M. McNicholas, 2003. Mutations in *Aspergillus fumigatus* resulting in reduced susceptibility to posaconazole appear to be restricted to a single amino acid in the cytochrome P450 14alpha-demethylase. Antimicrobial Agents and Chemotherapy 47: 577-581.

Maschmeyer, G., A. Haas and O.A. Cornely, 2007. Invasive aspergillosis: epidemiology, diagnosis and management in immunocompromised patients. Drugs 67: 1567-1601.

Mellado, E., G. Garcia-Effron, L. Alcazar-Fuoli, M. Cuenca-Estrella and J.L. Rodriguez-Tudela, 2004. Substitutions at methionine 220 in the 14alpha-sterol demethylase (Cyp51A) of *Aspergillus fumigatus* are responsible for resistance in vitro to azole antifungal drugs. Antimicrobial Agents and Chemotherapy 48: 2747-2750.

Mellado, E., G. Garcia-Effron, L. Alcázar-Fuoli, W.J. Melchers, P.E. Verweij, M. Cuenca-Estrella and J.L. Rodríguez-Tudela, 2007. A new *Aspergillus fumigatus* resistance mechanism conferring in vitro cross-resistance to azole antifungals involves a combination of cyp51A alterations. Antimicrobial Agents and Chemotherapy 51: 1897-1904.

Mellado, E., G. Garcia-Effron, M.J. Buitrago, L. Alcazar-Fuoli, M. Cuenca-Estrella and J.L. Rodriguez-Tudela, 2005. Targeted gene disruption of the 14-alpha sterol demethylase (cyp51A) in *Aspergillus fumigatus* and its role in azole drug susceptibility. Antimicrobial Agents and Chemotherapy 49: 2536-2538.

Mellado, E., T.M. Diaz-Guerra, M. Cuenca-Estrella and J.L. Rodriguez-Tudela, 2001. Identification of two different 14-alpha sterol demethylase-related genes (cyp51A and cyp51B) in *Aspergillus fumigatus* and other *Aspergillus* species. Journal of Clinical Microbiology 39: 2431-2438.

Meneau, I. and D. Sanglard, 2005. Azole and fungicide resistance in clinical and environmental *Aspergillus fumigatus* isolates. Medical Mycology 43 (Suppl 1): S307-S311.

Müller, F.M., A. Staudigel, S. Salvenmoser, A. Tredup, R. Miltenberger and J.V. Herrmann, 2007. Cross-resistance to medical and agricultural azole drugs in yeasts from the oropharynx of human immunodeficiency virus patients and from environmental Bavarian vine grapes. Antimicrobial Agents and Chemotherapy 51: 3014-3016.

Nascimento, A.M., G.H. Goldman, S. Park, S.A. Marras, G. Delmas, U. Oza, K. Lolans, M.N. Dudley, P.A. Mann and D.S. Perlin, 2003. Multiple resistance mechanisms among *Aspergillus fumigatus* mutants with high-level resistance to itraconazole. Antimicrobial Agents and Chemotherapy 47: 1719-1726.

Patterson, T.F., 2007. Treatment guidelines for invasive aspergillosis. 47th InterScience Conference on Antimicrobial Agents and Chemotherapy, Chicago 2007; abstract 1406.

Pfaller, M.A. and D.J. Diekema, 2007. Epidemiology of invasive candidiasis: a persistent public health problem. Clinical Microbiology Reviews 20: 133-163.

Sanglard, D. and F.C. Odds, 2002. Resistance of *Candida* species to antifungal agents: molecular mechanisms and clinical consequences. Lancet Infectious Diseases 2: 73-85.

Scott, L.J. and D. Simpson, 2007. Voriconazole: a review of its use in the management of invasive fungal infections. Drugs 67: 269-298.

Sobel, J.D., S.E. Ohmit, P. Schuman, R.S. Klein, K. Mayer, A. Duerr, J.A. Vazquez and A. Rampalo, 2001. The evolution of *Candida* species and fluconazole susceptibility among oral and vaginal isolates recovered from human immunodeficiency virus (HIV)-seropositive and at-risk HIV-seronegative women. Journal of Infectious Diseases 183: 286-293.

Ullmann, A.J., J.H. Lipton, D.H. Vesole, P. Chandrasekar, A. Langston, S.R. Tarantolo, H. Greinix, W. Morais de Azevedo, V. Reddy, N. Boparai, L. Pedicone, H. Patino and S. Durrant, 2007. Posaconazole or fluconazole for prophylaxis in severe graft-versus-host disease. New England Journal of Medicine 356: 335-347.

Upton, A., K.A. Kirby, P. Carpenter, M. Boeckh and K.A. Marr, 2007. Invasive aspergillosis following hematopoietic cell transplantation: outcomes and prognostic factors associated with mortality. Clinical Infectious Diseases 44: 531-540.

Van Leer-Buter, C., R.P. Takes, K.M. Hebeda, W.J. Melchers and P.E. Verweij, 2007. Aspergillosis--and a misleading sensitivity result. Lancet 370: 102.

Verweij, P.E., E. Mellado and W.J. Melchers, 2007a. Multiple-triazole-resistant aspergillosis. New England Journal of Medicine 356: 1481-1483.

Verweij, P.E., H.A.L. van der Lee, J. Kuijpers, E. Snelders, A.J.M.M. Rijs and W.J.G. Melchers, 2007b. Epidemiology of multiple-triazole-resistance in *Aspergillus fumigatus*. 47th InterScience Conference on Antimicrobial Agents and Chemotherapy, Chicago 2007; abstract M-2018.

Viscoli, C., R. Herbrecht, H. Akan, L. Baila, C. Doyen, A. Gallamini, A. Giagounidis, O. Marchetti, R. Martino, L. Meerts, M. Paesmans, M. Shivaprakash, A. Ullmann and J. Maertens, 2007. Caspofungin as first-line therapy of invasive aspergillosis in haematological patients: a study of the EORTC Infectious Diseases Group. 3rd Trends in Medical Mycology, Torino, 2007: abstract O12.

Walsh, T.J., V. Petraitis, R. Petraitiene, A. Field-Ridley, D. Sutton, M. Ghannoum, T. Sein, R. Schaufele, J. Peter, J. Bacher, H. Casler, D. Armstrong, A. Espinel-Ingroff, M.G. Rinaldi and C.A. Lyman, 2003. Experimental pulmonary aspergillosis due to *Aspergillus terreus*: pathogenesis and treatment of an emerging fungal pathogen resistant to amphotericin B. Journal of Infectious Diseases 188: 305-319.

Aspergillus species in human keratomycosis

Palanisamy Manikandan[1], *Ilona Dóczi*[2], *Sándor Kocsubé*[3], *János Varga*[3,4], *Tibor Mihály Németh*[3], *Zsuzsanna Antal*[3], *Csaba Vágvölgyi*[3], *Madhavan Bhaskar*[5] *and László Kredics*[3]
[1]*Aravind Eye Hospital and Postgraduate Institute of Ophthalmology, Avinashi road, Civil Aerodrome post, Coimbatore, 641014 Tamilnadu, India*
[2]*Department of Clinical Microbiology, Faculty of Medicine, University of Szeged, Somogyi Béla tér 1, H-6725 Szeged, Hungary*
[3]*Department of Microbiology, Faculty of Science and Informatics, University of Szeged, Közép fasor 52, H-6726 Szeged, Hungary*
[4]*CBS Fungal Biodiversity Centre, P.O.Box 85167, 3508 AD Utrecht, The Netherlands*
[5]*Department of Microbiology, Coimbatore Medical College, Coimbatore, Tamilnadu, India*

Abstract

Members of the filamentous fungal genus *Aspergillus* are among the agents frequently causing keratomycosis in humans. *Aspergillus* keratitis is most common among agricultural workers in geographical regions with hot, humid, tropical or semi-tropical climates. Trauma by vegetable matter during agricultural activities is the main predisposing factor of the infection. A persistent infiltrate at the site of superficial injury is often present, which gradually increases in size and density. If untreated, a full thickness corneal ulcer may develop, which leads to perforation and may progress to endophthalmitis. Early and accurate diagnosis coupled with appropriate antifungal therapy is crucial for improving the chances of complete recovery. The main causative agents of *Aspergillus* keratitis are *A. flavus*, *A. fumigatus*, *A. terreus* and *A. niger*, while *A. glaucus*, *A. ochraceus*, *A. fischerianus*, *A. nidulans* and *A. tamarii* are rarely occurring. Morphology-based diagnosis of *Aspergillus* keratitis can be made by means of smear, staining, fungal culture and confocal microscopy. Molecular methods based on simplified PCR kits could have the potential to provide highly sensitive and specific diagnostic capabilities for detecting *Aspergillus* species in corneal samples. Rabbits are frequently used in model systems of experimental keratitis for studying the possible virulence factors and testing potential antifungal agents. Potential virulence factors of *Aspergillus* species causing corneal infections include extracellular proteinases, mycotoxins and fungal adhesins. The treatment of *Aspergillus* keratitis involves both medical and surgical management. The disease is usually treated with a topical antifungal agent, but surgical interventions may be needed for patients whose corneal infections do not resolve. Possible surgical interventions include regular debridement at the base of the ulcer, superficial lamellar keratectomy, excimer laser photo ablation of superficial stromal corneal infiltrates, temporary or permanent amniotic membrane transplantation and penetrating keratoplasty.

Keywords: keratitis, keratomycosis, cornea, corneal ulcer, epidemiology, antifungal susceptibility, biodiversity

1. Introduction

Keratomycosis – also known as mycotic, or fungal keratitis – is a suppurative, usually ulcerative corneal disease (Thomas, 2003). Infection is exogenous in most of the cases with the organism entering through the corneal epithelium. Filamentous fungi have replaced bacteria as the predominant cause of infectious keratitis in some developing countries. Ocular fungal infections are being increasingly recognised as important causes of serious vision loss and blindness worldwide. Certain types of ocular mycoses may even be life threatening. Corneal infections of fungal aetiology are very common and represent 30% to 40% of all cases of culture positive infectious keratitis in India. The most common fungi involved in mycotic keratitis are *Fusarium* species (Dóczi *et al.*, 2004), *Candida* species (Tanure *et al.*, 2000) and members of the genus *Aspergillus.*

Aspergillus species have long been regarded as important pathogens in eye infections, especially keratitis and endogenous endophthalmitis (Klotz *et al.*, 2000; Thomas, 2003). In developing countries, where *Aspergillus* species are isolated from fungal keratitis in large numbers, the limited availability of antifungal drugs and the lack of response to therapy lead to blindness in a high number of patients. The aim of this review is to summarise the information available about fungal keratitis caused by *Aspergillus* species.

2. Epidemiology of *Aspergillus* keratitis

Mycotic keratitis occurs frequently in agrarian tropical countries, but it is relatively uncommon in the western world (O'Day, 1996; Srinivasan *et al.*, 1997; Bharathi *et al.*, 2002, 2003, 2006). The preponderance of a particular genus is likely in a given geographic area. Hence the species of fungal pathogens may vary from country to country (Table 1) and even within the same country (Table 2). In India, *Fusarium* is the predominant fungal pathogen isolated from mycotic keratitis cases in the southern part of the country (Srinivasan *et al.*, 1997; Bharathi *et al.* 2002, 2003, 2006), whereas *Aspergillus* is predominantly observed in north India (Vajpayee *et al.*, 1990; Panda *et al.*, 1997; Vajpayee *et al.*, 2000; Chowdhary and Singh, 2005; Saha and Das,

Table 1. Worldwide incidence of Aspergillus spp. among culture-proven cases of fungal keratitis.

Country	Period of study	Number of Aspergilli/ all fungal isolates (%)	Reference
Australia, Melbourne	July 1996 - May 2004	6/35 (17.1)	Bhartiya *et al.*, 2007
Bangladesh, Chittagong	11 months	19/51 (37.3)	Dunlop *et al.*, 1994
Bangladesh, Chittagong	NA	2/7 (28.6)	Williams *et al.*, 1987
Brasil	1975 - 2003	28/293 (9.6)	Hofling-Lima *et al.*, 2005
Brasil	1983 - 1997	8/49 (16.3)	Alvarez-de-Carvalho *et al.*, 2001
Brasil	January 1994 - December 1999	6/20 (30.0)	Salera *et al.*, 2002
China, Beijing (North)	January 1995 - October 2000	68/498 (13.6)	Zhang *et al.*, 2002
China, Beijing	2001 - 2004	116/681 (17.0)	Xuguang *et al.*, 2007
China	January 1975 - June 1997	NA/615	Wang *et al.*, 2000
China, Zhengzhou (Central)	January 1975 - June 1997	NA (20.5)	Wang *et al.*, 2000
China, Shijiazhuang (North)	January 1975 - June 1997	NA (21.7)	Wang *et al.*, 2000
China, Guangzhou (South)	January 1975 - June 1997	NA (30.7)	Wang *et al.*, 2000
China, Quingdao (North)	January 1996 - December 1999	14/97 (14.4)	Xie *et al.*, 2001
China, North	January 1999 - December 2004	72/596 (12.1)	Xie *et al.*, 2006
China	1989 - 2000	130/775 (16.8)	Sun *et al.*, 2004
France, Paris	January 1993 - January 2001	4/19 (21.0)	Rondeau *et al.*, 2002
Ghana	June 1999 - May 2001	19/109 (17.4)	Leck *et al.*, 2002
Ghana, Accra	NA	10/128 (7.8)	Hagan *et al.*, 1995
Great Britain, London	1994-2007	7/66 (10.6)	Galarreta *et al.*, 2007
Iran	May 2004 - March 2005	1/2 (50.0)	Shokohi *et al.*, 2006
Iran, Mashhad	1982 - 2001	15/27 (55.6)	Berenji *et al.*, 2003
Iran	NA	8/19 (42.1)	Javadi *et al.*, 1996
Nepal	1985 - 1987	32/68 (47.0)	Upadhyay *et al.*, 1991
Nepal	August 1998 - July 2001	78/145 (53.8)	Khanal *et al.*, 2005
Nigeria	NA	5/21 (23.8)	Gugnani *et al.*, 1976
Nigeria	1974 - 1977	5/42 (11.9)	Gugnani *et al.*, 1978
Paraguay	April 1988 - April 1989	5/26 (19.2)	Mino de Kaspar *et al.*, 1991

Table 1. Continued.

Country	Period of study	Number of Aspergilli/ all fungal isolates (%)	Reference
Paraguay	1988 - 2001	37/136 (27.2)	Laspina *et al.*, 2004
Paraguay	January 1997 - December 2000	2/23 (8.7)	Sonego-Krone *et al.*, 2006
Saudi Arabia	NA	11/27 (40.7)	Khairallah *et al.*, 1992
Singapore	January 1991 - December 1995	5/29 (17.2)	Wong *et al.*, 1997
Sri Lanka	1976 - 1977 and 1980 - 1981	4/22 (18.2)	Gonawardena *et al.*, 1994
Thailand (Central)	January 1988 - December 2000	7/35 (20.0)	Boonpasart *et al.*, 2002
USA, Florida	January 2004 - December 2005	19/122 (15.6)	Alfonso *et al.*, 2006
USA, Florida	January 1980 - January 2002	30/419 (7.1)	Marangon *et al.*, 2004
USA, South Florida	January 1982 - January 1992	5/125 (4.0)	Rosa *et al.*, 1994
USA, South Florida	January 1969 - December 1977	6/134 (4.4)	Liesegang and Forster 1980
USA, Minneapolis	January 1971 - January 1981	6/19 (31.6)	Doughman *et al.*, 1982
USA, Pennsylvania	January 1991 - March 1999	1/24 (4.2)	Tanure *et al.*, 2000

NA: No data available.

2006). These variations are considered to be due to (1) environmental factors such as humidity, rainfall and wind; (2) occupational factors such as agricultural work, construction work, etc.; and (3) populations like rural agrarian population, urban industrial workers, etc. (Thomas, 2003).

Seasonal variation has also been observed in mycotic keratitis. The peak occurrence is observed in November in South Florida (Rosa *et al.*, 1994). The isolation rate is increased during seasons of intense agricultural activities. In developing countries, increased incidence of *Aspergillus* keratitis is usually observed during winter (November to January) and around the harvesting season (June & September) (Bharathi *et al.*, 2003).

Trauma by paddy grains or vegetable matter during agricultural activities is the most common predisposing factor of *Aspergillus* keratitis (Singh *et al.*, 2006; Srinivasan

Table 2. Incidence of Aspergillus *spp. in India among culture-proven cases of fungal keratitis.*

Location	Period of study	Number of Aspergilli/ all fungal isolates (%)	Reference
Chennai (South)	December 1999 - May 2002	76/130 (58.5)	Therese *et al.*, 2006
Chidambaram (South)	July 2002 - June 2005	78/230 (34)	Vasudevan *et al.*, 2006
Coimbatore (South, children)	February 1997 - January 2004	11/37 (29.7)	Singh *et al.*, 2006
Hyderabad (South)	January 1991 - December 2000	417/1360 (30.7)	Gopinathan *et al.*, 2002
Hyderabad	January 1991 - December 1996	170/557 (30.5)	Garg *et al.*, 2000
Hyderabad	February 1991 - June 1995	7/21 (33.3)	Kunimoto *et al.*, 2000
Karnataka	October 1985 - September 1988	23/67 (34.3)	Kotigadde *et al.*, 1992
Madurai (South)	January 1994 - March 1994	25/155 (16.1)	Srinivasan *et al.*, 1997
Madurai	December 2002 - June 2003	26/100 (26.0)	Prajna *et al.*, 2004
Madurai	NA	10/58 (17.2)	Rahman *et al.*, 1998
Madras (South)	1980 - 1982	36/68 (52.9)	Sundaram *et al.*, 1989
Madras	NA	138/322 (42.9)	Venugopal *et al.*, 1989
Tiruchirapalli (South)	June 1999 - May 2001	76/353 (21.5)	Leck *et al.*, 2002
Tiruchirapalli	NA	74/248 (29.8)	Thomas *et al.*, 1986
Tirunelveli (South)	September 1999 - March 2001	135/554 (24.37)	Bharathi *et al.*, 2002
Tirunelveli	September 1999 - August 2002	286/1100 (26.0)	Bharathi *et al.*, 2003
Tirunelveli	September 1999 - September 2002	305/1226 (24.8)	Bharathi *et al.*, 2006
Vellore (South)	NA	3/7 (42.9)	Panhalkar *et al.*, 1985
Amristar (North)	NA	21/30 (70)	Usha *et al.*, 2006
Chandigarh (North)	6 years	25/61 (41.0)	Chander and Sharma 1994
New Delhi (North)	January 1999 - December 2001	85/215 (39.5)	Panda *et al.*, 1997
New Delhi	January 2000 - December 2004	37/77 (48.1)	Saha and Das 2006
New Delhi	NA	9/31 (29)	Vajpayee *et al.*, 1990
New Delhi	NA	2/12 (16.7)	Vajpayee *et al.*, 2000
New Delhi	January 1999 - June 2001	78/191 (40.8)	Chowdhary and Singh 2005

Table 2. Continued.

Location	Period of study	Number of Aspergilli/ all fungal isolates (%)	Reference
Calcutta (East)	June 1999 - September 2000	22/44 (50.0)	Agarwal *et al.*, 2001
Calcutta	January 2001 - December 2003	373/623 (59.9)	Basak *et al.*, 2005
Patna (East)	2 years	42/76 (55.3)	Kumari *et al.*, 2002
Goa (West)	February 1993 - January 1994	9/16 (56.3)	Verenkar *et al.*, 1998
Mumbai (West)	1988 - 1996	219/367 (59.7)	Deshpande and Koppikar 1999

NA: No data available.

et al., 1997; Bharathi *et al.*, 2002, 2003, 2006). Other predisposing factors include prolonged use of corticosteroids, inappropriate use of antibiotics, diabetes mellitus, other ocular diseases and contact lens wear (Thomas, 2003). Injury of the eye during harvesting is the most frequently described risk factor. The traumatizing agents were vegetable matter, mud, sand, dust particles, paddy grains, tree branches, twigs and metals (Srinivasan *et al.*, 1997; Bharathi *et al.*, 2003; Thomas, 2003). The wearing of contact lenses has been observed as a common risk factor for mycotic keratitis in industrialised countries, however, in developing countries it is uncommon. Ocular problems like corneal surface disorders, dry eye, bullous keratopathy and exposure keratitis, as well as therapeutic procedures like photorefractive keratectomy (PRK), radial keratotomy (Heidemann *et al.*, 1995; Panda *et al.*, 1998) and laser-assisted *in situ* keratomileusis (LASIK) (Sridhar *et al.*, 2000; Kuo *et al.*, 2001) may also predispose *Aspergillus* keratitis.

Aspergillus keratitis more commonly occurs in adult male agricultural workers in the age group of 21-60, as they are the commonest population exposed to agriculture related injuries (Srinivasan *et al.*, 1997; Bharathi *et al.*, 2003). Others affected are mostly women of the same occupation and male construction workers. The environmental and corneal isolates of *Aspergillus* do not differ in cultural characteristics. One thought is that the commensal fungi in the conjunctival sac become virulent under circumstances like trauma or the administration of corticosteroids, but this is less significant, than direct implantation during trauma.

We performed a retrospective study on 26 *Aspergillus* strains isolated from keratomycosis in the Aravind Eye Hospital between August 2005 and February 2006 (Manikandan *et al.*, 2007; Kredics *et al.*, 2007b). The 26 patients included 17 males and 9 females. Rural, semiurban and urban populations were represented among the patients with 23, 50 and 27%, respectively. Farmer was the most frequent occupation among the male patients, while most of the female patients were housewives. Corneal trauma was reported as the potential predisposing factor of the infection for 57.7% of the patients, the traumatising agents were dust or iron particles, insects or oil and mustard seeds. Among the further possible predisposing conditions, systemic diseases like diabetes mellitus and hypertension proved to be frequent.

3. Clinical features of *Aspergillus* keratitis

The onset of *Aspergillus* keratitis is almost always insidious. The patients usually present for examination during the first week of their illness. In others it may last weeks to recognise supervening infection. During this period, the patients do not feel any discomfort, and the infection may appear to respond to topically applied antibiotics. However, afterwards the patients may present with photophobia and discharge from the eyes. A persistent infiltrate at the site of superficial injury is often present, which gradually increases in size and density. The infiltrate appears grayish-white or yellowish-white and the base of the ulcer is filled with soft, creamy, raised exudates. The cornea becomes slightly thickened and satellite lesions may develop peripheral to the focal area of infiltrations. The signs of inflammation are minimal in comparison with bacterial keratitis. The absence of lid edema is a common feature. Feathery borders or hyphate edges and hypopyon (Figure 1A) are frequently present. However, some authors observed, that these descriptions do not always correlate with fungal ulcers (Srinivasan, 2004). Treated corneal ulcers show signs of improvements like lessening of pain, decrease in size of the infiltrate, disappearance of satellite lesions, and rounding out of the feathery margins of the ulcer (Figure 1B). If untreated, the ulceration leads to the formation of a descemetocele. A full thickness corneal ulcer (Figure 1C) may lead to perforation and rarely progress to endophthalmitis. Keratomycotic malignant glaucoma is a rare complication of severe corneal ulcers. It can be resulted in late perforation and rupture of the eyeball. Fungus penetrates into the anterior chamber and forms a fungal mass at the pupillary area, thereby causing the diversion of aqueous, leading to high intraocular pressure. It is important to recognise this syndrome, because the eye will be lost, unless the malignant glaucoma is managed by surgical intervention. Most reported cases developed from *Fusarium* keratitis (Thomas and Thomas 1991), with one case known to be caused by *A. flavus* (Jain *et al.*, 2007).

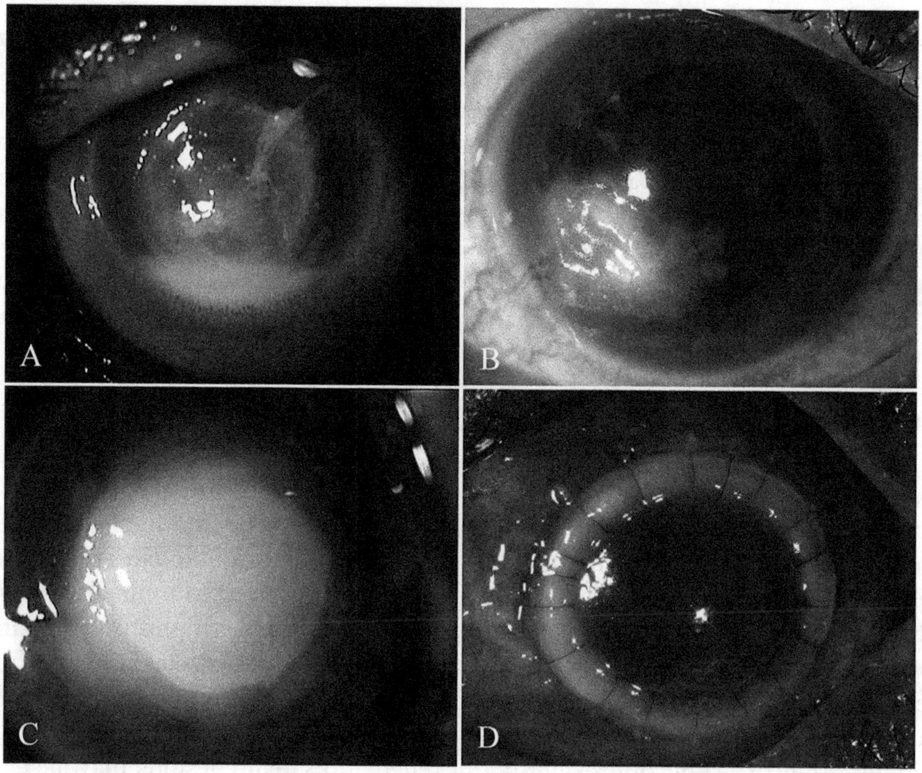

Figure 1. Slit-lamp biomicroscopic diffuse view of corneal ulcers caused by A. flavus. *(A) ulcer with hypopyon, (B) Healing ulcer, (C) Full thickness ulcer, (D) Therapeutic Keratoplasty (TKP) for non healing* Aspergillus *ulcer. Photos were taken at the Aravind Eye Hospital and Postgraduate Institute of Ophthalmology, Coimbatore, India.*

4. Morphological and molecular diagnosis of Aspergilli from human keratomycosis

4.1. The spectrum of Aspergillus species known from cases of fungal keratitis

Certain *Aspergillus* species, mainly *A. flavus* (Saha and Das, 2006; Thomas 2003), *A. fumigatus* (Saha and Das, 2006; Thomas, 2003), *A. terreus* (Thomas, 2003), and *A. niger* (Chowdhary and Singh, 2005.) have long been regarded as important pathogens in eye infections especially keratitis. Other members of the genus less frequently occurring in keratitis include *A. glaucus* (Venugopal *et al.*, 1989), *A. ochraceus* (Williams *et al.*, 1987), *A. fischerianus* (Coriglione *et al.*, 1990) and *A. nidulans* (Leck *et al.*, 2002). *A. tamarii* has also been recently reported from a case of keratitis after an ocular injury (Kredics *et al.*, 2007a). Since the description of this case we have found further corneal isolates of *A. tamarii* during a retrospective study (unpublished), suggesting that the lack of reports about *A. tamarii* keratitis may not be due to the lack of occurrence rather to the lack of recognition. *A. ustus* and *A. versicolor* have also been reported from eye infections, but not from keratitis. They are uncommon causative agents of endophthalmitis (Saracli *et al.*, 2007; Bifrare and Wolfensberger, 2007), therefore not discussed in this review.

4.2. Morphological identification

A standard management protocol should be followed in every case of suspected mycotic keratitis, as on clinical basis alone it is impossible to determine the causative agent of the infection with full certainty. Morphology-based diagnosis of *Aspergillus* keratitis can be made by means of smear, staining, fungal culture and confocal microscopy. Adequately collected scrapings from the base and edges of the corneal ulcer are important for direct microscopic examination as well as for culturing of *Aspergillus*. Platinum spatula Beaver blade, Bard-Parker knife and blunt cataract knife may be used to collect the scrapings. Where this is not possible, calcium alginate swabs premoistened with tryptone soy broth may facilitate the recovery of the pathogen. If corneal scrapings originated from patients with suspicion of fungal keratitis fail to yield positive results, corneal biopsy by superficial keratectomy may be helpful.

Direct microscopic evaluation is the most valuable and rapid diagnostic tool for detecting fungal filaments. Presumptive diagnosis of fungal keratitis in general can be made by 10% KOH mount and Gram stained smear of the corneal scrapings. Gram staining and KOH mount helps to identify fungal species in 45 to 80% of cases while Giemsa in 66%. Lactophenol cotton blue has a sensitivity of 70 to 80%, Grocott methenamine silver staining of as much as 89%, and calcofluor white about 80 to 90%. Though the sensitivity may vary between different studies when individual staining techniques are used, a combination of these methods definitely increases the chances of diagnosis. The hyphal elements of *Aspergillus* can be easily visualised by KOH and Gram's staining techniques in combination to make a presumptive diagnosis of fungal keratitis. Fluorescent microscopy is an excellent technique in presumptive diagnosis and may take the place of the conventional microscopic evaluation in the future.

Prompt and accurate diagnosis is stressed for all mycotic keratitis cases in order to be able to start an appropriate therapy, which increases the chances of complete recovery. This is more important in the case of developing countries where decreased awareness and poor socio-economic status complicates the issue. This requires a detailed clinical history record including possible risk factors, examinations, as well as the use of conventional and modern microbiological diagnostic techniques. Tandem screening by confocal microscopy is a relatively new, noninvasive technique for imaging the cornea *in vivo* in normal and infected states (Winchester *et al.*, 1997). Fungal hyphae from two, culture-proven cases of clinical keratitis due to *Aspergillus* spp. were imaged as high-contrast filaments 60 to 400 µm long and 6 µm wide. This technique may be useful for the identification of corneal pathogens in the early stages of infection with great implications for management and outcome. A detailed history record and the clinical signs may be suggestive of mycotic keratitis in general rather than suggestive of *Aspergillus* keratitis. Confocal microscopy may help the diagnosis of *Aspergillus* keratitis, however, it has limitations like the lack of distinctive morphology of some pathogens.

Conventional culture methods used in mycology are still preferred for the diagnosis of *Aspergillus* keratitis. After preliminary microscopic examination, the scrapped material is inoculated in SDA/PDA and incubated at 25 °C, which yields better growth of *Aspergillus*. The use of SDA/PDA with 5% sheep blood agar is sufficient for isolation of different *Aspergillus* species. Fifty-four percentage of *Aspergillus* isolates growing on these media can be identified within 2 days, which improves to 83% and 97% after 3 days and one week of culturing, respectively. As *Aspergillus* is known as a common lab contaminant, growth in more than one media or repeated isolation of the same species from the patient is mandatory to consider it as the causative agent.

Aspergillus colonies are usually fast growing, white, yellow, yellowish-brown, black, or shades of green, and they mostly consist of a dense felt of erect conidiophores. Conidiophores end in vesicles covered with either a single palisade-like layer of phialides (uniseriate) or a layer of subtending cells (metulae), which wear small whorls of phialides (biseriate structure). The vesicle, the metulae and the conidia form the conidial head. Conidia are one celled, smooth or rough walled, hyaline or pigmented and are basocatenate forming long dry chains, which may be divergent (radiate) or aggregated in compact columns (columnar). Some species produce Hülle-cells or sclerotia. Most species sporulate within 7 days. Identification is mainly based on colony pigmentation and conidial head. Lactophenol cotton blue mount is generally used in the identification of the cultured isolates. Table 3 compares the culture and micromorphological characteristics of the *Aspergillus* species reported to date from keratomycosis cases.

We performed morphological examinations, including studies on culture characteristics and microscopy for a total of 25 *Aspergillus* strains isolated from keratomycosis patients in the Aravind Eye Hospital, Tamilnadu, India between

Table 3. Characters aiding the morphological identification of Aspergillus *species reported from corneal infections. Based on the Atlas of Clinical Fungi (de Hoog* et al., *2004).*

	Colony characteristics	Micromorphology			
		Conidial heads/conidiogeneous cells	**Conidiophore stipes**	**Vesicles**	**Conidia**
A. flavus	colonies (CzA) yellowish-green, consisting of a dense felt of conidiophores	Radiate/uni- and biseriate	rough-walled, hyaline	spherical, 25-45 µm diam.	echinulate, (sub)spherical, 3.5 µm diam.
A. terreus	colonies (CzA) yellowish-brown to cinnamon-brown, consisting of a dense felt of conidiophores	densely columnar/ biseriate	smooth-walled, hyaline	subspherical, 10-20 µm diam.	smooth-walled, striate with SEM, spherical to broadly ellipsoidal, 1.5-2.5 µm, hyaline
A. fumigatus	colonies (CzA) dark blue-green, consisting of a dense felt of conidiophores, intermingled with aerial hyphae	Columnar/ uniseriate	smooth-walled, often green in the upper part	subclavate, 20-30 µm wide	verrucose, (sub)spherical, 2.5-3.0 µm diam.
A. niger	colonies (CzA) black, consisting of a dense felt of conidiophores	Radiate/ biseriate	smooth-walled, hyaline or pigmented	subspherical, 50-100 µm diam.	Brown, ornamented with warts and ridges, subspherical, 3.5-5.0 µm diam.
A. glaucus	colonies (CzA) spreading broadly, flat, dull green to grey-green	sparse, radiate/ uniseriate	700-800 × 2-3 µm, hyaline, smooth-walled	spherical, metulae twice as long as the phialides	ovoidal or aculeate, echinulate, hyaline, 4.5-7.5 µm diam.

Table 3. Continued.

	Colony characteristics	Micromorphology			
		Conidial heads/conidiogeneous cells	Conidiophore stipes	Vesicles	Conidia
A. ochraceus	colonies (CzA) with restricted growth, yellow-orange, ochraceous or buff	radiate splitting into several columns with age/ biseriate	brownish, commonly 1.0-1.5 mm in length with roughened walls	spherical, thin-walled, hyaline, 35-50 µm diam.	spherical to subspherical, 2.5-3.5 µm diam., smooth-walled to finally roughened
A. nidulans	colonies (CzA) growing rapidly, green, cream-buff or honey-yellow, reverse dark purplish	short, columnar/ biseriate	brownish, 60-130 × 2.5-3.0 µm	hemispherical, 8-10 µm diam.	Spherical, rugulose, subhyaline, green in mass, 3-4 µm diam.
A. fisherianus (Neosartorya fisheri)	colonies (OA) growing rapidly, farinose to felty, whitish with greyish-blue shades, somewhat pinkish around ascomata, reverse off-white to pinkish	columnar, terminally somewhat radiate/ uniseriate	smooth-walled, hyaline to pale green, 300-500 µm	flask shaped, up to 18 µm diam.	subsphaerical, verruculose, 2-3 × 2.0-2.5 µm, pale greyish-green
A. tamarii	colonies (CzA) growing rapidly, dark brown	compact and spherical, loosely radiate/ uni- and biseriate	usually 1-2 mm in length, hyaline, usually roughened	spherical, 10-50 µm diam.	echinulate to tuberculate, subspherical, 5-8 µm diam.

CzA: Czapek Dox agar, OA: oatmeal agar, SEM: scanning electron microscopy.

September 2005 and March 2006 (Kocsubé *et al.*, 2006). Most of the isolates proved to belong to *A. flavus*, however, other species of the genus including *A. terreus* and *A. fumigatus* were also represented in the samples, although at much lower frequencies. The initial, morphology-based identification of 24 out of the 25 isolates could be confirmed by sequence analysis of the internal transcribed spacer (ITS) region performed by nucleotide-nucleotide BLAST analysis (Kocsubé *et al.*, 2006), suggesting that successful identification can be carried out based on culture characteristics and micromorphology alone. The only exception was an isolate originally identified as *A. flavus* but later reidentified as *A. tamarii*, a species of *Aspergillus* section *Flavi* not hitherto reported from keratitis (Kredics *et al.*, 2007a). Some conidia of this isolate were not ornamented with tubercules and warts like the typical wild *A. tamarii* isolates, but were smooth-walled and hyaline. This case indicated that even the routinely performed, morphology-based identification may be misleading in the cases of *Aspergillus* species that are not expected as keratitis pathogens due to the lack of literature data about their involvement in corneal infections.

4.3. Molecular identification and genetic diversity

Most of the *Aspergillus* strains isolated from keratomycosis are being identified and reported at the genus level only, although their accurate identification at the species level would be of great importance, as the pathogenic potential may vary between different species of the genus.

The use of molecular techniques based on polymerase chain reaction (PCR) offers a significant reduction in time required for an accurate diagnosis, when compared with culture-based morphological identification. Anand *et al.* (2001) evaluated the usefulness of PCR in the diagnosis of fungal endophthalmitis. They used the universal fungal primers U1 and U2, previously designed for the variable region of the 28S rRNA gene (Sandhu *et al.*, 1995). The method proved to be more sensitive in the diagnosis of fungal infections when compared with the conventional mycological methods, but it is not capable for diagnosis at the genus and species level. A positive result available early enough by such panfungal primer-based PCR techniques can, however, justify the use of antifungal therapy. Panfungal primers for the 18S rRNA sequences were also designed and a nested PCR method was developed for the detection of three fungal species (*A. fumigatus, Fusarium solani* and *Candida albicans*) in ocular samples (Jaeger *et al.*, 2000). A similar, nested PCR-strategy was used to establish a detection system for *A. fumigatus, F. oxysporum* and *C. albicans* from corneal scrapings of keratitis patients (Gaudio *et al.*, 2002). PCR-RFLP analysis of the amplified ITS region using the restriction enzyme *Cfo*I has also been used successfully to detect *A. fumigatus* in a patient with fungal keratitis (Ferrer *et al.*,

2000). Based on a study of Kumar and Shukla (2005), single-stranded conformation polymorphism analysis of sequence variation in different regions of rRNA genes was successfully tested for the detection of fungal keratitis caused by *A. fumigatus.*

The previous studies involved ocular samples with *A. fumigatus* only, however, as mentioned before, other members of the genus, e.g. *A. flavus* and *A. niger* are also occurring frequently as causative agents of fungal keratitis. Direct detection of fungi in corneal samples by PCR amplification and sequencing of the ITS region was used successfully by Mancini *et al.* (2006) for fast (24 h) identification of the fungal pathogens, including *A. flavus* (in comparison, culture-based identification took for 4 days). Ferrer *et al.* (2001) developed a typing method based on seminested PCR amplification of the ITS region of fungal pathogens, and used successfully to detect various pathogens including *A. niger* and *A. fumigatus* in ocular samples. Wang *et al.* (2007) developed a multi-PCR system for the fast diagnosis of causative agents of mycotic keratitis, including *A. flavus, A. fumigatus* and *A. niger*, while Zhang *et al.* (2007) developed a gene microarray carrying oligonucleotide probes homologous to species-specific regions of ITS for the detection of 12 species causing keratitis in China (including *A. terreus, A. niger, A. flavus* and *A. fumigatus*).

Kanbe *et al.* (2002) applied nested PCR with a mixture of specific primers to the DNA topoisomerase II gene for the specific identification of five medically important *Aspergillus* species (*A. flavus, A. fumigatus, A. niger, A. nidulans* and *A. terreus*). In another study, a PCR identification system was described for the three major human pathogenic *Aspergillus* species, *A. flavus, A. fumigatus* and *A. niger*, based on the amplification of ITS1 and subsequent, separate PCR reactions or nested PCR with species-specific primers (Sugita *et al.*, 2004). De Aguirre *et al.* (2004) successfully designed DNA probes directed to the ITS2 region of rDNA to differentiate seven medically important *Aspergillus* species, *A. flavus, A. fumigatus, A. nidulans, A. niger, A. terreus, A. ustus* and *A. versicolor* from one another and from other opportunistic molds and yeasts by using an amplification and detection system based on PCR-enzyme immunoassay. Such identification systems that are capable for the detection of a wider range of species – including most of the Aspergilli causing keratitis – have the potential to be adapted for the fast identification of the causative agents of *Aspergillus* keratitis at the species level.

These studies suggest that simplified PCR kits or gene arrays could provide highly sensitive and specific diagnostic capabilities for detecting ocular pathogens in the laboratory. Real-time PCR may also be promising for this purpose (Goebes *et al.*, 2007). However, the main limitation of the widespread use of PCR-based techniques is the cost of PCR, which is much higher than that of classical culture-based methods.

Besides the need of fast and accurate diagnosis of fungal keratitis, it is also very important to study the genetic diversity of the causative agents as the pathogenic potential may vary not only between different species of the genus, but also between different strains belonging to the same species. Molecular strain typing methods for studying the genetic diversity of *Aspergillus* isolates may become powerful tools for the examination of the epidemiology of keratitis. Bagyalakshmi *et al.* (2007) analysed the genetic similarity among ocular isolates of *A. flavus* by PCR-RFLP and DNA sequencing of the ITS region, and reported a novel pattern of nucleotide polymorphisms due to inversions and substitutions as well as both single-nucleotide differences and short lengths of sequence diversity due to insertions or deletions.

Based on studies performed with non-clinical as well as clinical, non-keratitis isolates from the genus *Aspergillus*, further possibilities for molecular diagnosis of *Aspergillus* keratitis and strain typing of the causative agents include sequence analysis of the D1 and D2 regions of the 28S RNA, the IGS regions, the 18S RNA gene, and of genes encoding for the second largest RNA polymerase subunit (RPB2), mitochondrial cytochrome B, β-tubulin and calmodulin, as well as multiplex PCR based on amplicon melting point analysis, random amplified polymorphic DNA (RAPD) analysis, restriction fragment length polymorphism (RFLP) analysis, DNA fingerprinting with repetitive DNA sequences, amplification fragment length polymorphism (AFLP) analysis, microsatellite marker analysis, heteroduplex panel analysis and DNA-microarrays (reviewed by Balajee and Marr, 2006).

5. Animal models for studying *Aspergillus* keratitis

The increasing frequency of ocular fungal infections caused by *Aspergillus* species has focused attention on the development of an effective modelling system (Clemons and Stevens, 2005, 2006). Animal models of *Aspergillus* keratitis provide the opportunity to study miscellaneous aspects of pathogenesis, innate and acquired host-response, disease transmission as well as the effect of therapeutic agents.

Several factors should be considered before any particular model of infection is employed. The model has to be carefully defined and the parameters of infection should be controllable in order to mimic the clinical situation as close as possible. In addition, ethical questions arise concerning the human care and use of animals. Animal models are useful for studying diagnostic assays, pharmacology of antifungals as well as toxicity. Thanks to the well-standardised methods, it is possible to address issues *in vivo*, that cannot be answered by *in vitro* tests. Generally, animal models are crucial for the advancement of the therapy, for the investigation of drugs prior to a clinical trial and for the improvement of our current understanding of pathogenesis and host-resistance (Clemons and Stevens, 2005, 2006).

Rabbits are more expensive than mice and require greater efforts from animal care personnel. However, rabbit models do have the benefits of allowing multiple sampling from the same animal, better clinical evaluations and opportunity to obtain body fluids in sufficient quantity. Rabbits have been frequently used to study fungal keratitis. To establish the disease, animals are highly immunosuppressed and the choice of immunosuppressive regimen alters the type of the disease manifested (Clemons and Stevens, 2005). Rabbit models of keratitis have been used primarily for studies of therapy, showing various antifungals to have some therapeutic effects. The models vary somewhat in the manner of establishment, but often the route of infection is anaesthetising the rabbit with an intramuscular injection of xylazine and ketamine, which is followed by the injection of conidia into the corneal stroma. Samples are usually collected by scraping the cornea. The importance of immunosuppression in the model is in question, as the use of glucocorticoids, usually given subconjunctivally, alters the course of disease and facilitates the establishment of infection. In non-suppressed animals, the fungal burden in the cornea declines to undetectable by day 7, whereas if corticoids are given, the fungal burden remains stable through 15 days of infection and the inflammatory response in the cornea is worsening. During studies to establish a rabbit model of keratitis due to *Aspergillus*, O'Day *et al.* (1979) found that the usual effective duration of the model is about 5-10 days.

Inoculation of the corneas of Dutch-belted rabbits with *A. fumigatus* (strain VE148) produced an infection that persisted for at least 10 days (O'Day *et al.,* 1979). The observable clinical patterns were the following: in most cases the earliest sign was the appearance of an infiltrate at the site of injection followed by filamentous projections around the stroma from the central infiltrate. During the next days, the intensity of the inflammation was increasing slowly in most animals as the size of the infiltrate enlarged and the filamentous projections became more apparent. In some cases occasional hypopyon and ulceration could also be observed (O'Day *et al.,* 1979).

6. Antimycotics for the treatment of *Aspergillus* keratitis: antifungal susceptibilities and effectiveness for the therapy of experimental and clinical infections

Medical therapy consists of specific antifungal therapy against *Aspergillus* and nonspecific, supportive therapy. In developing countries, various antifungals have been tried for the treatment of *Aspergillus* keratitis with varying success rates. Factors like species of the fungal pathogen, cost effectiveness of the treatment, easy availability of the drug, extent of the ulcer and the concentration achieved on the cornea decide for the drug to be chosen. The choice of antifungal agents depends on host defense mechanisms to eradicate the fungal pathogen, as most of them - except for amphotericin B - reach an effective concentration only for the inhibition of the pathogens. The selective effect of most antifungals is based on the sterol difference (ergosterol in fungal and cholesterol in mammalian cell membrane).

The most frequently used antifungal agents against *Aspergillus* keratitis include polyenes and azole compounds. Combination therapy is often being tried because of the lower efficacy of antifungal agents when used individually. Due to the high number of cases unresponsive to the antifungal therapy applied in the practice, further studies are needed for the evaluation of new agents effective against *Aspergillus* for the treatment of fungal keratitis.

The effective treatment of fungal keratitis is often hindered by the lack of routine antifungal susceptibility testing of the isolates. The recent explosion in the rates of opportunistic fungal infections, the increasing number of reports of resistance to the available antifungal agents and the limited number of antifungal susceptibility data for ocular fungal isolates have propelled interest in clinically relevant methods for antifungal susceptibility testing (Therese *et al.*, 2006). Standardised susceptibility testing methods that have been tested for clinical *Aspergillus* isolates include the CLSI M-38A method (Clinical and Laboratory Standard Institute, 2002), the Sensititre YeastOne method and the Etest method modified for molds (Etest technical guide 10, 1999). The CLSI M38-A protocol is a broth microdilution method standardised for filamentous fungi, which can be applied for the susceptibility testing to any antifungal drug. The Sensititre YeastOne method (Trek Diagnostic Systems Ltd., East Grinsted, Sussex, England) is a commercial colorimetric microdilution method based on the colour change of the oxidation-reduction indicator Alamar blue, and it is available for amphotericin B, fluconazole, itraconazole, ketoconazole, voriconazole, posaconazole, 5-flucytosine and caspofungin. The Etest method (AB Biodisk, Solna, Sweden) is an agar diffusion-dilution method commercially available for amphotericin B, fluconazole, itraconazole, ketoconazole, voriconazole, posaconazole, caspofungin and 5-flucytosine, which is based on the application of strips with a concentration gradient of the tested antimicrobial agent (Figure 2).

Melatiadis *et al.* (2002) compared the Etest and the Sensititre YeastOne method with the CLSI M-38P method for susceptibility testing of 5 medically important *Aspergillus* species, *A. fumigatus*, *A. flavus*, *A. terreus*, *A. nidulans* and *A. ustus* to itraconazole and amphotericin B. In this study, broader ranges of MICs and higher values were obtained by the Etest method than by the other methods, while the

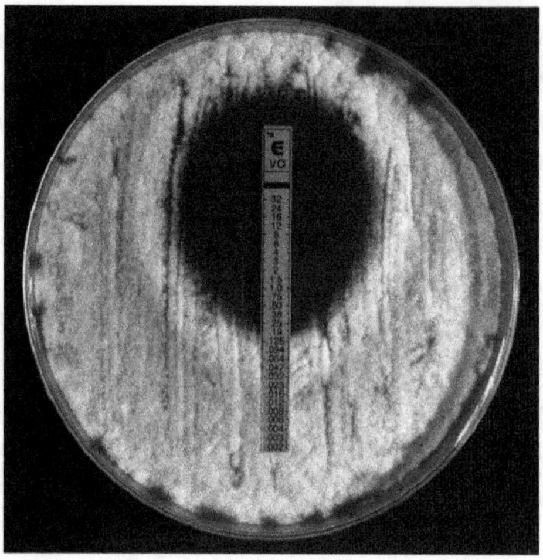

Figure 2. Susceptibility of an A. flavus *isolate from fungal keratitis to voriconazole examined by the E-test method for molds. The MIC-value is read as the drug concentration at which the elliptical inhibition zone intersects the scale of the Etest strip (MIC=0.25 μg/ml).*

Sensititre method yielded the lowest MICs. The authors concluded that - although the Etest and Sensititre methods were found promising - both of them require further optimalisation for filamentous fungi, particularly for *A. ustus* and *A. flavus*.

6.1. Natamycin

Natamycin, also known as pimaricin, is a naturally occurring macrolide polyene antifungal agent produced by *Streptomyces natalensis*. It is reported to have a broad spectrum of activity against various fungi, including *Aspergillus* species (Agarwal *et al.*, 2001; Thomas, 2003). Natamycin, available as 5% suspension, is considered as the drug of choice for the topical treatment of corneal infections. It may be stored at room temperature or refrigerated, but exposure to light or extreme temperature should be avoided. Unfortunately, there is a lack of information in the literature about the antifungal susceptibilities of *Aspergillus* isolates to natamycin.

Natamycin often adheres to areas of corneal ulceration increasing the drug contact. It is mainly useful in superficial ulcers as it is poorly absorbed by the cornea. The optimal dosing schedule varies according to the severity of the ulcer. Natamycin in conjunction with oral and topical ketoconazole appeared to be effective against *A. fumigatus* in a rabbit model of fungal keratitis (Komadina *et al.*, 1985). In human cases of keratitis due to *Aspergillus* spp., favourable responses were noted in 6 (54.5%)

of 11 patients receiving 1% natamycin (Kalavathy *et al.*, 2005). In a randomised trial of 2.5% natamycin for fungal keratitis, the zone diameters of 22 *Aspergillus* isolates were between 0 and 21 mm determined by the well diffusion method (Rahman *et al.*, 1998). Seven of these cases showed no response to 5% natamycin in various combinations with clotrimazole and econazole, even though the fungi from 4 of these cases showed sensitivity to 2.5% natamycin *in vitro*. Natamycin proved to be useful by topical drops at 5% and less useful at 10% as well as by subconjunctival injections (Garcia de Lomas *et al.*, 1985). Favourable responses were noted in 6 (54.5%) of 11 patients with *Aspergillus* keratitis receiving natamycin (Kalavathy *et al.*, 2005).

6.2. Amphotericin B

Amphotericin B is a polyene antifungal drug originally extracted from *Streptomyces nodosus*. It can be used for the topical treatment of *Aspergillus* keratitis, but its subconjunctival administration has also been proposed. It is available as topical solution in the concentration of 0.15%, however, the drug was found to be effective even at a reduced concentration of 0.05% in a small series of cases (O'Day, 1996), thus the optimal rate of administration is not precisely established. Amphotericin B has antifungal activity against Aspergilli, however, resistant ocular *Aspergillus* isolates including *A. flavus*, *A. terreus* and *A. niger* have also been reported in the literature (Hahn *et al.*, 1993; Therese *et al.*, 2006). In a study by Qiu *et al.* (2005) only 4 of 9 examined *Aspergillus* keratitis isolates were found to be sensitive to amphotericin B. The minimal inhibitory concentration (MIC) values of 4 ocular *Aspergillus* isolates to amphotericin B proved to be 1-2 μg/ml by the Sensitire YeastOne microdilution method (Marangon *et al.*, 2004). MIC-values of 26 corneal *Aspergillus* isolates for amphotericin B determined by the Etest method were between 0.064 and 4 μg/ml (Manikandan *et al.*, 2006).

In an experimental keratitis study of *A. fumigatus* performed in 130 rabbits, amphotericin B was found to be useful at 5 mg/ml (0.5%) concentration by topical drops but less useful at 2 mg/ml (Garcia de Lomas *et al.*, 1985). Results of a study involving three patients of culture proven *A. flavus* corneal ulcer with hypopyon not responding to 5% topical natamycin suggested that intracameral amphotericin B (7.5 to 10 μg in 0.1 ml) may be a useful modality in the treatment of severe keratomycosis enabling to avoid surgical intervention in the acute stage of the disease (Kaushik *et al.*, 2001). It has been used to treat three patients with keratitis and hypopyon due to *A. flavus*, with minimal toxicity being reported. Three patients with *Aspergillus* keratomycosis were treated with collagen shields soaked in amphotericin B (0.5%) for 2 hours at 25 degrees C before application, in conjunction with amphotericin B (0.25%) eye drops applied every 2 hours (Mendicute *et al.*, 1995). Results of a retrospective study by the same authors comprising 8 patients suggested that collagen shields soaked in amphotericin B may be an effective therapeutic option also in cases of *Aspergillus* keratomycosis that developed after cataract surgery (Mendicute *et al.*, 2000).

6.3. Miconazole

Miconazole is an imidazole antifungal agent, developed by Janssen Pharmaceutica, and commonly applied topically. Miconazole was suggested for the therapy of *Aspergillus* keratitis in the 1980s. In an experimental keratitis study of *A. fumigatus* performed in 130 rabbits, miconazole was found to be useful at 10 mg/ml (1%) concentration by topical drops and subconjunctival injections, but was less useful at 5 mg/ml (Garcia de Lomas *et al.*, 1985). The comparative efficacy of 1% miconazole in ointment base and drops has also been evaluated in a rabbit model of keratomycosis against fungal pathogens including *Aspergillus* species (Gupta, 1986). The ointment proved to be more effective in healing the fungal ulcers and associated lesions, it required much less frequent application than drops, furthermore it was well tolerated by the rabbit eye and produced no ocular or systemic toxicity. Miconazole has been applied for the therapy of mycotic keratitis caused by different *Aspergillus* species including *A. flavus* (Foster, 1981), *A. fumigatus* (Foster, 1981; Fahad *et al.*, 2004) and *A. niger* (Rondeau *et al.*, 2002). Topical and subconjunctival miconazole therapy resulted in resolution of all lesions in two mycotic keratitis cases due to *A. fumigatus*, and one due to *A. flavus* (Foster, 1981).

6.4. Ketoconazole

Ketoconazole is a synthetic antifungal drug used to prevent and treat fungal infections, especially in immunocompromised patients. It can be used topically and/or orally for the treatment of *Aspergillus* keratitis. The ketoconazole MIC-values for 4 ocular *Aspergillus* isolates proved to be 2-4 μg/ml (sensitive) by the Sensititre YeastOne microdilution test (Marangon *et al.*, 2004). *In vitro* susceptibility testing by the agar dilution method revealed that corneal *Aspergillus* isolates resistant to ketoconazole could be found in *A. flavus, A. terreus* and *A. niger* (Therese *et al.*, 2006). Ketoconazole MIC-values for 26 corneal *Aspergillus* isolates were between 0.25 and 1 μg/ml by the Etest method (Manikandan *et al.*, 2006).

The topical application of 1% ketoconazole in arachis oil was found to have a significant prophylactic potential in inhibiting the development of corneal stromal fungal lesions when administered to the cornea of New Zealand white male rabbits before the inoculation of their cornea with an ocular pathogenic *A. flavus* strain (Oji *et al.*, 1982).

The same ketoconazole solution also had a therapeutic effect in the reduction of well-established *A. flavus* keratitis in rabbits, all the corneal lesions could be cured in about sixteen days. Data from another rabbit model of *A. fumigatus* keratitis revealed that neither oral nor topical ketoconazole was effective despite moderate *in vitro* sensitivity of the fungus to this antifungal agent, while its oral and topical forms used in conjunction with natamycin appeared to augment elimination of this pathogenic mold (Komadina *et al.*, 1985). Ketoconazole with MIC of 30 μg/ml (agar dilution method) after 11 days of incubation was found to be effective against an *A. fumigatus* strain isolated from a human case of keratomycosis (Singh *et al.*, 1989). In the case of experimental keratitis in rabbits caused by the same isolate, oral ketoconazole therapy exhibited partial response followed by relapse. Laboratory-proven corneal *Aspergillus* infections could also be successfully treated with topical ketoconazole, no signs of progression of the corneal infection were seen after the antifungal therapy was initiated and the infection disappeared after three weeks of ketoconazole therapy without any biomicroscopic signs of ocular surface toxicity (Torres *et al.*, 1985).

6.5. Itraconazole

Itraconazole is a broad-spectrum triazole antifungal drug with a high degree of efficacy against filamentous fungi including *Aspergillus* species. *In vitro* susceptibility testing results performed on 3 *Aspergillus* isolates (two isolates of *A. flavus*, one from a human case of recalcitrant mycotic keratitis, and an environmental isolate of *A. fumigatus*) suggested that the use of itraconazole could be a primary consideration in the treatment of *Aspergillus* keratitis (Hahn *et al.*, 1993). All nine isolates of *Aspergillus* tested by Qiu *et al.* (2005) using the Etest method for the susceptibility to itraconazole proved to be sensitive. By the Sensititre YeastOne Method, the MIC-values of 4 tested *Aspergillus* isolates were between 0.256 and 1 μg/ml (Marangon *et al.*, 2004). MIC-values of 26 corneal *Aspergillus* isolates for itraconazole were between 0.064 and 2 μg/ml by the Etest method with the single exception of an *A. terreus* isolate showing complete resistance (MIC=32 μg/ml) (Manikandan *et al.*, 2006).

Itraconazole was found to be effective in experimental keratitis due to *Aspergillus* spp. (Van Cutsem and Van Gerven, 1991). It was also found to be effective for the treatment of patients, however, unresponsive cases were also reported in the literature (Kaushik *et al.*, 2001; Thomas *et al.*, 1988; Heidemann *et al.*, 1995). According to Rajasekaran *et al.* (1987), 69 out of 110 human cases of keratomycosis due to *Aspergillus* responded well to itraconazole, and a combination of oral and topical itraconazole yielded the highest percentage of excellent or moderate responses. One-percent-suspension of itraconazole prepared in a commercial isotonic eye drop formulation was found to be well tolerated when used for therapy of *Aspergillus* keratitis but was not very effective in treating severe infections (Thomas and Rajasekaran, 1988). In the case of oral itraconazole (200 mg/day) therapy, favourable response was observed in 10 out of 15 patients (9 with *A. flavus* keratitis, 1 with *A. fumigatus* keratitis) (Thomas *et al.*, 1988). In a preliminary study on the efficacy of topical and systemic itraconazole as a broad-spectrum antifungal agent against keratomycosis, an 80% cure rate was achieved in *Aspergillus* keratitis without any significant side effects (Agarwal *et al.*, 2001). In another study, favourable responses were noted in 5 of 10 patients receiving itraconazole (Kalavathy *et al.*, 2005).

6.6. Fluconazole

Fluconazole is a fungistatic bis-triazole evaluated as a topical and systemic agent for the treatment of fungal keratitis. According to a series of studies, fluconazole is not very effective against *Aspergillus* spp. All 4 *Aspergillus* isolates tested by the Sensititre YeastOne Method proved to be completely resistant to fluconazol with MIC-values >256 μg/ml (Marangon *et al.*, 2004). Therese *et al.* (2006) observed high percentage of resistance to fluconazole among *A. flavus*, *A. niger* and *A. terreus* isolates. In another study, only 2 out of 9 *Aspergillus* isolates from keratomycosis were sensitive (Qiu *et al.*, 2005). In our experience, all of 26 *Aspergillus* strains examined by the Etest method proved to be resistant to fluconazole with MIC-values >256 μg/ml (Manikandan *et al.*, 2006).

Both topical and oral fluconazole were effective for the treatment of *A. fumigatus* keratitis in a rabbit model, suggesting that fluconazole could be considered as an effective alternative in the therapy of keratitis caused by this species (Avunduk *et al.*, 2003). Some human cases of *Aspergillus* keratitis had favourable outcome when treated with 0.2% topical fluconazole alone (Sonego-Krone *et al.*, 2006), however, fluconazole generally does not show encouraging results against *Aspergillus* species (Rao *et al.*, 1997).

6.7. Further therapeutic possibilities

Marangon *et al.* (2004) suggested that voriconazole, a derivative of fluconazole with a broader spectrum of antifungal activity may have a role in the therapeutic management of ocular infections caused by *Aspergillus* species. The MIC-values of 4 tested isolates to voriconazole determined by the Sensititre YeastOne Method were between 0.128 and 0.5 μg/ml. MIC-values of 26 corneal *Aspergillus* isolates for voriconazole were between 0.064 and 1 μg/ml by the Etest method (Manikandan *et al.*, 2006). Thiel *et al.* (2007) determined the voriconazole concentrations in human aqueous humor and plasma during topical or combined topical and systemic administration for fungal keratitis. The results of this study suggested that both topical and combined voriconazole therapy could be effective in treating of fungal keratitis. A case of *A. flavus* keratitis was successfully treated with oral voriconazole and penetrating keratoplasty after conventional antifungal medical therapy failure (Freda, 2006). Recently, the combination of topical (1%) and oral voriconazole (800 mg twice a day) with topical amphotericin B (0.15%) proved successful for the treatment of *A. fumigatus* keratitis (Stewart *et al.*, 2007).

In vitro susceptibility tests indicated good antifungal activity of clotrimazole and econazole against *A. fumigatus* (Gugnani *et al.*, 1978). In the Aravind Eye Hospitals in South India, preparations of 2% econazole (Aurozole) and 1% clotrimazole (Auroclot) – both manufactured by Aurolab (Madurai, India) – are routinely used for the therapy of corneal infections caused by filamentous fungi including *Aspergillus* species. Econazole is an antifungal drug from the imidazole class. All except one of the 22 examined ocular *Aspergillus* isolates were sensitive to 1% econazole by the well diffusion method, the zone diameters were between 0 and 28 mm (Rahman *et al.*, 1998). An *A. tamarii* isolate recently described from fungal keratitis had MIC-value of 0.064 μg/ml for econazole by the Etest method (Kredics *et al.*, 2007a). Saperconazole, an experimental lypophilic triazole also appears to be highly effective against *Aspergillus* species *in-vitro* (O'Day, 1996).

Lalitha *et al.* (2007) characterised the susceptibility of 90 filamentous fungal isolates – including 41 *Aspergillus* strains – from keratitis to various antifungal agents such as caspofungin and posaconazole. Besides triazoles, caspofungin had the lowest MICs against *Aspergillus* species.

Based on a study involving 22 cases of *Aspergillus* keratitis as well as 22 cases caused by *Fusarium* species, chlorhexidine was suggested to have potential as an inexpensive topical agent for fungal keratitis (Rahman *et al.*, 1998). However, treatment of patients with this compound in two locations in Africa has not had encouraging results (Johnson, 1998).

Mohan *et al.* (1988) compared the efficacy of a 1% silver sulfadiazine ointment with that of 1% miconazole in therapy of clinical mycotic keratitis. The response of *Aspergillus* keratitis was comparable for the two types of treatment. However, silver sulfadiazine was not found effective in therapy of culture-proven mycotic keratitis in a study in southern India (Thomas, 2003). In experimental *Aspergillus* keratitis, polyhexamethylene biguanide (0.02%) was found to be a moderately effective drug, while 1% povidone iodine was not effective (Panda *et al.*, 2003). Fluorinated pyrimidines may not have a significant role in the management of *Aspergillus* keratitis.

7. Surgical interventions

Surgical intervention is required only if the mycotic keratitis responds poorly or fails to respond to antimycotic therapy. Though the necessity for surgical intervention is considerably reduced at present due to the availability of specific antimycotic therapy, it is still much higher than in the case of bacterial keratitis. Since the antimycotic therapy makes the fungus nonviable and surgical debridement helps better penetration of drugs, combined therapy may be advantageous in some cases. Thomas (2003) summarised the data about the therapy of 61 patients with *Aspergillus* keratitis. In the case of superficial ulcers caused by *A. flavus*, *A. fumigatus*, and other *Aspergillus* species, more than 80% of the patients responded to the medical therapy with a variety of topical or systemic antifungals. For the rest, combined surgical and medical management was advocated. On the other hand, in the case of deep corneal lesions due to *Aspergillus* species, almost 60% of the patients do not respond to medical therapy alone, particularly if natamycin is not used. In such cases surgery is required to control the infection (Thomas, 2003). Besides poor response to drug therapy, impending perforation and desmatocele formation are further indications for surgery. The possible surgical interventions are (1) regular debridement at the base of the ulcer, which removes the necrotic material and facilitates drug penetration (O'Day *et al.*, 1984; Agrawal *et al.*, 1994); (2) superficial lamellar keratectomy, which removes fungal filaments and increases drug penetration (Garg *et al.*, 2000; Wilhelmus and Jones, 2001); and (3) excimer laser photo ablation of superficial stromal corneal infiltrates, which can be used cautiously for superficial, localised infections, but it may not be useful in cases of advanced infections with deep stromal involvement and suppuration (Gottsch *et al.*, 1991).

The other modes of surgical interventions are temporary or permanent amniotic membrane transplantation (Kim *et al.*, 2001) and use of cyanoacrylate tissue adhesives (Forster, 1994), which are bacteriostatic (Agrawal *et al.*, 1994). Penetrating keratoplasty is the commonly used surgery, which is a full thickness corneal grafting. It is indicated whenever an impending perforation (a perforation exceeding 2 mm) exists and/or there is no response to the medical therapy. The lens is extracted if there is a large perforation exposing it, otherwise left undisturbed to prevent spread of the infection to the posterior segment (Portnoy *et al.*, 1989; Thomas, 1994). Donor grafts of 8 mm or less in diameter give a better prognosis (Killingsworth *et al.*, 1993). Antimycotics should be continued postoperatively and the topical corticosteroids (0.5% cyclosporin A) may be used as an adjuvant therapy to avoid graft rejection (Perry *et al.*, 2002).

8. Pathogenesis of *Aspergillus* keratitis

Future research needs to focus on the pathogenesis of ophthalmic mycoses including *Aspergillus* keratitis in order to be able to identify new targets for the therapy (Thomas, 2003). Information about the pathogenic mechanism of fungal keratitis could be obtained from animal experiments, impression debridement of corneal ulcers, diagnostic corneal biopsy, or from corneal buttons removed during penetrating keratoplasty (Vemuganti *et al.*, 2002). In a retrospective study of 148 patients including 55 culture proven cases of *Aspergillus* keratitis, the correlation of host and agent factors in the progression of mycotic keratitis were examined through the microbiologic evaluation and histologic study of human corneal buttons obtained at the time of therapeutic keratoplasty (Figure 1D) (Vemuganti *et al.*, 2002). Based on the results the authors suggested that both agent factors (heavy fungal load with deep penetration) and host factors (insufficient inflammatory response) influence the progression of the disease in the early stages.

A. flavus strains isolated from patients with keratitis are known to be able to secrete proteinases (Zhu *et al.*, 1990; Gopinathan *et al.*, 2001). To get a better understanding of the possible role of proteainases in the pathogenesis of fungal keratitis, the extracellular proteinases of a clinical *A. flavus* isolate from a severe case of keratitis were identified and partially characterised (Zhu *et al.*, 1990). The isolate secreted primarily a metalloproteinase when grown on minimal medium with milk protein as the nitrogen source, and a mixture of a serine-, a cysteine and a metalloproteinase when insoluble collagen or elastin was used as the nitrogen source. The authors suggested that the collagenase activity might have been involved in the severe corneal destruction caused by the examined *A. flavus* isolate (Zhu *et al.*, 1990). Gopinathan *et al.* (2001) characterised the extracellular proteases produced *in vitro* by corneal fungal pathogens including *A. flavus* when collagen was provided as the sole nitrogen source. The fungal isolates produced predominantly serine proteinase activity, and, to a lesser extent, metalloproteinase activity. However, homogenates of infected rabbit corneas showed the presence of metalloproteinase activity alone, the enzymatic activities entirely being sensitive to ethylene diamine tetraacetate (EDTA), a metalloprotease inhibitor. The results of this study also suggested that activated resident corneal cells or inflammatory cells may largely contribute to the increased proteolytic activities in fungal infected corneas resulting in tissue matrix degradation in fungal keratitis, and that the possible role of fungal serine proteases in the activation of corneal matrix metalloproteinases (MMPs) cannot be ruled out (Gopinathan *et al.*, 2001). Further studies are therefore needed to answer the question about the contribution of *Aspergillus* proteinases to the pathogenesis of mycotic keratitis.

After inoculation of fungal conidia of *A. fumigatus, C. albicans, F. solani* and *Penicillium citreo-viride* into rabbits to induce fungal keratitis, the adherence ability, chemotaxis to neutrophils and metalloproteinase expression levels differed in eyes with different fungal pathogens, suggesting that these factors may contribute to the different growth patterns of fungi in cornea (Joseph *et al.*, 2006). During studies to establish a rabbit model of keratitis due to *A. fumigatus*, O'Day *et al.* (1979) noted even substantial intraspecies variation in virulence.

Other putative virulence factors of *Aspergillus* species causing corneal infections include mycotoxins, fungal adhesins with potential binding sites (laminin, fibronectin, collagen) in the corneal tissue (Thomas, 2003) as well as the binding of human fibrinogen to the surface of hyphae and conidia (Bouchara *et al.*, 1988). The possible role of these factors in *Aspergillus* keratitis requires clarification. The availability of complete, annotated genome sequences for *Aspergillus* species involved in fungal keratitis will provide solid basis for future studies in this field.

9. Conclusions

Aspergillus keratitis is frequent in agriculture based geographical regions with hot, humid, tropical or subtropical climates. Those at highest risk are young and middle-aged male agricultural workers, as they are more exposed to the possibility of corneal trauma with infected material. *A. flavus* is the most commonly isolated fungal pathogen in *Aspergillus* keratitis cases followed by *A. fumigatus, A. terreus* and *A. niger*.

Early and accurate microbiological diagnosis coupled with appropriate treatment is crucial for increasing the chances of complete recovery. For distinctive diagnosis of *Aspergillus* keratitis, one has to depend on a combination of microscopy, colony morphology as well as molecular techniques.

In developing countries, the knowledge on antifungal susceptibilities is mainly based on the western literature. Because of the magnitude of the problem in India, surveys of the local susceptibility patterns would be vital. The results of such surveys including a large number of *Aspergillus* isolates from keratitis patients could provide significant contribution to the solution of the fungal keratitis problem.

Multivariate analyses including morphological and molecular identification as well as studies on the genetic diversity, antifungal susceptibilities and potential virulence factors of *Aspergillus* strains isolated from corneal ulcers could provide further important informations about *Aspergillus* keratitis as well as ideas for ophthalmologists regarding the apparent initiation of treatment and the selection of the appropriate antifungal agents.

Acknowledgments

This work was supported by the Indian National Science Academy and the Hungarian Academy of Sciences within the frames of the Indo-Hungarian bilateral exchange programme (No.IA/INSA-HAS Project/2007). L. K. is a grantee of the János Bolyai Research Scholarship.

References

Agarwal, P.K., P. Roy, A. Das, A. Banerjee, P.K. Maity and A.R. Banerjee, 2001. Efficacy of topical and systemic itraconazole as a broad-spectrum antifungal agent in mycotic corneal ulcer. A preliminary study. Indian Journal of Ophthalmology 49: 173-176.

Agrawal, V., J. Biswas, H.N. Madhavan, G. Mangat, M.K. Reddy, J.S. Saini, S. Sharma and M. Srinivasan, 1994. Current perspectives in infectious keratitis. Indian Journal of Ophthalmology 42: 171-191.

Alfonso, E.C., D. Miller, J. Cantu-Dibildox, T.P. O'brien and O.D. Schein, 2006. Fungal keratitis associated with non-therapeutic soft contact lenses. American Journal of Ophthalmology 142: 154-155.

Alvarez-de-Carvalho, A.C., H. Iran-Ruthes, M. Maia, D. Yana, M. Teruo-Sato, H. Moreira, G. Fernandes-Bordignon and F. de-Queiroz-Telles, 2001. [Fungal keratitis in the State of Parana-Brazil: clinical, epidemiological and diagnostic findings] Revista Iberoamericana de Micologia 18: 76-78. In Portuguese.

Anand, A.R., H.N. Madhavan, N.V. Sudha and K.L. Therese, 2001. Polymerase chain reaction in the diagnosis of *Aspergillus* endophthalmitis. Indian Journal of Medical Research 114: 133-140.

Avunduk, A.M., R.W. Beuerman, E.D. Warnel, H.E. Kaufman and D. Greer, 2003. Comparison of efficacy of topical and oral fluconazole treatment in experimental *Aspergillus* keratitis. Current Eye Research 26: 113-117.

Bagyalakhsmi, R., K.L. Therese and H.N. Madhavan, 2007. Nucleotide polymorphisms associated with Internal Transcribed Spacer (ITS) regions of ocular isolates of *Aspergillus flavus*. Journal of Microbiological Methods 68: 1-10.

Balajee, S.A. and K.A. Marr, 2006. Phenotypic and genotypic identification of human pathogenic aspergilli. Future Microbiology 1: 435-445.

Basak S.K., S. Basak, A. Mohanta and A. Bhowmick, 2005. Epidemiological and microbiological diagnosis of suppurative keratitis in Gangetic West Bengal, eastern India. Indian Journal of Ophthalmology 53: 17-22.

Berenji, F., S.R. Elahi, A.M. Fata, H. Khakshour and A. Darakhshan, 2003. Fungal keratitis in patients at Imam Reza Hospital, Mashhad (1982 – 2001) [in Persian]. Medical Journal of Mashhad University of Medical Sciences 45: 49-54.

Bharathi, M.J., R.R. Ramakrishnan, S. Vasu, R. Meenakshi and R. Palaniappan, 2002. Aetiological diagnosis of microbial keratitis in South India - A study of 1618 cases. Indian Journal of Medical Microbiology 20: 19-24.

Bharathi, M.J., R.R. Ramakrishnan, R.R. Meenakshi, S.S. Mittal, C.S. Kumar and M.M. Srinivasan, 2006. Microbiological Diagnosis of Infective Keratitis: Comparative Evaluation of Direct Microscopy and Culture Results. British Journal of Ophthalmology 90: 1271-1276.

Bharathi, M.J., R. Ramakrishnan, S. Vasu, R. Meenakshi and R. Palaniappan, 2003. Epidemiological characteristics and laboratory diagnosis of fungal keratitis. A three-year study. Indian Journal of Ophthalmology 51: 315-321.

Bhartiya, P., M. Daniell, M. Constantinou, F.M. Islam and H.R. Taylor, 2007. Fungal keratitis in Melbourne. Clinical and Experimental Ophthalmology 35: 124-130.

Bifrare, Y.-D. and T.J. Wolfensberger, 2007. Protracted *Aspergillus versicolor* endophthalmitis caused by corneal microperforation. Klinische Monatsblätter für Augenheilkunde 224: 314-316.

Boonpasart, S., N. Kasetsuwan, V. Puangsricharern, L. Pariyakanok and T. Jittpoonkusol, 2002. Infectious keratitis at King Chulalongkorn Memorial Hospital: a 12-year retrospective study of 391 cases. Journal of the Medical Association of Thailand 85: S217-S230.

Bouchara, J.P., A. Bouali, G. Tronchin, R. Robert, D. Chabasse and J.M. Senet, 1988. Binding of fibrinogen to the pathogenic *Aspergillus* species. Journal of Medical and Veterinary Mycology 26: 327-334.

Chander, J. and A. Sharma, 1994. Prevalence of fungal corneal ulcers in northern India. Infection. 22: 207-209.

Chowdhary, A. and K. Singh, 2005. Spectrum of fungal keratitis in North India. Cornea 24: 8-15.

Clemons, K.V. and D.A. Stevens, 2005. The contribution of animal models of aspergillosis to understanding pathogenesis, therapy and virulence. Medical Mycology 43: S101-S110.

Clemons, K.V. and D.A. Stevens, 2006. Animal models of *Aspergillus* infection in preclinical trials, diagnostics and pharmacodynamics: What can we learn from them? Medical Mycology 44: S119-S126.

Coriglione, G., G. Stella, L. Gafa, G. Spata, S. Oliveri, A. A. Padhye, and L. Ajello, 1990. *Neosartorya fischeri var. fischeri* (Wehmer) Malloch and Cain 1972 (anamorph: *Aspergillus fischerianus* Samson and Gams, 1985) as a cause of mycotic keratitis. European Journal of Epidemiology 6: 382-385.

De Aguirre, L., S.F. Hurst, J.S. Choi, J.H. Shin, H.P. Hinrikson and C.J. Morrison, 2004. Rapid differentiation of *Aspergillus* species from other medically important opportunistic molds and yeasts by PCR-enzyme immunoassay. Journal of Clinical Microbiology 42: 3495-3504.

De Hoog, G.S., J. Guarro, J. Gene and M.J. Figueras (eds.), 2004. Atlas of Clinical Fungi, Centraalbureau voor Schimmelcultures, Utrecht, The Netherlands.

Deshpande, S.D. and G.V. Koppikar, 1999. A study of mycotic keratitis in Mumbai. Indian Journal of Pathology and Microbiology 42: 81-87.

Dóczi, I., T. Gyetvai, L. Kredics and E. Nagy, 2004. Involvement of *Fusarium* spp. in fungal keratitis. Clinical Microbiology and Infection 10: 773-776.

Doughman, D.J., N.M. Leavenworth, R.C. Campbell and R.L. Lindstrom, 1982. Fungal keratitis at the University of Minnesota: 1971-1981. Transactions of the American Ophthalmological Society 80: 235-247.

Dunlop, A.A., E.D.Wright, S.A. Howlader, I. Nazrul, R. Husain, K. McClellan and F.A. Billson, 1994. Suppurative corneal ulceration in Bangladesh. A study of 142 cases examining the microbiological diagnosis, clinical and epidemiological features of bacterial and fungal keratitis. Australian and New Zealand Journal of Ophthalmology 22: 105-110.

Etest technical guide 10, 1999. Antifungal susceptibility testing of moulds. AB Biodisk, Solna, Sweden.

Fahad, B., M. McKellar, M. Armstrong, D. Denning and A. Tullom, 2004. *Aspergillus* keratitis following corneal foreign body. British Journal of Ophthalmology 88: 847-848.

Ferrer, C., F. Colom, J.L. Abad, J.J. Perez-Santonja and J.L. Alio, 2000. Detection and identification of *Aspergillus fumigatus* in a patient with keratitis by amplification and RFLP of 5.8 S rDNA. Revista Iberoamericana de Micologia 17: S134.

Ferrer, C., F. Colom, S. Frases, E. Mulet, J.L. Abad and J.L. Alio, 2001. Detection and identification of fungal pathogens by PCR and by ITS2 and 5.8S ribosomal DNA typing in ocular infections. Journal of Clinical Microbiology 39: 2873-2879.

Forster, R.K. 1994. Fungal keratitis and conjunctivitis: clinical aspects. In: G. Smolin and R.A. Thoft (eds.) The cornea: scientific foundations and clinical practice. Little, Brown & Co., Boston, USA, pp. 239-252.

Foster, C.S. 1981. Miconazole therapy for keratomycosis. American Journal of Ophthalmology 91: 622-629.

Freda, R. 2006. Use of oral voriconazole as adjunctive treatment of severe cornea fungal infection: case report. Arquivos Brasileiros de Oftalmologia 69: 431-434.

Galarreta, D.J., S.J. Tuft, A. Ramsay and J.K. Dart, 2007. Fungal keratitis in London: microbiological and clinical evaluation. Cornea 26: 1082-1086.

Garcia de Lomas, J., M.A. Fons, J.M. Nogueira, F. Rustom, R. Borrás and F.J. Buesa, 1985. Chemotherapy of *Aspergillus fumigatus* keratitis: an experimental study. Mycopathologia 89: 135-138.

Garg, P., U. Gopinathan, K. Choudhary and G.N. Rao, 2000. Keratomycosis: clinical and microbiologic experience with dematiaceous fungi. Ophthalmology 107: 574-580.

Gaudio, P.A., U. Gopinathan, V. Sangwan and T.E. Hughes, 2002. Polymerase chain reaction based detection of fungi in infected corneas. British Journal of Ophthalmology 86: 755-760.

Goebes, M.D., L. M. Hildemann, E. Kujundzic and M. Hernandez, 2007. Real-time PCR for detection of the *Aspergillus* genus. Journal of Environmental Monitoring 9: 599-609.

Gonawardena, S.A., K.P. Ranasinghe, S.N. Arseculeratne, C.R. Seymon and R. Ajello, 1994. Survey of mycotic and bacterial keratitis in Sri Lanka. Mycopathologia 127: 77-81.

Gopinathan, U., P. Garg, M. Fernandes, S. Sharma, S. Athmanathan and G.N. Rao, 2002. The epidemiological features and laboratory results of fungal keratitis: a 10-year review at a referral eye care center in South India. Cornea 21: 555-559.

Gopinathan, U., T. Ramakrishna, M. Willcox, C.M. Rao, D. Balasubramanian, A. Kulkarni, G.K. Vemuganti and G.N. Rao, 2001. Enzymatic, clinical and histologic evaluation of corneal tissues in experimental fungal keratitis in rabbits. Experimental Eye Research 72: 433-442.

Gottsch, J.D., M.L. Gilbert, D.F. Goodman, M.E. Sulewski, J.D. Dick and W.J. Stark, 1991. Excimer laser ablation of microbial keratitis. Ophthalmology 98: 146-149.

Gugnani, H.C., R.S. Talwar, A.N. Njoku-Obi and H.C. Kodilinye, 1976. Mycotic keratitis in Nigeria. A study of 21 cases. British Journal of Ophthalmology 60: 607-613.

Gugnani, H.C., S. Gupta and R.S. Talwar, 1978. Role of opportunistic fungi in ocular infections in Nigeria. Mycopathologia 65: 155-166.

Gupta, S.K. 1986. Efficacy of miconazole in experimental keratomycosis. Australian and New Zealand Journal of Ophthalmology 14: 373-376.

Hagan, M., E. Wright, M. Newman, P. Dolin and G. Johnson, 1995. Causes of suppurative keratitis in Ghana. British Journal of Ophthalmology 79: 1024-1028.

Hahn, Y.H., D.G. Ahearn and L.A. Wilson, 1993. Comparative efficacy of amphotericin B, clotrimazole and itraconazole against *Aspergillus* spp. An *in vitro* study. Mycopathologia 123: 135-140.

Heidemann, D.G., S. P. Dunn and J. C. Watts, 1995. *Aspergillus* keratitis after radial keratotomy. American Journal of Ophthalmology 120: 254-256.

Hofling-Lima, A.L., A. Forseto, J.P. Duprat, A. Andrade, L.B. Souza, P. Godoy and D. Freitas, 2005. [Laboratory study of the mycotic infectious eye diseases and factors associated with keratitis]. Arquivos Brasileiros de Oftalmologia 68: 21-27. In Portuguese.

Jaeger, E.E., N.M. Carroll, S. Choudhury, A.A. Dunlop, H.M. Towler, M.M. Matheson, P. Adamson, N. Okhravi and S. Lightman, 2000. Rapid detection and identification of *Candida, Aspergillus* and *Fusarium* species in ocular samples using nested PCR. Journal of Clinical Microbiology 38: 2902-2908.

Jain, V., A. Maiti, D. Shome, N. Borse and S. Natarajan, 2007. *Aspergillus*-induced malignant glaucoma. Cornea 26: 762-763.

Javadi, M.A., R. Hemati, M.M. Mohammadi, A. Farsi, F. Karimian and B. Einolahi, 1996. [Causes of fungal keratitis and its management: review of 23 cases from Labbafinejad Medical Center (LMC)]. Bina 2: 38-54. In Persian.

Johnson, G. J. 1998. External eye infections. Journal of Community Eye Health 12: 1-2.

Joseph, J., S. Savitri, P. Garg, U. Gopinathan and G. Vemuganti, 2006. Roles of adherence and matrix metalloproteinases in growth patterns of fungal pathogens in cornea. Current Eye Research 31: 467-468.

Kalavathy, C.M., P. Parmar, J. Kaliamurthy, V.R. Philip, M.D. Ramalingam, C.A. Jesudasan and P.A. Thomas, 2005. Comparison of topical itraconazole 1% with topical natamycin 5% for the treatment of filamentous fungal keratitis. Cornea 24: 449-452.

Kanbe, T., K. Yamaki and A. Kikuchi, 2002. Identification of the pathogenic *Aspergillus* species by nested PCR using a mixture of specific primers to DNA topoisomerase II gene. Microbiology and Immunology 46: 841-848.

Kaushik, S., J. Ram, G.S. Brar, A.K. Jain, A. Chakraborti and A. Gupta, 2001. Intracameral amphotericin B: initial experience in severe keratomycosis. Cornea 20: 715-719.

Khairallah, S.H., K.A. Byrne and K.F. Tabbara, 1992. Fungal keratitis in Saudi Arabia. Documenta Ophthalmologica 79: 269-276.

Khanal, B., M. Deb, A. Panda and H.S. Sethi, 2005. Laboratory diagnosis in ulcerative keratitis. Ophthalmic Research 37: 123-127.

Killingsworth, D.W., G.A. Stern, W.T. Driebe, A. Knapp and D.M. Dragon, 1993. Results of therapeutic penetrating keratoplasty. Ophthalmology 100: 534-541.

Kim, J.S., J.C. Kim, T.W. Hahn and W.C. Park, 2001 Amniotic membrane transplantation in infectious corneal ulcer. Cornea 20: 720-726.

Klotz, S.A., C.C. Penn, G.J. Negvesky and S. I. Butrus, 2000. Fungal and parasitic infections of the eye. Clinical Microbiology Reviews 13: 662-685.

Kocsubé, S., P. Manikandan, J. Varga, V. Narendran, Z. Antal, R. Revathi, C. Vágvölgyi, E. Nagy and L. Kredics, 2006. Identification and genetic variability of *Aspergillus* strains isolated from patients with keratomycosis in South-India. Acta Microbiologica et Immunologica Hungarica 53: 298-299.

Komadina, T.G., T.D. Wilkes, J.P. Shock, W.C. Ulmer, J. Jackson and R.W. Bradsher, 1985. Treatment of *Aspergillus fumigatus* keratitis in rabbits with oral and topical ketoconazole. American Journal of Ophthalmology 99: 476-479.

Kotigadde, S., M. Ballal, Jyothirlatha, A. Kumar, R. Srinivasa and P.G. Shivananda, 1992. Mycotic keratitis: a study in coastal Karnataka. Indian Journal of Ophthalmology 40: 31-33.

Kredics, L., J. Varga, S. Kocsubé, I. Dóczi, R.A. Samson, R. Revathi, V. Narendran, M. Bhaskar, C. Vágvölgyi and P. Manikandan, 2007a. Case of keratitis caused by *Aspergillus tamarii*. Journal of Clinical Microbiology 45: 3464-3467.

Kredics, L., P. Manikandan, S. Kocsubé, I. Dóczi, V. Narendran, J. Varga, Z. Antal, C. Vágvölgyi, R. Revathi and E. Nagy, 2007b. *Aspergillus* keratomycosis at an eye hospital in south India: a retrospective study. International Journal of Antimicrobial Agents 29: S356.

Kumar, M. and P.K. Shukla, 2005. Use of PCR targeting of internal transcribed spacer regions and single-stranded conformation polymorphism analysis of sequence variation in different regions of rRNA genes in fungi for rapid diagnosis of mycotic keratitis. Journal of Clinical Microbiology 43: 662-668.

Kumari, N., A. Xess and S.K. Shahi, 2002. A study of keratomycosis: our experience. Indian Journal of Pathology and Microbiology 45: 299-302.

Kunimoto, D.Y., S. Sharma, P. Garg, U. Gopinathan, D. Miller and G.N. Rao, 2000. Corneal ulceration in the elderly in Hyderabad, south India. British Journal of Ophthalmology 84: 54-59.

Kuo, I.C., T.P. Margolis, V. Cevallos and D.G. Hwang, 2001. *Aspergillus fumigatus* keratitis after laser *in situ* keratomileusis. Cornea 20: 342-344.

Lalitha, P., B.L. Shapiro, M. Srinivasan, N.V. Prajna, N.R. Acharya, A.W. Fothergill, J. Ruiz, J.D. Chidambaram, K.J. Maxey, K.C. Hong, S.D. McLeod and T.M. Lietman, 2007. Antimicrobial susceptibility of *Fusarium*, *Aspergillus*, and other filamentous fungi isolated from keratitis. Archives in Ophthalmology 125: 789-793.

Laspina, F., M. Samudio, D. Cibils, C.N. Ta, N. Farina, R. Sanabria, V. Klauss, H. Mino de Kaspar, 2004. Epidemiological characteristics of microbiological results on patients with infectious corneal ulcers: a 13-year survey in Paraguay. Graefe's Archive for Clinical and Experimental Ophthalmology 242: 204-209

Leck, A.K., P.A. Thomas, M. Hagan, J. Kaliamurthy, E. Ackuaku, M. John, M.J. Newman, F.S. Codjoe, J.A. Opintan, C.M. Kalavathy, V. Essuman, C.A. Jesudasan and G.J. Johnson, 2002. Aetiology of suppurative corneal ulcers in Ghana and south India, and epidemiology of fungal keratitis. British Journal of Ophthalmology 86: 1211-1215.

Liesegang, T.J. and R.K. Forster, 1980. Spectrum of microbial keratitis in South Florida. American Journal of Ophthalmology 90: 38-47.

Mancini, N., M. Perotti, C.M. Ossi, A. Cavallaro, S. Matuska, G. Paganoni, R. Buroni, P. Rama and M. Clementi, 2006. Rapid molecular identification of fungal pathogens in corneal samples from suspected keratomycosis cases. Journal of Medical Microbiology 55: 1505-1509.

Manikandan, P., I. Dóczi, S. Kocsubé, V. Narendran, J. Varga, R. Revathi, C. Vágvölgyi, E. Nagy and L. Kredics, 2006. Epidemiology and antifungal susceptibilities of *Aspergillus* strains isolated from patients with keratomycosis in South India. Acta Microbiologica et Immunologica Hungarica 53: 313-314.

Manikandan, P., S. Kocsubé, I. Dóczi, V. Narendran, M. Bhaskar, J. Varga, Z. Antal, C. Vágvölgyi, R. Revathi, E. Nagy and L. Kredics, 2007. *Aspergillus* keratitis: epidemiological features, molecular identification, antifungal susceptibility and clinical outcome at the Aravind Eye Hospital in South India. Acta Microbiologica et Immunologica Hungarica 54: 80-81.

Marangon, F.B., D. Miller, J.A. Giaconi and E.C. Alfonso, 2004. *In vitro* investigation of voriconazole susceptibility for keratitis and endophthalmitis fungal pathogens. American Journal of Ophthalmology 137: 820-825.

Melatiadis, J., J.W. Mouton, J.F.G.M. Meis, B.A. Bouman, P.E. Verweij and the EUROFUNG Network, 2002. Comparison of the Etest and the Sensititre Colorimetric Methods with the NCCLS proposed standard for antifungal susceptibility testing of *Aspergillus* species. Journal of Clinical Microbiology 40: 2876-2885.

Mendicute, J., A. Ondarra, F. Eder, J.I. Ostolaza, M. Salaberria and J.M. Lamsfus, 1995. The use of collagen shields impregnated with amphotericin B to treat *Aspergillus* keratomycosis. CLAO Journal 21: 252-255.

Mendicute, J., J. Orbegozo, M. Ruiz, A. Sáiz, F. Eder and J. Aramberri, 2000. Keratomycosis after cataract surgery. Journal of Cataract and Refractive Surgery 26: 1660-1666.

Mino de Kaspar, H., G. Zoulek, M.E. Paredes, R. Alborno, D. Medina, M. Centurion de Morinigo, M. Ortiz de Fresco and F. Aguero, 1991. Mycotic keratitis in Paraguay. Mycoses. 34: 251-254.

Mohan, M., S.K. Gupta, V.K. Kalra, R.B. Vajpayee and M.S. Sachdev, 1988. Topical silver sulphadiazine: a new drug for ocular keratomycosis. British Journal of Ophthalmology 72: 192-195.

National Committee for Clinical Laboratory Standards (Clinical and Laboratory Standard Institute), 2002. Reference method for broth dilution antifungal susceptibility testing of filamentous fungi; approved standard. NCCLS document M38-A. NCCLS (CLSI), Pennsylvania, USA.

O'Day, D.M 1996. Fungal Keratitis. In: J.S. Pepose, H.N. Holland and K.R. Wilhelmus (eds.) Ocular infection and immunity. Mosby, St. Louis, USA, pp. 1048-1061.

O'Day, D.M., P.L. Akrabawi, L.D. Richmond, B.R. Jones and Y. Clayton, 1979. An animal model of *Fusarium solani* endophthalmitis. British Journal of Ophthalmology 63: 277-280.

O'Day, D.M., W.A. Ray, W.S. Head and R.D. Robinson, 1984. Influence of the corneal epithelium on the efficacy of topical antifungal agents. Investigative Ophthalmology and Visual Science 25: 855-859.

Oji, E.O. 1982. Ketoconazole: a new imidazole antifungal agent has both prohylactic potential and therapeutic efficacy in keratomycosis of rabbits. International Ophthalmology 5: 163-167.

Panda, A., G.K. Das, M. Vanathi and A. Kumar, 1998. Corneal infection after radial keratotomy. Journal of Cataract and Refractive Surgery 24: 331-334.

Panda, A., N. Sharma, G. Das, N. Kumar and G. Satpathy, 1997. Mycotic keratitis in children: epidemiologic and microbiologic evaluation. Cornea 16: 295-299.

Panda, A., R. Ahuja, N.R. Biswas, G. Satpathy and S. Khokhar, 2003. Role of 0.02% polyhexamethylene biguanide and 1% povidone iodine in experimental *Aspergillus* keratitis. Cornea 22: 138-141.

Panhalkar, S., A. Thomas, T.A. Alexander and G. Koshi, 1985. Bacterial and mycotic agents of corneal ulcers in Vellore. Indian Journal of Ophthalmology 33: 289-293.

Perry, H.D., S.J. Doshi, E.D. Donnenfeld and G.S. Bai, 2002. Topical cyclosporin A in the management of therapeutic keratoplasty for mycotic keratitis. Cornea 21: 161-163.

Portnoy, S.L., M.S. Insler and H.E. Kaufman, 1989. Surgical management of corneal ulceration and perforation. Survey of Ophthalmology 34: 47-58.

Prajna, N.V., P.K. Nirmalan, R. Mahalakshmi, P. Lalitha and M. Srinivasan, 2004. Concurrent use of 5% natamycin and 2% econazole for the management of fungal keratitis. Cornea 23: 793-796.

Qiu, W.Y., Y.F. Yao, Y.F. Zhu, Y.M. Zhang, P. Zhou, Y.Q. Jin and B. Zhang, 2005. Fungal spectrum identified by a new slide culture and *in vitro* drug susceptibility using Etest in fungal keratitis. Current Eye Research 30: 1113-1120.

Rahman, M.R., G.J. Johnson, R. Husain, S. Howlader and D. Minassian, 1998. Randomised trial of 0.2% chlorhexidine gluconate and 2.5% natamycin for fungal keratitis in Bangladesh. British Journal of Ophthalmology 82: 919-925.

Rao, S.K., H.N. Madhavan, G. Rao and P. Padmanabhan, 1997. Fluconazole in filamentous fungal keratitis. Cornea 16: 700.

Rajasekaran, J., P.A. Thomas, C.M. Kalavathy, P.C. Joseph and D.J. Abraham, 1987. Itraconazole therapy for fungal keratitis. Indian Journal of Ophthalmology 35: 157-160.

Rondeau, N., T. Bourcier, C. Chaumeil, V. Borderie, O. Touzeau, Y. Scat, F. Thomas, C. Baudouin, J.P. Nordmann and L. Laroche, 2002. [Fungal keratitis at the Centre Hospitalier National d'Ophtalmologie des Quinze-Vingts: retrospective study of 19 cases]. Journal Francaise d'Ophtalmologie 25: 890-896. In French.

Rosa, R.H. Jr, D. Miller and E.C. Alfonso, 1994. The changing spectrum of fungal keratitis in south Florida. Ophthalmology. 101: 1005-1013.

Saha, R. and S. Das, 2006. Mycological profile of infectious Keratitis from Delhi. Indian Journal of Medical Research 123: 159-164.

Salera, M.C., M.A. Guarino Tanure, W.T. Montenegro Lima, C.M. Campos, F.C. Trindade and J. dos Anjos Moreira, 2002. [Spectrum of fungal keratitis at the São Geraldo Hospital – Belo Horizonte – MG]. Arquivos Brasileiros de Oftalmologia 65: 9-13. In Portuguese.

Sandhu, G.S., B.C. Kline, L. Stockman and G.D. Roberts, 1995. Molecular probes for diagnosis of fungal infections. Journal of Clinical Microbiology 33: 2913-2919.

Saracli, M.A., F.M. Mutlu, S.T. Yildiran, A.E. Kurekci, A. Gonlum, Y. Uysal, U. Erdem, A.C. Basustaoglu and D.A. Sutton, 2007. Clustering of invasive *Aspergillus ustus* eye infections in a tertiary care hospital: a molecular epidemiologic study of an uncommon species. Medical Mycology 45: 377-384.

Shokohi, T., K. Nowroozpoor-Dailami and T. Moaddel-Haghighi, 2006. Fungal keratitis in patients with corneal ulcer in Sari, Northern Iran. Archives of Iranian Medicine 9: 222-227.

Singh, G., P. Manikandan, M. Bhaskar, R. Revathi, K. Narendran, A. Kour and V. Narendran, 2006. Multivariate analysis of childhood microbial keratitis in South India. Annals of the Academy of Medicine, Singapore 35: 185-189.

Singh, S.M., R. Khan, S. Sharma and P.K. Chatterjee, 1989. Clinical and experimental mycotic corneal ulcer caused by *Aspergillus fumigatus* and the effect of oral ketoconazole in the treatment. Mycopathologia 106: 133-141.

Sonego-Krone, S., D. Sanchez-Di Martino, R. Ayala-Lugo, G. Torres-Alvariza, C.N. Ta, L. Barbosa and H.M. de Kaspar, 2006. Clinical results of topical fluconazole for the treatment of filamentous fungal keratitis. Graefe's Archive for Clinical and Experimental Ophthalmology 244: 782-787.

Sridhar, M.S., P. Garg, A.K. Bansal and U. Gopinathan, 2000. *Aspergillus flavus* keratitis after laser in situ keratomileusis. American Journal of Ophthalmology 129: 802-804.

Srinivasan, M. 2004. Fungal keratitis. Current Opinion in Ophthalmology 15: 321-327.

Srinivasan, M., C.A. Gonzales, C. George, V. Cevallos, J.M. Mascarenhas, B. Asokan, J. Wilkins, G. Smolin and J.P. Whitcher, 1997. Epidemiology and aetiological diagnosis of corneal ulceration in Madurai, south India. British Journal of Ophthalmology 81: 965-971.

Stewart, R.M., S.A. Quah, T.J. Neal, S.B. Kaye, 2007. The use of voriconazole in the treatment of *Aspergillus fumigatus* keratitis. British Journal of Ophthalmology 91: 1094-1096.

Sugita, T., K. Kamimura, K. Uchida, H. Yamaguchi and A. Nagai, 2004. PCR identification system for the genus *Aspergillus* and three major pathogenic species: *Aspergillus fumigatus, Aspergillus flavus* and *Aspergillus niger*. Medical Mycology 42: 433-437.

Sun, X.G., Y. Zhang, R. Li, Z.Q. Wang, S.Y. Luo, X.Y. Jin and W.H. Zhang, 2004. Etiological analysis on ocular fungal infection in the period of 1989 - 2000. Chinese Medical Journal 117: 598-600.

Sundaram, B.M., S. Badrinath and S. Subramanian, 1989. Studies on mycotic keratitis. Mycoses 32: 568-572.

Tanure, M.A., E.J. Cohen, S. Sudesh, C.J. Rapuano and P.R. Laibson, 2000. Spectrum of fungal keratitis at Wills Eye Hospital, Philadelphia, Pennsylvania. Cornea 19:307-312.

Therese, K.L., R. Bagyalakshmi, H.N. Madhavan and P. Deepa, 2006. *In-vitro* susceptibility testing by agar dilution method to determine the minimum inhibitory concentrations of amphotericin B, fluconazole and ketoconazole against ocular fungal isolates. Indian Journal of Medical Microbiology 24: 273-279.

Thiel, M.A., A.S. Zinkernagel, J. Burhenne, C. Kaufmann and W.E. Haefeli, 2007. Voriconazole concentration in human aqueous humor and plasma during topical or combined topical and systemic administration for fungal keratitis. Antimicrobial Agents and Chemotherapy 51: 239-244.

Thomas, P.A. 1994. Mycotic keratitis: an underestimated mycosis. Journal of Medical and Veterinary Mycology 32: 235-254.

Thomas, P.A. 2003. Current perspectives on ophthalmic mycoses. Clinical Microbiology Reviews 16: 730-797.

Thomas, P.A. and J. Rajasekaran, 1988. Treatment of *Aspergillus* keratitis with imidazoles and related compounds. In: H. van den Bossche, D.W.R. Mackenzie and G. Cauwenbergh (eds.) *Aspergillus* and aspergillosis. Plenum Press, New York, USA, pp. 267-279.

Thomas, K. and P.A. Thomas, 1991. Keratomycotic malignant glaucoma. Indian Journal of Ophthalmology 39: 118–121.

Thomas, P.A., C.M. Kalavathy and J. Rajasekaran, 1986. Microbial keratitis - a study of 774 cases and review of the literature. Journal of Madras State Ophthalmic Association 23: 13-21.

Thomas, P.A., D.J. Abraham, C.M. Kalavathy and J. Rajasekaran, 1988. Oral itraconazole therapy for mycotic keratitis. Mycoses 31: 271-279.

Torres, M.A., J. Mohamed, H. Cavazos-Adame and L.A. Martinez, 1985. Topical ketoconazole for fungal keratitis. American Journal of Ophthalmology 100: 293-298.

Upadhyay, M.P., P.C. Karmacharya, S. Koirala, N.R. Tuladhar, L.E. Bryan, G. Smolin and J.P. Whitcher, 1991. Epidemiologic characteristics, predisposing factors, and etiologic diagnosis of corneal ulceration in Nepal. American Journal of Ophthalmology 111: 92-99.

Usha, A., A. Aruna and J. Vijay, 2006. Fungal profile and susceptibility pattern in cases of keratomycosis. Journal of Medical Education 8: 39-41.

Vajpayee, R.B., S.K. Gupta, U. Bareja and K. Kishore, 1990. Ocular atopy and mycotic keratitis. Annals of Ophthalmology 22: 369-372.

Vajpayee, R.B., T. Dada, R. Saxena, M. Vajpayee, H.R. Taylor, P. Venkatesh and N. Sharma, 2000. Study of the first contact management profile of cases of infectious keratitis: a hospital-based study. Cornea 19: 52-56.

Van Cutsem, J. and F. van Gerven, 1991. Activite' antifongique in vitro de l'itraconazole sur les champignons filamenteux opportunistes: traitement de la keratomycose et de la penicilliose experimentales. Journal de Mycologie Médicale 1: 10-15.

Vasudevan, R., M. Jeya and V. Udhaya, 2006. Prevalence of bacterial and fungal keratitis in and around Chidambaram. Journal of Tamil Nadu Ophthalmic Association 44: 25-29.

Vemuganti, G.K., P. Garg, V. Gopinathan, T.J. Naduvilath, R.K. John, R. Buddi and G.N. Rao, 2002. Evaluation of agent and host factors in progression of mycotic keratitis - A histologic and microbiologic study of 167 corneal buttons. Ophthalmology 109: 1538-1546.

Venugopal, P.L., T.L. Venugopal, A. Gomathi, E.S. Ramakrishna and S. Ilavarasi, 1989. Mycotic keratitis in Madras. Indian Journal of Pathology and Microbiology 32: 190-197.

Verenkar, M.P., B. Shubhangi, M.J.W. Pinto, N. Pradeep, 1998. Study of mycotic keratitis in Goa. Indian Journal of Medical Microbiology 16: 58-60.

Wang, L., Y. Zhang, Y. Wang, G. Wang, J. Lu and J. Deng, 2000. [Spectrum of mycotic keratitis in China]. Zhonghua Yan Ke Za Zhi 36: 138-140. In Chinese.

Wang, L.Y., Z.J. Yang, X.Y.Yang and S.T. Sun, 2007. Experimental study of fast diagnosis of mycotic keratitis using molecular biology technical. Chinese Journal of Ophtalmology 43: 256-259.

Wilhelmus, K.R. and D.B. Jones, 2001. *Curvularia* keratitis. Transactions of the American Ophthalmological Society 99: 111-132.

Williams, G., F. Billson, R. Husain, S.A. Howlader, N. Islam and K. McClellan, 1987. Microbiological diagnosis of suppurative keratitis in Bangladesh. British Journal of Ophthalmology 71: 315-321.

Winchester, K., W.D. Mathers and J.E. Sutphin, 1997. Diagnosis of *Aspergillus* keratitis *in vivo* with confocal microscopy. Cornea 16: 27-31.

Wong, T.Y., K.S. Fong and D.T. Tan, 1997. Clinical and microbial spectrum of fungal keratitis in Singapore: a 5-year retrospective study. International Ophthalmology 21: 127-130.

Xie, L., W. Zhong, W. Shi and S. Sun, 2006. Spectrum of Fungal Keratitis in North China. Ophthalmology 113: 1943-1948.

Xie, L., X. Dong and W. Shi, 2001. Treatment of fungal keratitis by penetrating keratoplasty. British Journal of Ophthalmology 85: 1070-1074.

Xuguang, S., W. Zhixin, W. Zhiqun, L. Shiyun and L. Ran, 2007. Ocular fungal isolates and antifungal susceptibility in northern China. American Journal of Ophthalmology 143: 131-133.

Zhang, W., Z. Pan, Z. Wang, X. Jin, S. Luo, Y. Zou, Y. Wu and R. Li, 2002. [The variance of pathogenic organisms of purulent ulcerative keratitis] Zhonghua Yan Ke Za Zhi 38: 8-12. In Chinese.

Zhang, Y.L., L.Y. Wang, Z.T. Li and S. T. Sun, 2007. Gene microarray for the detection and identification of pathogenic species in fungal keratitis. Journal of Clinical Rehabilitative Tissue Engineering Research 11: 1873-1877.

Zhu, W.S., K. Wojdyla, K. Donlon, P.A. Thomas and H.I. Eberle, 1990. Extracellular proteases of *Aspergillus flavus*. Fungal keratitis, proteases, and pathogenesis. Diagnostic Microbiology and Infectious Disease 13: 491-497.

Index

A

B

C

O

P

www.ingramcontent.com/pod-product-compliance
Lightning Source LLC
LaVergne TN
LVHW080848300926
849162LV00003B/373
* 9 7 8 9 0 8 6 8 6 0 6 5 4 *